Honey Bees, Beekeeping and Bee Products

Honey bees are social insects; they live together in large, well-organized family groups comprising three castes: queen (fertile female), workers (sterile females) and drones (males). During honey flow season, there is a considerable increase in the foraging activity of the workers and in the rate of egg laying by the queen. Sex determination in honey bees involves a multi-allelic locus, such that homozygotes develop as males and heterozygotes as females, whereas diet quality influences the caste determination in honey bees. Like all living organisms, honey bees can be infested with diseases and pests. Some of these are more deleterious to bee colonies than others, but it is important for the beekeeper to be able to recognize conditions that might be disease or pest-related and respond accordingly so as to improve the quality of honey and honey bee by-products.

The best-known primary products of beekeeping are honey and wax, but pollen, propolis, royal jelly, venom, queens, bees and their larvae are also marketable primary bee products. The purpose of this book is to make available information on bee biology and beekeeping as well as to provide comprehensive information on the manufacturing, processing and marketing of value-added bee products.

This book has been designed as a useful tool for the many diverse professionals who characterize and market honey bee products, including beekeepers, non-beekeepers, small entrepreneurs, extension officers and those involved in small business development. This edited book will be the first of its kind to contain comprehensive information on both bees and bee products.

Key Features:

- Contains comprehensive information on beekeeping.
- Discusses the recent advances in beekeeping.
- Sheds light on bee colony integration and organization.
- Contains brief information on honey bee products.

Honey Bees, Beekeeping and Bee Products

Edited by
Aijaz Ahmad Wachkoo, Gulzar Ahmad Nayik, Jalal Uddin
and Mohammad Javed Ansari

CRC Press
Taylor & Francis Group
Boca Raton London New York

CRC Press is an imprint of the
Taylor & Francis Group, an **informa** business

Designed cover image: Shutterstock

First edition published 2024
by CRC Press
2385 NW Executive Center Drive, Suite 320, Boca Raton FL 33431

and by CRC Press
4 Park Square, Milton Park, Abingdon, Oxon, OX14 4RN

CRC Press is an imprint of Taylor & Francis Group, LLC

ISBN: 9781032406503 (hbk)
ISBN: 9781032406510 (pbk)
ISBN: 9781003354116 (ebk)

DOI: 10.1201/9781003354116

Typeset in Times
by Deanta Global Publishing Services, Chennai, India

The editors dedicate this book to their beloved families

Contents

Preface

Beekeeping (or apiculture) is the practice of intentional maintenance of bees by humans, and mainly refers to maintenance of colonies of honey bees. Honey bees are social insects; they live together in large, well-organized family groups comprising three castes: queen (fertile female), workers (sterile females) and drones (males). Honey bee workers perform a complex age-dependent division of labour, with the most pronounced behavioural differences occurring between in-hive bees and foragers. During honey flow season, there is a considerable increase in the foraging activity of the workers and in the rate of egg laying by the queen. The additional space necessary for these is provided through the supply of new, clean, yellow combs or comb foundation sheets. Sex determination in honey bees involves a multi-allelic locus, such that homozygotes develop as males and heterozygotes as females, whereas diet quality influences the caste determination in honey bees. Honey bees exhibit striking caste-specific differences in longevity, and their life cycle comprises four stages: egg, larva, pupa and adult. The combined behaviours of individuals within insect societies determine the survival and development of the colony. Individual behaviours include nest building, foraging, storing and ripening food, nursing the brood, temperature regulation, hygiene and defence. Like all living organisms, honey bees can be infested with diseases and pests. Some of these are more deleterious to bee colonies than others, but it is important for the beekeeper to be able to recognize conditions that might be disease or pest-related and respond accordingly so as to improve the quality of honey and honey bee by-products.

The best-known primary products of beekeeping are honey and wax, but pollen, propolis, royal jelly, venom, queens, bees and their larvae are also marketable primary bee products. While most of these products can be consumed or used in the state in which they were produced by the bees, there are many additional uses where these products form only a part of all the ingredients of another product. Because of the quality and sometimes almost mystical reputation and characteristics of most primary bee products, their addition to other products usually enhances the value or quality of these secondary products. For this reason, the secondary products, which can be made up partially or wholly of primary bee products, are referred to here as "value-added" products from beekeeping.

Honey is a sweet, nutritious and stable food produced by honey bees from the nectar of plants or excretions of plant-sucking insects found on living parts of plants. It has a unique taste—sweet—which has been accessible to ancient human beings for ages in most parts of the world. Scientific research on the composition of honey has led to an ever-increasing understanding of the quality factors associated with beekeeping and has had a profound influence on current guidelines for honey. Wax has played a considerable role in only a few parts of the world, and propolis is even less known. However, with increasing knowledge about beekeeping and an awareness of the beneficial aspects of many bee products, the use and demand for other products are increasing. The inclusion of "natural" bee products in cosmetics, medicines and foods has improved consumer appeal. While such appeal is not always based on scientific evidence, more and more studies are confirming at least some of the traditionally claimed benefits of primary bee products.

The purpose of this book is to make available information on bee biology and beekeeping as well as to provide comprehensive information on the manufacturing, processing and marketing of value-added bee products. It has been designed as a useful tool for the many diverse professionals who characterize and market honey bee products, including beekeepers, non-beekeepers, small entrepreneurs, extension officers and those involved in small business development. It will enable the reader to properly store, process, package and market the primary products, as well as the value-added products derived from them. The reader will be able to draw conclusions about the myriad of sometimes miraculous effects and cures claimed for bee products. This edited book will be the first of its kind to contain comprehensive information on both bee and bee products.

This book will serve as a comprehensive guide for students, scholars, researchers and scientists working in apiculture. All the contributing authors are acknowledged for their constant dedication and contribution to the editorial guidelines and timeline. We are fortunate to have contributors of international repute from Australia, India, Turkey, Pakistan, Saudi Arabia, Romania, Bangladesh, Morocco, Algeria and Poland. We would like to thank colleagues from the production team of the Taylor & Francis group for their constant help during the editing and production process. Finally, we as editors have a message to all readers: this book may contain minor errors or gaps. Suggestions, criticism and comments are always welcome, so please do not hesitate to contact us about any relevant issue.

Aijaz Ahmad Wachkoo
Gulzar Ahmad Nayik
Jalal Uddin
Mohammad Javed Ansari

About the Editors

Aijaz Ahmad Wachkoo is currently posted as an Assistant Professor (Zoology) at Government Degree College Shopian, Jammu & Kashmir, India. He obtained his Master's Degree in Zoology from the University of Kashmir, Jammu & Kashmir, India, and his PhD from Punjabi University, Patiala, India. He has also been a DS Kothari Postdoctoral Fellow (PDF) at University of Kashmir, Jammu & Kashmir, India, besides having the distinction of qualifying in the National Eligibility Test (Council of Scientific & Industrial Research/University Grants Commission), the National Eligibility Test (Indian Council of Agricultural Research/Agricultural Scientists Recruitment Board), and the Graduate Aptitude Test in Engineering. He has also qualified at the Union Public Service Commission for the post of Assistant Professor. To date, he has discovered 31 new species of insects: mostly ants (27 species), 2 beetle species, 1 dipteran fly and 1 braconid wasp. He has also reported 33 insect taxa across Diptera, Hemiptera and Hymenoptera for the first time from India. To stabilize the nomenclature, he has designated neotypes and lectotypes of several species. He has proposed new combinations and raised many infraspecific taxa to full species status. Other than systematics, he is actively engaged in ecological and conservation studies. He has already published over 60 research articles, 1 book and 3 book chapters. He has also attended 8 international conferences and 35 national/regional conferences, symposia and workshops.

Gulzar Ahmad Nayik completed his Master's Degree in Food Technology from Islamic University of Science & Technology, Awantipora, Jammu & Kashmir, India, and his PhD from Sant Longowal Institute of Engineering & Technology, Longowal, Sangrur, Punjab, India. He has published over 90 peer-reviewed research and review papers, authored/co-authored 90 book chapters, and edited 13 books with Springer, Elsevier, Royal Society of Chemistry and Taylor & Francis. He has also delivered a number of presentations at various national and international conferences, seminars, workshops and webinars. He was shortlisted twice, in 2017 and 2018, for the prestigious Inspire-Faculty Award from the Indian National Science Academy, New Delhi, India. He was nominated for India's prestigious National Award (Indian National Science Academy Medal for Young Scientists 2019–20). He also fulfils the roles of editor, associate editor, assistant editor and reviewer for many food science and technology journals. He has received many awards, appreciations and recognitions and is a member of various international societies and organizations. He is currently editing several book projects for Elsevier, Taylor & Francis, Springer Nature and the Royal Society of Chemistry.

Jalal Uddin completed his Master's Degree in Chemistry from University of Karachi, Pakistan, and his PhD in Food Analytical Chemistry from University of Padua, Italy. He has also completed a postdoctoral qualification from Montana State University, USA. He is currently working as Assistant Professor in Pharmaceutical Analytical Chemistry in King Khalid University, Abha, Saudi Arabia. He has specialized in food safety, authenticity, and quality evaluation and also in nuclear magnetic resonance and the mass-based metabolomic/metabonomic approach in nutritional and epidemiological studies. He has published over 60 peer-reviewed research and review papers, and many book chapters, and has delivered a number of presentations at various national and international conferences, seminars, workshops and webinars. He manages the honey quality control laboratory in the Pharmaceutical Analytical Chemistry department, King Khalid University, Abha, Saudi Arabia.

Mohammad Javed Ansari is currently working as Assistant Professor in the Department of Botany, Hindu College Moradabad, Uttar Pradesh, India. He has also worked as Assistant Professor in the Department of Plant Protection, College of Food & Agricultural Sciences, King

Saud University, Riyadh, Saudi Arabia, from 2010 to 2017. He obtained his PhD in Biotechnology from Indian Institute of Technology, Roorkee, India, and his MSc in Botany from Ch. Charan Singh University, Meerut, India. He has published over 150 research/review articles in high impact factor journals. He has worked as Principal Investigator in many funded projects in King Saud University, Riyadh, Saudi Arabia. He has participated in many national and international conferences (Abu Dhabi, Belgium, Indonesia, Turkey, the USA, Italy, Romania, Germany, Spain, Cuba, Australia, Egypt, Ukraine, Ethiopia, Portugal). He is an editorial board member of many renowned journals and also a member of many international scientific associations and societies. He has supervised many students at master's level and doctorate level.

Contributors

Ulas Acaroz
Department of Food Hygiene and Technology
Faculty of Veterinary Medicine
Afyon Kocatepe University
Afyonkarahisar, Turkey
ACR Bio Food and Biochemistry Research & Development
Afyonkarahisar, Turkey

Hina Alim
Department of Life Sciences
University of Mumbai
Vidyanagari
Santacruz (East)
Mumbai, India

Mohammad Javed Ansari
Department of Botany
Hindu College Moradabad
Mahatma Jyotiba Phule Rohilkhand University
Bareilly, India

Uzmai Arifie
Krishi Vigyan Kendra/Extension Training Centre Malangpora Pulwama
Sher-e-Kashmir University of Agricultural Sciences and Technology
Srinagar, India

Damla Arslan-Acaroz
ACR Bio Food and Biochemistry Research & Development
Afyonkarahisar, Turkey
Department of Biochemistry
Faculty of Veterinary Medicine
Afyon Kocatepe University
Afyonkarahisar, Turkey

Aneela Ashiq
Department of Biochemistry
Khwaja Fareed University of Engineering and Information Technology
Rahim Yar Khan, Punjab, Pakistan

Syeda Tasmia Asma
Department of Food Hygiene and Technology
Faculty of Veterinary Medicine
Afyon Kocatepe University
Afyonkarahisar, Turkey

Meryem Bakour
Laboratory of Natural Substances, Pharmacology, Environment, Modeling, Health, and Quality of Life
Department of Biology
Faculty of Sciences Dhar Mehraz
Sidi Mohamed Ben Abdellah University
Fez, Morocco
The Higher Institute of Nursing Professions and Health Techniques
Fez, Morocco

Muhammad Amjad Bashir
Department of Plant Protection
Faculty of Agriculture Sciences
Ghazi University
Dera Ghazi Khan, Punjab, Pakistan

Umer Bin Farook
Krishi Vigyan Kendra/Extension Training Centre Malangpora Pulwama
Sher-e-Kashmir University of Agricultural Sciences and Technology
Srinagar, India

Otilia Bobiş
Department of Beekeeping and Sericulture
Faculty of Animal Science and Biotechnology
University of Agricultural Sciences and Veterinary Medicine Cluj-Napoca
Cluj-Napoca, Romania

Imran Bodlah
Department of Entomology
Pir Mehr Ali Shah Arid Agriculture University
Rawalpindi, Pakistan

Muhammad Adnan Bodlah
Fareed Biodiversity Conservation Centre
Department of Agricultural Engineering
Khwaja Fareed University of Engineering and Information Technology
Rahim Yar Khan, Punjab, Pakistan

Tridip Boruah
P.G. Department of Botany
Madhab Choudhury College
Barpeta, Assam, India

Shahnawaz Ahmad Dar
Krishi Vigyan Kendra/Extension Training Centre Malangpora Pulwama
Sher-e-Kashmir University of Agricultural Sciences and Technology
Srinagar, India

Asmae El Ghouizi
Laboratory of Natural Substances, Pharmacology, Environment, Modeling, Health, and Quality of Life
Department of Biology, Faculty of Sciences Dhar Mehraz
Sidi Mohamed Ben Abdellah University
Fez, Morocco

Hamed A. Ghramh
Unit of Bee Research and Honey Production
King Khalid University
Abha, Saudi Arabia

Ammara Gul-E-Fareen
Department of Entomology
Pir Mehr Ali Shah Arid Agriculture University
Rawalpindi, Pakistan

Ali Hamza
Department of Food Science & Technology
Muhammad Nawaz Shareef University of Agriculture
Multan, Pakistan

Shabbir Hussain
Institute of Chemistry
Khwaja Fareed University of Engineering and Information Technology
Rahim Yar Khan, Pakistan

Javid Iqbal
Department of Plant Protection
College of Food and Agriculture Sciences
King Saud University
Riyadh, Saudi Arabia

Ruqqaya Javid
Division of Entomology
Sher-e-Kashmir University of Agricultural Sciences and Technology
Srinagar, India

Md. Fahad Jubayer
Department of Food Engineering and Technology
Sylhet Agricultural University
Sylhet, Bangladesh

Md. Shahidullah Kayshar
Department of Food Engineering and Technology
Sylhet Agricultural University
Sylhet, Bangladesh

Nadeya Khaliq
High Mountain Arid Agricultural Research Institute
Leh, Ladakh

Khalid Ali Khan
Unit of Bee Research and Honey Production
King Khalid University
Abha, Saudi Arabia

Shumaila Khan
Fareed Biodiversity Conservation Centre
Department of Agricultural Engineering
Khwaja Fareed University of Engineering and Information Technology
Rahim Yar Khan, Punjab, Pakistan

Sevgi Kolayli
Department of Chemistry
Karadeniz Technical University
Trabzon, Turkey

Anna Kurek-Górecka
Department of Community Pharmacy
Faculty of Pharmaceutical Sciences in Sosnowiec
Medical University of Silesia in Katowice
Sosnowiec, Poland

Hassan Laaroussi
Laboratory of Natural Substances, Pharmacology, Environment, Modeling, Health, and Quality of Life
Department of Biology
Faculty of Sciences Dhar Mehraz
Sidi Mohamed Ben Abdellah University
Fez, Morocco

Badiaa Lyoussi
Laboratory of Natural Substances, Pharmacology, Environment, Modeling, Health, and Quality of Life
Department of Biology
Faculty of Sciences Dhar Mehraz
Sidi Mohamed Ben Abdellah University
Fez, Morocco

A.R. Malik
Division of Fruit Science
Sher-e-Kashmir University of Agricultural Sciences and Technology
Srinagar, India

Mujahid Manzoor
Faculty of Agricultural Sciences
University of the Punjab
Lahore, Pakistan

Md. Anisur Rahman Mazumder
Department of Food Technology and Rural Industries
Bangladesh Agricultural University
Mymeningh, Bangladesh

Alishbah Mohsin
Fareed Biodiversity Conservation Centre
Department of Agricultural Engineering
Khwaja Fareed University of Engineering and Information Technology
Rahim Yar Khan, Punjab, Pakistan

Shamas Murtaza
Department of Food Science & Technology
Muhammad Nawaz Shareef University of Agriculture
Multan, Pakistan

Hammad Naeem
Department of Food Science & Technology
Muhammad Nawaz Shareef University of Agriculture
Multan, Pakistan

Vikas Nanda
Department of Food Engineering & Technology
Sant Longowal Institute of Engineering and Technology
Longowal, Sangrur, Punjab, India

Gulzar Ahmad Nayik
Department of Food Science & Technology
Government Degree College Shopian
Jammu & Kashmir, India

Noor-ul-Huda
Department of Food Science & Technology
Muhammad Nawaz Shareef University of Agriculture
Multan, Pakistan

Paweł Olczyk
Department of Community Pharmacy
Faculty of Pharmaceutical Sciences in Sosnowiec
Medical University of Silesia in Katowice
Sosnowiec, Poland

Driss Ousaaid
Laboratory of Natural Substances, Pharmacology, Environment, Modeling, Health, and Quality of Life
Department of Biology
Faculty of Sciences Dhar Mehraz
Sidi Mohamed Ben Abdellah University
Fez, Morocco

Bilal Ahmad Pandit
Division of Fruit Science
Sher-e-Kashmir University of Agricultural Sciences and Technology
Srinagar, India

Jessica Pandohee
Sydney Mass Spectrometry
University of Sydney
Camperdown, New South Wales, Australia

Dayanand Peter
Department of Food Processing Technology
School of Agriculture and Bioscience
Karunya Institute of Technology and Sciences
Coimbatore, Tamil Nadu, India

Afzal Rahman
Department of Food Technology and Rural Industries
Bangladesh Agricultural University
Mymeningh, Bangladesh

Thottiam Vasudevan Ranganathan
Department of Food Processing Technology
School of Agriculture and Bioscience
Karunya Institute of Technology and Sciences
Coimbatore, Tamil Nadu, India

Tariq Rasheed
Department of Entomology
Pir Mehr Ali Shah Arid Agriculture University
Rawalpindi, Pakistan

Narimane Segueni
Faculty of Medicine
University Salah Boubnider Constantine
Constantine, Algeria
Department of Chemistry
University Mentouri-Constantine
Constantine, Algeria

Syed Rizwan Ali Shah
Department of Animal Nutrition and Nutritional Diseases
Faculty of Veterinary Medicine
Afyon Kocatepe University
Afyonkarahisar, Turkey

Muhammad Shahbaz
Department of Food Science & Technology
Muhammad Nawaz Shareef University of Agriculture
Multan, Pakistan

Rattan Singh
Department of Food Technology
Guru Jambeshwar University of Science & Technology
Hisar, Haryana, India

Soumaya Touzani
Laboratory of Natural Substances, Pharmacology, Environment, Modeling, Health, and Quality of Life
Department of Biology
Faculty of Sciences Dhar Mehraz
Sidi Mohamed Ben Abdellah University
Fez, Morocco

Jalal Uddin
Department of Pharmaceutical Chemistry
College of Pharmacy
King Khalid University
Asir, Saudi Arabia

Aijaz Ahmad Wachkoo
Department of Zoology
Government Degree College
Shopian, Jammu & Kashmir, India

Sajad H. Wani
Department of Biotechnology
Government Degree College
Shopian, Jammu & Kashmir, India

Ayesha Younas
Fareed Biodiversity Conservation Centre
Department of Agricultural Engineering
Khwaja Fareed University of Engineering and Information Technology
Rahim Yar Khan, Punjab, Pakistan

1 The Origin and Evolution of Beekeeping

Syeda Tasmia Asma, Ulas Acaroz, Hina Alim and Jessica Pandohee

1.1 INTRODUCTION

Insects are the most diversified group of organisms and form a major part of the Earth's living biomass (Bar-On et al., 2018). Considering their historical profusion and distribution, as well as numerous cases of remarkable resistance to natural or purposeful extinction, certain insects have historically been regarded as the greatest survivors among most apocalyptic circumstances. Plant pollination is one of the ecosystem functions that have been dependent on insects as the primary carriers of pollen for flowering plant species. The most significant group is a lineage that contains over 20,000 known species, belonging to the superfamily Apoidea: bees (Zattara and Aizen, 2021). *Osmiini*, *Lasioglossum*, *Andrena*, and *Apis* bees are well-known pollinators to a range of species (Ballantyne et al., 2017; Culliney, 1983). As the reproduction of the large diversity of plant populations is dependent on the richness of bee populations, the distribution of these pollinators is crucial for plant survival. Bees are also of great importance for human development; our interactions have been traced back to ancient human history. Beeswaxes, honey and minute quantities of royal jelly were of great importance in trade for our ancestors, not limited to household purposes but also of regional economic importance (Michener, 2000). The economic value of beeswax and honey prompted practical endeavors to study bees' behavior and social structure, with the highly concrete objective of enhancing the ease and efficacy with which these precious resources might be harvested. Maintaining a pure bee culture to achieve commodities of enhanced quality is of great advantage for regional prosperity. Hence, to meet the increasing demand, our ancestors used to practice the science of beekeeping, also known as apiculture (bees are denoted as *apis* in Latin). The initial attraction for beekeeping was focused on one bee species, *Apis mellifera*, called the honey bee today. This was due to *Apis mellifera*'s role in the pollination of diverse plant species (Hung et al., 2018). In recent years, scientists have observed a global decline of bee species every year since the 1990s, as shown in data provided by the Global Biodiversity Information Facility (GBIF). There has been a 25% fall in the number of bee species in 2006 and 2015 in comparison to the year 1990 (Zattara and Aizen, 2021). This alarming decline has motivated beekeepers to advance their technology and equipment to scale up the bee population and prevent the overexploitation of existing bee species. In this chapter, the history of beekeeping and advancements in beekeeping are extensively discussed.

1.2 HISTORY OF BEEKEEPING

Beekeeping is an intrinsic human activity, which dates back more than 10,000 years to the Paleolithic period (Eroğlu and Yuksel, 2020). In the 1980s, scientists believed that honey bees were an exemplar for inculcating the most tangible programming found in nature (Gould, 2002). Beekeeping goes back a long way and predates human civilization. It has been utilized for various objectives due to the medicinal properties of bee products. Honey hunting—the opportunistic theft of honey from

DOI: 10.1201/9781003354116-1

wild honey bee nests—was the foundation of beekeeping (Sanford and Tew, 2004). In the Middle Ages, simple structures were used as hives; however, honey was mainly consumed from natural and wild hives (Graham et al., 1992). True beekeeping started when humans began to provide artificial holes for the bees to build comb in order for the queen to lay her eggs and the workers to process honey. The Sumerians, who lived in Mesopotamia in approximately 3000 BCE, were also known to use honey as a remedy. Dry honey aged 3200 years has been discovered in the graves of Egyptian pharaohs, and records from plaques several thousand years old indicate that in those times, Egyptians utilized honey for sustenance, medicinal, and traditional practices. Supporting information about beekeeping may be found in hieroglyphs and pictures dating back to the times of the Hittites and Sumerians from North Africa, Europe, the Middle East, and Asia. Honey was also used extensively in Etruscan and Roman food and medicine (Fontana and Fontana, 2020). By around 2000 BCE, the Egyptians had established advanced beekeeping. They subsequently shared their innovations with the rest of the Mediterranean over the next two millennia. Honey and wax were key commodities for commerce during the Middle Ages in Europe, and beekeeping in skep, log, box, and tree hives grew to supply the need. Beginning in approximately 300 BCE, other honey bee species contributed to the establishment and spread of beekeeping in Asia. Meanwhile, with the stingless bee *Melipona beecheii*, beekeeping evolved independently throughout Mesoamerica, as evidenced by archaeological findings and literary descriptions that survived the Spanish invasion (Kritsky, 2017).

In South Africa, honey was collected largely from natural hives. Following the discovery of the first modern hive, honey gathering evolved into an animal-producing activity managed by human hands. South Africa is still well known for the presence of a high population of wild honey bee colonies (Grout, 1949). Ceramic beehives are categorized according to their layout; this can be either vertical or horizontal. Horizontal beehives are tubular vessels, commonly used in antiquity in the Mediterranean region. The world's most ancient horizontal beehive, from the tenth century BC, was discovered in Israel (Topal et al., 2021).

Stories from Sumerian-Akkadian scriptures portray the adoration of the "bee goddess," and the bee appears on seals of the Egyptian pharaohs; the honey bee was regarded as a holy insect by the Sumerians and Egyptians. It is thought that the Egyptians believed that bees originated from the tears of the deity Ra; they are claimed to have served as symbolic representations of the Egyptian state for thousands of years, alongside water and the common reed (*Phragmites australis*). Aristaeus was a demi-god in Mediterranean mythology who taught humanity beekeeping, whereas Aphrodite and Bacchus were associated with honey bees in mythology. Common uses and ways to consume honey include use in religious rites, in cooking, and as an essential ointment. In the kitchen, honey was combined with olive oil to form a marinade to enhance the flavor of fried meat. Honey was utilized by the Persians in sacrificial rites to the deities. Honey was commonly utilized in marinating meat, sauces, dishes, and sweets in the classical Greek and Roman eras (Topal et al., 2021). A shrine to the goddess of fertility and other cultic items near the hives symbolized the importance and the link between Israelite cultic rituals, bees, honey, and beeswax (Topal et al., 2021).

Anton Jana (1734–1773) was a beekeeping pioneer and one of the world's foremost bee specialists. Maria Theresa, the Holy Roman Empress, appointed him as a professor at the Vienna Beekeeping School, and his lessons were followed by the state beekeeping teachers. Anton Jana was an artist who painted the frontal panels of the beehive. The origins of this folk art may be traced back to the mid-nineteenth century (Topal et al., 2021).

1.3 RECENT ADVANCEMENTS IN BEEKEEPING EQUIPMENT

Beekeeping practices originated long ago in ancient Greece and Egypt. Hives, gloves, protective clothing, hive tools, and smokers were the traditionally used basic equipment for beekeeping. Hives are manmade constructions used for the maintenance of honey bee colonies. Over time, different types of hives have been invented. Nowadays, many beekeepers are using Langstroth or mobile

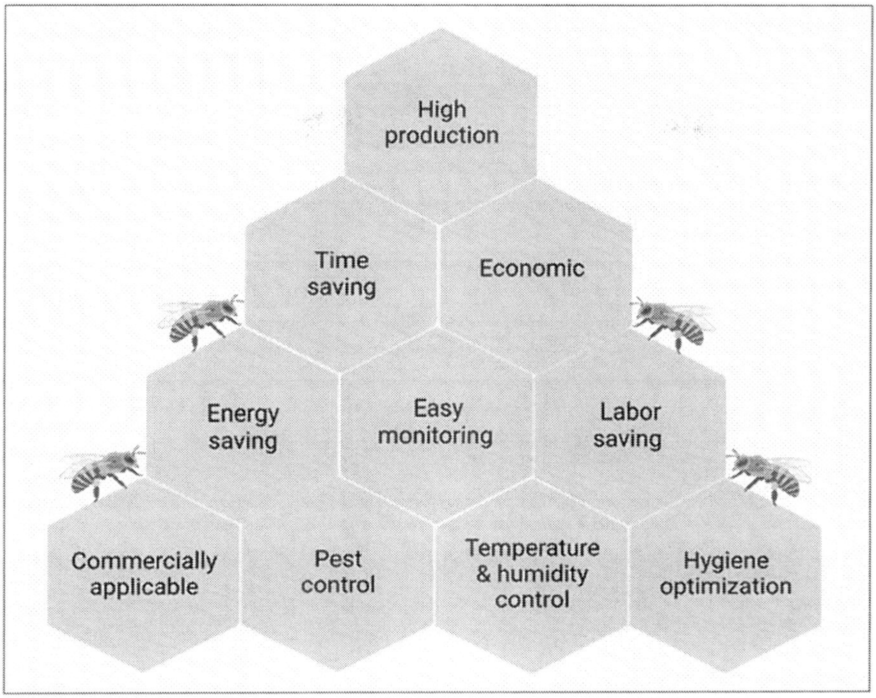

FIGURE 1.1 Advantages of modern beekeeping over traditional beekeeping.

frames. The progression of modern or advanced beekeeping originated with the development of the Langstroth beehive by a beekeeper, Langstroth, from the USA. Langstroth's beehive design was based on portable and inter-changeable bee frames. Since 1861, Langstroth's beehive has become popular in the USA and European countries (Gupta et al., 2014).

One of the significant benefits of modern beekeeping is the invention of mobile frames. Mobile apiary mediates the process of pollination and enhances the production of honey. Women and even children can also handle them due to their lightweight (Yap and Devlin, 2015). Modern beekeeping is rapidly spreading worldwide due to the invention of mobile beekeeping. Modern beekeeping provides the following advantages over traditional beekeeping (Figure 1.1):

1. Hive delivery to fields for pollen collection saves the energy and time of honey bees for producing honey.
2. A higher production rate can be achieved (Ignjatijević, Milojević, and Andžić, 2018).
3. Better-quality honey can be obtained.
4. Frames can be easily inspected.
5. The disease rate is lower.

The developments in beekeeping equipment and techniques have influenced beekeeping practices worldwide (Table 1.1). Mobile apiaries have ended the traditionally used beekeeping practices, allowing beekeepers to harvest honey throughout all seasons of the year (Kang et al., 2017). High expenses, the need for highly skilled beekeepers, lack of wax, predators, and pests are the few manageable shortcomings of modern beekeeping (Jebesa, 2017).

Franz introduced honey-picking equipment based on a centrifugal mechanism for honey extraction without any damage to the nest (Jung, 2014). This honey extractor was based on a simple configuration, was easy to handle, and enabled the beekeepers to extract honey without causing

TABLE 1.1
Some Developments in Beekeeping Equipment and Techniques in Recent Years

Objective	Method/equipment	Description	Reference
Beehive monitoring	Vibration and radar sensors	A piezoelectric transducer inside the beehive and doppler radar outside the hive were used to monitor bee activity and swarming.	(Aumann, Aumann, and Emanetoglu, 2021)
	Image classification models	The models were used for classifying the beehive images to accurately detect any abnormal conditions with a high accuracy rate of about 99.07%.	(Berkaya, Gunal, and Gunal, 2021)
	Remote-controlled wireless weight sensor	A set of four sensors was customized for beehive monitoring.	(Bratek and Dziurdzia, 2021)
	Microclimatic and auditory monitoring	An auditory approach was used to monitor the sound of honey bees during swarming, and the microclimatic approach was used to monitor the temperature, humidity, and weight of beehives.	(Anand et al., 2018)
	Automated control and data-acquisition equipment	Used to monitor temperature, weight, and acoustic sounds produced by bees. Can also detect swarming.	(Rybin et al., 2017)
	Sensor	A wireless sensor-based monitoring system was used to analyze the stress conditions in beehives that can lead to absconding.	(Kridi, de Carvalho, and Gomes, 2016)
	BeePi (multi-sensor monitoring)	An electronic multi-sensor monitoring device was used to monitor the beehive's health and production status by tracking continuous datasets of weather and beehive images, audios, and videos.	(Kulyukin and Reka, 2016)
	Acoustic sensor	Analyzed the operational productivity by monitoring the flying sounds of bees near the entrance of the beehive.	(Brundage, 2012)
Queen bee detection	MFCC-based descriptor	The mel-frequency cepstrum coefficients-based descriptor can distinguish the presence or absence of the queen bee.	(Soares et al., 2022)
	Sounds analysis	Hardware-based infrastructures were employed for monitoring the presence of the queen bee.	(Cejrowski et al., 2018)
	Radio frequency identification–based detection device	Aimed to detect the presence of the queen bee and also to locate the queen bee in the beehive.	(Berry, 2011)
Temperature and humidity monitoring	Sensor	A system was used to control the microclimate of the beehive by monitoring the inside temperature and humidity.	(Ghamdi and Al Khazim, 2016)
	Sensor	The sensor development was based on a moisture and climate control system that can monitor humidity and temperature in the beehive. The electronic ventilation fan and the heating element were also connected with the sensor system to optimize the humidity and temperature.	(Bulanyy, 2017)
Pest detection and control	Anti-predator device	This device was invented particularly to provide anti-predator activity with enhanced defense against different pests, such as predator bees and wasps.	(Atkinson, 2022)
	Var-Gor system	The smart technology–based Var-Gor system was designed to detect varroa mites by imaging and monitoring beehives.	(Sevin, Tutun, and Mutlu, 2021)
	Airborne ultrasound system	High-frequency airborne ultrasound (without chemicals) was used for controlling varroa mites in beehives.	(Barry et al., 2018)
	Small hive beetle trap	An attractant and a trap system were used in beehives to destroy small hive beetles.	(Teal et al., 2011)

any damage to the bee frames. It is still in active use by numerous beekeepers worldwide and has undergone modifications, such as the switch from crank handling to electrical handling. This modification progressively enhanced the productivity of beekeepers (Choi et al., 2021). Some other beekeeping equipment, including electronic honey capping technology, smart beehives, unmanned remote control harvesters for honey extraction, and beeswax separators, has been designed and employed with several kinds of beekeeping hives (Muthusamy et al., 2021) (Archibong, 2021). Muthusamy et al. designed and developed drone-based remote-controlled technology for honey extraction. Remote-controlled harvesting technology replaced the traditionally used manual harvesting of honey. Manual honey extraction is not as preferable an approach due to several limitations, such as labor budgets, time and energy conservation, and difficulty of access to higher places. The invention of remote-controlled machines for honey extraction has significantly reduced or overcome the shortcomings of manual extraction. Remote-controlled extraction enables extractors to easily collect honey from inaccessible areas, such as mountains or tall trees, in less time. As the manual collection of honey from hills or higher places was not easy or safe, this technology not only ensured the safety of the extractor but also saved the time and energy of the personnel involved (Muthusamy et al., 2021).

Ajamian introduced a manmade device named a *drone bee* for pollination. Bees perform pollination naturally by shifting pollen from the anthers of a plant to the stigma. Weather issues such as gales, storms, and rain are unfavorable for natural pollination. Ignored/missed plant fields may also disturb production. Better consistency, improved accuracy, and precision are desired features to enhance agricultural yield and bring financial paybacks to farmers or beekeepers. Over recent decades, researchers have observed an evident decline in bee pollination in North America. The collapse or drop in bee colonies adversely affects pollination and may cause the extinction of endangered bee species. The manmade drone bee was designed to overcome the shortcomings of natural bee pollination. This device was designed and equipped to fly over the desired plants to execute the process of pollination similarly to a bee. It was prepared with pattern identification ability to recognize a designated type of crop or plant and transfer pollen from the anthers to the stigma, executing pollination in a strategic time and area (Ajamian, 2022).

1.4 BEEHIVE MONITORING

Several factors can affect the productivity of beehives, such as temperature and humidity inside the beehive. Optimization and monitoring of these factors are essential to obtain an ensured optimal yield. Several types of devices have been designed to facilitate hive monitoring. Beekeepers typically practiced physical inspection for monitoring beehives and perceiving production. Physical monitoring has the limitation of allowing only current snapshots. It was not easy to entirely understand the progression of problems associated with isolated assessments. Another technique employed for counting bees inside the hive was using sensors. The use of sensors also has some drawbacks; for example, it can determine the passage of the bees along a specific track only, and the bee's direction cannot be followed. However, recent advancements in beehive monitoring tools have overcome the problems associated with typical or traditional monitoring.

Engel introduced a system to continuously monitor the weight of beehives (Engel, 2014).

Chapa designed a monitoring device to track and screen the conditions inside beehives, such as activity, production, and health status. This system can provide information about the internal beehive conditions along with the related environmental conditions. The obtained values were processed via the control organization and compared with the expected baseline figures. Information was communicated by the control organization to the remote-based device to inform the beekeeper about the beehive's current and previous health and production status (Chapa, 2018).

Yildirim introduced demounted beehives with improved air circulation and heat isolation. Improper ventilation and heat problems are not favorable for bees' health and activity. In addition, beehives contain a large space and are not easy to transport; the invention of plastic beehives has overcome this issue (Yildirim, 2013). Schmitz introduced beehives with a movable entrance at the top. This invention can significantly reduce blockages of the entrance by growing vegetation, dead honey bees, and snow, among other things (Schmitz, 2002).

Qandour et al. (2014) developed a remote-controlled beehive monitoring device based on acoustic signals. Several studies have indicated that honey bee health and activity can be tracked and evaluated via acoustic determination. This system was designed to determine the presence of any pest invasion in honey bee colonies by making a comparative analysis of the acoustic hive fingerprint with a known fingerprint. It will facilitate an efficient monitoring program with a considerable reduction in labor, cost, and time spent organizing monitoring programs (Qandour et al., 2014).

Tashakkori et al. (2021) prepared a beehive monitoring device named *beemon*. Using the MQ-Telemetry Transport procedure, this device was designed to record the weight, humidity, and temperature data in the Thingsboard control panel (Tashakkori, Hamza, and Crawford, 2021). de Souza et al. introduced an electromechanical device for maintaining hygiene, humidity, and temperature conditions inside beehives to combat pathogens such as bacteria, fungi, molds, viruses, and mites that can cause detrimental effects to bees. The air convection system was used to dehumidify, sanitize, and warm up the circulating air inside beehives (de Souza, Sampaio, and Buzzatti, 2010).

1.5 PEST CONTROL

Sinanis and Lambropoulos introduced a frame and comb base to beehives for controlling and monitoring varroa mites. Varroa mites threaten the honey bee's survival and adversely affect the production rate by invading the hive's brood and repeating the cycle in young honey bees. This invention aimed to offer a significant, cost-effective, commercially available device for eliminating varroa mites (Sinanis and Lambropoulos, 2012). Mudd introduced an apparatus for controlling the entrance of varroa mites with the movement of honey bees in and out of the beehive entrance. The objective behind this invention was to improve the hive's health status and honey yield by reducing mite infestation and deformities in young bees. This mite control apparatus also provided some other advantages, such as ease of handling due to lighter weight, easy cleaning, cost-effectiveness, affordability for commercial beekeeping without requiring any chemicals or electricity, and more importantly, eco-friendliness (Mudd, 2008).

Brisson and Brisson designed a beehive equipped with a bottom board for detecting, controlling, and decreasing infestations by parasites of bees in the beehive. This apparatus can be used to reduce the parasitic infestations of honey bees (Brisson and Brisson, 2007).

Siefert et al. designed a device to observe parasitic infestations and separate the varroa mites from the honey bee population. A container configured with a basket containing several openings at the side wall and bottom of the basket and a closing lid on the container was the basis of the device. This invention was intended to measure the degree of varroa mite infestation in honey bee colonies (Siefert et al., 2021).

Bramante and Jain recently introduced an effective way to control parasite infestation in beehives. They designed a thymol-coated beehive entrance, allowing the honey bees to enter or exit the hive with the consistent discharge of thymol varroacide for prolonged time periods to generate miticidal effects (Bramante and Jain, 2022).

1.6 CONCLUSION

Bees may be found in practically all civilizations, in food, medicines, religious activities, cosmetics, and industrial items. Bee products are a valuable source of nourishment for human beings and are also frequently employed in cosmetics and apitherapy. Studies on the health benefits of bee products have recently gained popularity. Apitherapy, a type of alternative medicine, uses various bee products as therapeutic agents to prevent and slow the progression of illnesses, particularly in the treatment of Parkinson's disease. The hive air contains isoprenoids, carotenoids, terpenes, essential oils, phytohormones, volatile wax components, high-value alcohols, and aerosols that have been shown to be useful against respiratory tract illnesses.

In today's world, where there is a rising demand for natural living and natural goods, api-tourism is gaining popularity. Thus, while countries are conveying their beekeeping history, they can also showcase their natural beauty, customs, and cultural acquisitions. However, with the extinction of bee species, it is our responsibility to pass on the vestiges of the past to future generations by maintaining bees in a healthy manner using contemporary technologies. To satisfy current expectations, it is recommended to increase the contribution of beekeeping to commercial activity and economic potential. Moreover, it will be necessary to evaluate the risk of bees' extinction and to develop technology and methods for their preservation.

REFERENCES

Ajamian, Arnaud Z. 2022. *Autonomous Drone Bees*. Google Patents.

Al Ghamdi, and Ahmad Al Khazim. 2016. *Local Bee Hive with System to Control the Humidity and Temperature Inside It*. Google Patents.

Anand, Nirmal, V. Bharath Raj, M.S. Ullas, and Ashutosh Srivastava. 2018. "Swarm detection and beehive monitoring system using auditory and microclimatic analysis." 2018 3rd International Conference on Circuits, Control, Communication and Computing (I4C).

Archibong, Friday Nwankwo. 2021. "Design, construction and performance evaluation of a honey extracting Machine." *Agricultural Engineering International: CIGR Journal* 23(3), 279–289.

Atkinson, Kyle Blair Wayne. 2022. *Anti-Predator Device for Beehive*. Google Patents.

Aumann, Herbert M., Margery K. Aumann, and Nuri W. Emanetoglu. 2021. "Janus: A combined radar and vibration sensor for beehive monitoring." *IEEE Sensors Letters* 5(3):1–4.

Ballantyne, G., K.C. Baldock, L. Rendell, and P.G. Willmer. 2017. "Pollinator importance networks illustrate the crucial value of bees in a highly speciose plant community." *Scientific Reports* 7(1):1–13.

Bar-On, Y.M., R. Phillips, and R. Milo. 2018. "The biomass distribution on Earth." *Proceedings of the National Academy of Sciences* 115(25):6506–6511.

Barry, Brendan C., Lindy Verstraten, Fidelma T. Butler, Padraig M. Whelan, and William M.D. Wright. 2018. "The use of airborne ultrasound for Varroa destructor mite control in beehives." 2018 IEEE International Ultrasonics Symposium (IUS).

Berkaya, Selcan Kaplan, Efnan Sora Gunal, and Serkan Gunal. 2021. "Deep learning-based classification models for beehive monitoring." *Ecological Informatics* 64:101353.

Berry, Jennifer. 2011. *System to Detect the Presence of a Queen Bee in a Hive*. Google Patents.

Bramante, Andrew, and Raina Jain. 2022. *Miticidal Beehive Entranceway*. Google Patents.

Bratek, Piotr, and Piotr Dziurdzia. 2021. "Energy-efficient wireless weight sensor for remote beehive monitoring." *Sensors* 21(18):6032.

Brisson, Janet, and Michael Brisson. 2007. *Beehive Combination Bottom Board for Detection, Monitoring and/or Reduction of Parasite Infestations*. Google Patents.

Brundage, Trenton J. 2012. *Acoustic Sensor for Beehive Monitoring*. Google Patents.

Bulanyy, Igor. 2017. *System for Controlling Climate and Moisture in Beehive*. Google Patents.

Cejrowski, Tymoteusz, Julian Szymański, Higinio Mora, and David Gil. 2018. "Detection of the bee queen presence using sound analysis." Asian Conference on Intelligent Information and Database Systems.

Chapa, Mario. 2018. *Beehive Monitoring System*. Google Patents.

Choi, Hong-Min, Kim, Se-Gun, Kim, Hyo-Young, Woo, Soon-Ok, Kim, Seon-Mi, Moon, Hyo-Jung, and Han, Sang-Mi. 2021. "Comparative evaluation of productivity increase through improvement of royal jelly production technique." *Journal of Apiculture*, 36(3): 221–228.

Culliney, T.W. 1983. "Origin and evolutionary history of the HoneybeesApis." *Bee World* 64(1):29–38.

de Souza, Marcio Junqueira, Marcelo Regattieri Sampaio, and Marlo Golin Buzzatti. 2010. *Device for Hygienizing, Warming, and Dehumidifying a Beehive*. Google Patents.

Engel, Stephen Adam. 2014. *Apparatus for Continuous Weight Monitoring of Beehives*. Google Patents.

Eroglu, O., and S. Yuksel. 2020. "Historical development and current status of beekeeping in Turkey and the world." *Atlas Journal* 6(27):345–354

Fontana, C., and P. Fontana. 2020. "The first evidence." In I. Floris (Ed.), *Italian Apiculture: A Journey Through History and Honey Diversity; Part I*. Accademia Nazionale Italiana di Entomologia, 36–47.

Gould, J.L. 2002. "Can honey bees create cognitive maps." In Marc Bekoff, Colin Allen, and Gordon M. Burghardt (Eds), *The Cognitive Animal: Empirical and Theoretical Perspectives on Animal Cognition*, 41–45.

Graham, J.M., J.T. Ambrose, and L.L. Langstroth. 1992. *The Hive and the Honey Bee: A New Book on beekeeping Which Continues the Tradition of Langstroth on the Hive and the Honeybee*. Dadant and Sons, 1324.

Grout, R.A. 1949. *The Hive and Honey Bee*. Dadant and Sons.

Gupta, Rakesh Kumar, Wim Reybroeck, Johan W. van Veen, and Anuradha Gupta. 2014. *Beekeeping for Poverty Alleviation and Livelihood Security*. Springer Dordrecht Heidelberg.

Hung, K.L.J., J.M. Kingston, M. Albrecht, D.A. Holway, and J.R. Kohn. 2018. "The worldwide importance of honey bees as pollinators in natural habitats." *Proceedings of the Royal Society. Series B,: Biological Sciences* 285(1870):20172140.

Ignjatijević, Svetlana, Ivan Milojević, and Rosa Andžić. 2018. "Economic analysis of exporting Serbian honey." *International Food and Agribusiness Management Review* 21(7):929–944.

Jebesa, Sheleme Refera. 2017. "Assessment of factors affecting adoption of modern beehive in East Wolega Zone, Western Oromia." *International Journal of Engineering Research and Technology* 6(01):85–91.

Jung, Chuleui. 2014. "A note on the early publication of beekeeping of western honeybee, Apis mellifera in Korea: Yangbong Yoji (Abriss Bienenzucht) by P. Canisius Kugelgen." *Journal of Apiculture* 29(1):73–77.

Kang, D.Y., A. Seol, J. Oh, Y.K. Jung, H. Han, and J. Chung. 2017. "Analyzing the management characteristics of beekeeping households according to their beekeeping types." *Korean Journal of Apiculture* 32: 1–9.

Kridi, Douglas S., Carlos Giovanni N. de Carvalho, and Danielo G. Gomes. 2016. "Application of wireless sensor networks for beehive monitoring and in-hive thermal patterns detection." *Computers and Electronics in Agriculture* 127:221–235.

Kritsky, G. 2017. "Beekeeping from antiquity through the middle ages." *Annual Review of Entomology* 62:249–264.

Kulyukin, Vladimir A., and Sai Kiran Reka. 2016. "Toward sustainable electronic beehive monitoring: Algorithms for omnidirectional bee counting from images and harmonic analysis of buzzing signals." *Engineering Letters* 24(3):1–33.

Michener, C.D. 2000. *The Bees of the World* (Vol. 1). JHU Press.

Mudd, Karen Anne Wassmer. 2008. *Varroa Mites Control Entrance (VMCE)*. Google Patents.

Muthusamy, Senthil Kumar, S.V. Tresa Sangeetha, Aida Saud Al Busaidi, Khalid Sulaiman Al Moqbali, Muhannad Al Balushi, and Saba Ahmad Al Mawali. 2021. "Design and development of drone-based remote-controlled honey harvesting machine." 2021 Fifth International Conference on I-SMAC (IoT in Social, Mobile, Analytics and Cloud)(I-SMAC).

Qandour, Amro, Iftekhar Ahmad, Daryoush Habibi, and Mark Leppard. 2014. Remote beehive monitoring using acoustic signals. *Australian Acoustical Society* 42(3):204–209.

Rybin, Vyacheslav G., Denis N. Butusov, Timur I. Karimov, Dmitriy A. Belkin, and Maria N. Kozak. 2017. "Embedded data acquisition system for beehive monitoring." 2017 IEEE II International Conference on Control in Technical Systems (CTS).

Sanford, M.T., and J.E. Tew. 2004. "Apiculture (beekeeping)." In John L. Capinera (Eds), *Encyclopedia of Entomology*. Springer, 147–158.

Schmitz, Edmund P. 2002. *Beehive Movable Top Entrance*. Google Patents.

Sevin, Sedat, Hidayet Tutun, and Selime Mutlu. 2021. "Detection of Varroa mites from honey bee hives by smart technology Var-Gor: A hivemonitoring and image processing device." *Turkish Journal of Veterinary and Animal Sciences* 45(3):487–491.

Siefert, Benoit, Ludovic De Feraudy, Gael Charpentier, and Raphaele Massard. 2021. *Device for Separating Parasites*. Google Patents.

Sinanis, George, and George Lambropoulos. 2012. *Beehive Frame and Comb Foundation for Controlling Varroa Mites*. Google Patents.

Soares, Bianca Sousa, Jederson Sousa Luz, Valderlândia Francisca de Macêdo, Romuere Rodrigues Veloso e, Silva, Flávio Henrique Duarte de Araújo, and Deborah Maria Vieira Magalhães. 2022. "MFCC-based descriptor for bee queen presence detection." *Expert Systems with Applications* 201:117104.

Tashakkori, Rahman, Abdelbaset S. Hamza, and Michael B. Crawford. 2021. "Beemon: An IoT-based beehive monitoring system." *Computers and Electronics in Agriculture* 190:106427.

Teal, Peter E.A., Baldwyn Torto, James H. Tumlinson, and Drion G. Boucias. 2011. *In-Hive Trap and Lure for Control of the Small Hive Beetle, Aethina tumida*. Google Patents.

Topal, E., L. Adamchuk, I. Negri, M. Kösoğlu, G. Papa, M.S. Dârjan, M. Cornea-Cipcigan, and R. Mărgăoan. 2021. "Traces of honeybees, Api-tourism and beekeeping: From past to present." *Sustainability* 13(21):11659.

Yap, Nonita T., and John F. Devlin. 2015. "Beekeeping innovation for sustaining rural livelihoods. A success story." *International Journal of Innovation and Sustainable Development* 9(2):103–117.

Yildirim, Muzaffer. 2013. *Plastic Beehive*. Google Patents.

Zattara, E.E., and M.A. Aizen. 2021. "Worldwide occurrence records suggest a global decline in bee species richness." *One Earth* 4(1):114–123.

2 Honey Bee Social Structure

Khalid Ali Khan and Hamed A. Ghramh

2.1 THE COLONY AND ITS ORGANIZATION

Complex insect societies are categorized by multiple generations, in which offspring, members of the worker caste, stay inside the colonies and support the reproductive potential of their parents at the expense of their own reproduction. Therefore, the duration of reproduction was an essential factor in the evolution of social insects (Costa and Costa, 2006). Honey bees are highly social insects; they live in sizable, well-coordinated domestic units. Most solitary insects do not engage in the variety of complicated tasks that social insects do (Smith et al., 2008a). Social insects are highly developed invertebrates, and social insect colonies demonstrate complex group problem-solving under varying limitations (Winston, 1987; Charbonneau et al., 2013). In order to survive successfully in social colonies, honey bees have evolved a variety of activities, including division of labor, intricate nest construction, environmental control, communication, and defense (Robinson, 1992a; Smith et al., 2008a). Social insects are the most fascinating species on earth, especially honey bees due to their unique activities.

A bee colony normally comprises three different adult bee species: workers, drones, and a queen (Page and Peng, 2001). Together, thousands of worker bees construct nests, gather food, and care for the young (Streinzer et al., 2013). Concerning their level of adulthood, each member has a certain responsibility to complete (Streinzer et al., 2013). But, the entire colony must work together to survive and procreate. Individuals such as workers, drones, and queens cannot survive without the colony's support.

A colony typically has a single queen and a few hundred drones during the late spring and summer in addition to thousands of adult worker bees (Winston, 1987; Rangel and Fisher, 2019). The queen and workers keep the colony's social structure intact and rely on a strong communication network (Charbonneau et al., 2013). The actions required for the colony's existence are regulated by the exchange of chemical pheromones among members and communicative "dances." Worker bees' labor-related activities are mostly influenced by their age, although they also change depending on the demands of the colony (Winston, 1987). The queen, worker force size, and the quantity of food storage all influence reproduction and colony strength. The effectiveness of the colony decreases as its size grows, up to a maximum of 60,000 workers.

2.2 DIVISION OF LABOR

One of the topics in the study of animal performance that have received the most attention is the honey bee's division of labor (DOL). Even though studies have been undertaken since the 1800s from every aspect of biology, a dedicated experimental effort started in the 1930s and has continued to the present (Beekman et al., 2007; Smith et al., 2008b). Notably, tens of thousands of sterile female workers make up honey bee colonies, together with a few thousand seasonal haploid male drones and just one queen, capable of laying both fertilized and unfertilized eggs (Winston, 1987). In honey bees, castes with a distinct system of labor division serve as the greatest indicators of colony structure, as shown in Figure 2.1.

2.2.1 Honey Bee Queen

The well-being and production of the bee colony are significantly dependent on the queen, who is a crucial member of the colony. A queen's quality is often assessed by her reproductive

DOI: 10.1201/9781003354116-2

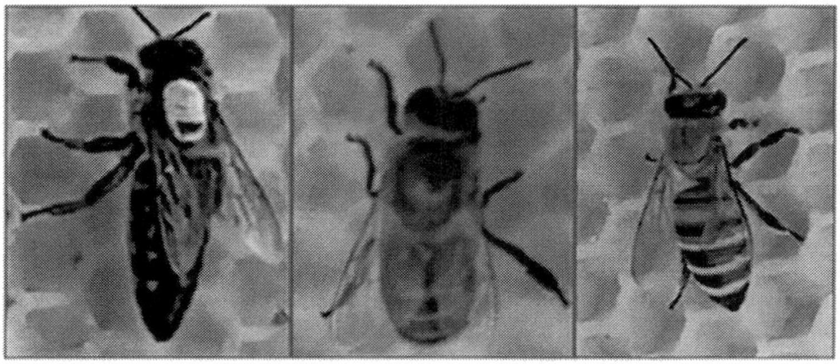

FIGURE 2.1 Caste and division of labor in bee colonies.

longevity, the amount of healthy offspring she might be able to produce, how many drones the queen has mated with, and the genetic variety of her mates (Amiri et al., 2017). Queen quality is correlated with a number of characteristics: overall weight (Szabo, 1974; Cassier et al, 1994; Tibor et al., 1987; Collins and Pettis, 2013), weight of ovaries (Gilley et al., 2003; Gregorc and Škerl, 2015), spermatheca weight, and the quantity of stored viable spermatozoa are all measured (Al-Lawati et al., 2009).

In addition, age and feeding materials such as royal jelly also affect the quality of the queen (Tarpy et al., 2000; Rangel et al., 2016). First instar larvae are preferred, although if the mother queen is suddenly misplaced, elder larvae may be raised into queens (Hatch et al., 1999; Tarpy et al., 2000). Genetic makeup, pesticide exposure, and parasite or pathogen load are additional factors that affect a queen's quality (Tarpy et al., 2012; Sandrock et al., 2014; Wu-Smart and Spivak, 2016).

Virgin queens fly on one to several occasions over several days about a week after they emerge (Winston, 1987). Approximately two weeks after emergence, drones that have reached sexual maturity mate with many honey bee queens, because honey bee queens are polyandrous (Moritz, 1989; Czekońska and Chuda-Mickiewicz, 2015; Czekońska et al., 2015). The queen will travel up to 3 kilometers during a nuptial trip to meet with several drones at a drone congregation area (DCA), which is situated 5 to 40 meters above the ground (Koeniger et al., 2014). According to earlier studies, queens typically mate with 12 drones (Winston, 1987; Kraus et al., 2005); however, a current study discovered that queens could mate with 34 to 77 (Withrow and Tarpy, 2018). The drone delivers his semen into the oviduct during copulation, everts his endophallus irrevocably into the female, and then falls to the earth (Woyke and Ruttner, 1958). The oviduct of the queen receives 10% of each male's ejaculation (Schlüns et al., 2005).

A queen begins accumulating sperm in her spermatheca, which is a unique organ found to enable spermatozoa storage, once she has finished her final nuptial flight and returned to the hive, and she also starts to lay eggs (Koeniger et al., 2014). The queen's spermatheca only holds about 3% to 5% of each drone's ejaculated spermatozoa for potential egg fertilization in the future (Schlüns et al., 2005; Koeniger et al., 2014). A queen's spermatheca can hold between five and six million spermatozoa (Koeniger and Koeniger, 2000). Queens are incredibly productive and fertilize each egg on average with two spermatozoa (Baer et al., 2016). Due to their parthenogenic nature, honey bees produce both fertilized and unfertilized eggs, which grow into haploid drones and female diploid workers or queens, respectively (Ellis et al., 2015). For queen egg laying, larger cells are allocated for drone eggs, while smaller cells are used for worker eggs (Ellis et al., 2015). The queen serves as the colony's main generative female and continuously secretes a mixture of pheromones that passively maintain the hive's social structure and other features of colony organization (Winston, 1987).

Furthermore, mating circumstances have an impact on queen reproduction (Keeling et al., 2003; Peso et al., 2013). A queen experiences a wide range of behavioral, physiological, and molecular changes when she mates with drones, including decreased sexual receptivity, altered ovarian development, ovulation, altered pheromone production, and transcriptional control. These modifications affect queen reproductive quality in ways that could have broad effects (Tanaka and Hartfelder, 2004; Kocher et al., 2008; Niño et al., 2013). Different studies using instrumental insemination have found that a significant portion of semen, including a variety of proteins and metabolites, triggers various post-mating alterations in queens and probably plays a significant role in determining queen quality (Kocher et al., 2009; Niño et al., 2012). The precise chemicals in semen and seminal liquid that start the post-mating modifications in the honey bee queens are still unknown. Additionally, due to their polyandrous nature, queens are more vulnerable to contracting sexually transmitted diseases like *Nosema* species or the Deformed Wing virus, which could endanger their fitness and general well-being (Yue et al., 2006; Amiri et al., 2016).

2.2.2 Drones

A drone in a honey bee colony is in charge of manufacturing semen and transferring it to the bee queen. In the course of drone growth, which lasts 24 or 25 days depending on the temperature of the rearing environment, spermatozoa are created (Czekońska et al., 2013). Spermatozoa are produced from the larval to pupal stages (Bishop, 1920). During the initial stages of a drone's imagined existence, spermatozoa are moved to the seminal vesicles (Woyke, 1958). Drones spend 10–12 days maturing in the nest after emerging. After 15–23 days (21 days on average), drones mate with the queen (Couvillon et al., 2010).

Drones are given royal jelly that has been pollen-enhanced during their ontogenetic development; the amount of pollen is adjusted to the age of the larvae when the royal jelly is prepared (Hrassnigg and Crailsheim, 2005). After emerging, workers continue to feed the young drones food that is high in protein and amino acids (Schmickl and Crailsheim, 2004). Additionally, during the first six days of their life, drones consume pollen, while older drones feed on carbohydrate-rich honey. In comparison to worker bee larvae, drone larvae require significantly more food with a more varied protein composition (Winston, 1987). Only colonies with abundant food resources can raise drones in large numbers (Hrassnigg and Crailsheim, 2005; Rueppell et al., 2006). Drones raised in higher pollen-accumulating colonies are highly energetic and start their flights earlier (Rueppell et al., 2006). During their sexual development, they don't need any more pollen nutrients (Stürup et al., 2013).

Drones exist in colonies in mild climates from May through August (Boes, 2010). Their proportion in a colony varies with the time of year, the number of individuals, the existence and age of the queen, the frequency of the drone brood, and the food availability (Boes, 2010; Taha and Al-Kahtani, 2013). During mating, some drones die, while the majority of drones, with an average lifespan of 30 days, die for some other reason, such as older age or biotic and abiotic factors (Wharton et al., 2007; Boes, 2010).

2.2.3 Honey Bee Worker

The majority of bees living in the colony are workers, who are the smallest bees. As immature females, they are unable to reproduce normally in the hive and do not lay eggs. The lifespan of a worker honey bee is 21 days: 3 days as an egg, the larval stage for 6 days, and 12 days for pre-pupa and pupa (Winston, 1987). The youngest bees, known as newly emerged bees, are typically involved in cleaning the nest and feed only on royal jelly. Nurse bees are involved in brood care and other nest events. After this, they perform various activities such as the building of comb and storing of honey, followed by activities outside the nest, including gardening and foraging (Yerushalmi et al., 2006).

Sterile diploid workers perform all other colony-related functions, whereas haploid males (drones) and queens tend to have reproductive responsibilities (Robinson, 1992b). Caste is a helpful concept to describe people who have a similar job repertoire when DOL is so ubiquitous that workers overlook most chores (Johnson, 2005). In social insects, there are two main types of DOL: between workers and queens, and among the workers for non-reproductive duties (Friedman et al., 2020). Sometimes, two organizational patterns for labor are seen in honeybees. Worker survival during the lean season is crucial during the winter, and bees become generalists. Labor division is employed to increase the growth rate and reserve accumulation in the spring and summer (Johnson, 2010). From the time they emerge as adults, workers go through a succession of activities and transformation responsibilities as they get older until they begin foraging, which is an age-related activity for workers, at around 21 days of age (Seeley, 1989). Due to the documented genetic, behavioral, and hormonal regulation of the pace of DOL in workers, individuals of similar ages can be observed executing a variety of jobs (Leoncini et al., 2004). The in-hive bees and foragers exhibit the most distinct behavioral distinctions in bees' complicated, age-dependent DOL. Outside the hive, foragers search for nectar and pollen for the entire colony, as opposed to nurse bees, who remain inside the hive and feed the larvae (Reim and Scheiner, 2014). In the colony, activities like caring for the brood, cleaning the cells, fanning, and foraging are carried out continuously. It has been questioned whether bees undertake these duties continuously or if they employ shift work techniques. Previous research looked into whether nurses took care of the brood using shift work or other comparable methods. According to Moore and co-researchers' (1998) marking and task performance observations of a bee (Moore et al., 1998), care for the brood is provided throughout the day without a specific timetable (Moore et al., 1998). In addition, researchers genotyped arriving pollen foragers, in contrast to caring for the brood, with microsatellite DNA markers for four days (Bernhard Kraus et al., 2011). This research implies that certain foragers for pollen go on foraging excursions in shifts, and that the genetic makeup of an individual may have an impact on this behavior (Bernhard Kraus et al., 2011). As was done for nursing, changing task arrangements for collecting pollen and possibly other tasks can be studied through direct observation.

2.3 COLONY INTEGRATION

A very important research question in the study of insect societies is how the conduct of individuals fits into a functioning colony (Chittka and Rossi, 2022). The DOL is one colony activity that relies on the fusion of worker behavior (Tanaka et al., 2022). Worker tasks alter due to changes in the environment and colony conditions, such as resource availability and population age demographics (Lee et al., 2022). It is crucial for colony organizations to be responsive to shifting circumstances within a structured labor system (Navas-Zuloaga et al., 2022).

How individual workers' tasks are integrated to allow colonies to grow and reproduce despite shifting internal and external environments is a key subject in insect sociobiology (Giraldo and Traniello, 2014). A high degree of colony integration is necessary to regulate the distribution of labor among workers based on age (Huang and Robinson, 1996). Honey bees typically spend their last three weeks of youth working in the nest before they die, but depending on how the colony or the environment is doing, they may accelerate, retard, or reverse this process (Robinson, 1992b). One of the greatest mysteries of a honey bee colony is how thousands of bees can cooperate so cohesively that the colony operates as a single, well-organized, and purposeful entity (Marshman, 2022). The ability of bees to precisely detect the demand for a certain worker activity is crucial for colony survival and reproduction, since the transition from nest chores to harvesting nectar needs intricate physiological changes (Akratanakul, 1986). As a result of variable individual birth and death rates, the colony age structure is one aspect of the ecosystem that changes throughout the course of the year (Hughes, 1984). For instance, the majority of honey bees in a colony in early spring are older bees that have survived the winter (Mattila et al., 2001). The functions often carried out by younger bees must now be handled by older bees. Fewer older bees are left to forage after they pass away a few weeks later (Huang et al., 1994), but an increasing number of younger bees must now take on an abnormally wide range of responsibilities, including precocious foraging. Reproductive colony splitting, predation on foragers, and brood illnesses are additional events that can change the age demographics of a colony (Robinson et al., 1989). The ability of honey bees to alter their regular age polyethism patterns according to colony needs is an example of the main cooperation between colony members that occurs in bee colonies in the absence of centralized control (Seeley, 1989; Anderson and McShea, 2001; Moritz and Southwick, 2012).

2.3.1 Chemical Communication and Labor Division Management

The influence of genetic diversity on the DOL among worker honey bees has been well documented (Smith et al., 2008b). Today, it is believed that honey bees still need their natural instincts, such as their capacity for communication and orientation. Individual communication is a key requirement for the integrity of swarms. Chemical communication is the primary source of communication (Trhlin and Rajchard, 2011), which is very helpful in maintaining colony integration.

The primary controlling force of colony operations is the honey bee queen. Pheromones, made by various glands and released as a multifaceted chemical brew called the "queen signal," play a significant role in this regulation (Bloch and Grozinger, 2011). The worker bees in the colony undergo numerous physiological and behavioral changes as a result of the queen signal, which primarily serves as a primer pheromone. By creating a social hierarchy and maintaining the queen's reproductive dominance, these alterations allow the colony to maintain homeostasis (Conte and Hefetz, 2008). The queen signal has three additional effects in addition to its primer effect: it directs the queen's retinue to form around her and calls on them to feed and groom her; and during the mating flights of immature premating queens, it attracts drones. Moreover, it keeps the swarm together while it is swarming (Bortolotti and Costa, 2014).

Results with queen mandibular pheromone (QMP) show that chemical communication can affect age polyethism (Vergoz et al., 2009). But, a signal coming from the workers acts as the main modulator of behavioral development in the workforce. This is due to the fact that modifying the age demography of colony workers while preserving the queen's consistent presence can significantly change age polyethism (Huang and Robinson, 1992). Additionally, there is little indication to suggest that queens' pheromone transmission or production varies other than in relation to colony density (Grozinger et al., 2014). The DOL is probably adjusted by QMP as an auxiliary inhibitor in response to changing colony needs such as eusociality. In lineages where eusociality has independently emerged, various forms of DOL have developed through alterations in the control of highly conserved molecular pathways linked to a number of fundamental life-history features, such as feeding, metabolism, and reproduction.

Eusociality has several benefits, one of which is the joint defense of the nest, young, and food reserves (Costa and Costa, 2006), as shown in Figure 2.2.

The biology of honey bees places a high value on nest defense. Pheromones can partially induce and control defensive behavior. These warning pheromones are produced by worker bees' mandibular glands and stinger (Pankiw, 2004). The Koschewnikow gland and the sting gland create the majority of the elements of the honey bee alarm pheromone (Cassier et al., 1994), and the recent finding that DOL is also influenced by genetics (Ben-Shahar et al., 2003; Lucas and Ben-Shahar, 2021; Hoekstra and Robinson, 2022) raises fresh concerns regarding how genetics and environment interact to integrate activity in insect societies. A key topic in all biological systems studies is the regulation of developmental plasticity, and analyzing the differences between developments in insect colonies and multicellular creatures may yield discoveries with widespread applications (Revely et al., 2021).

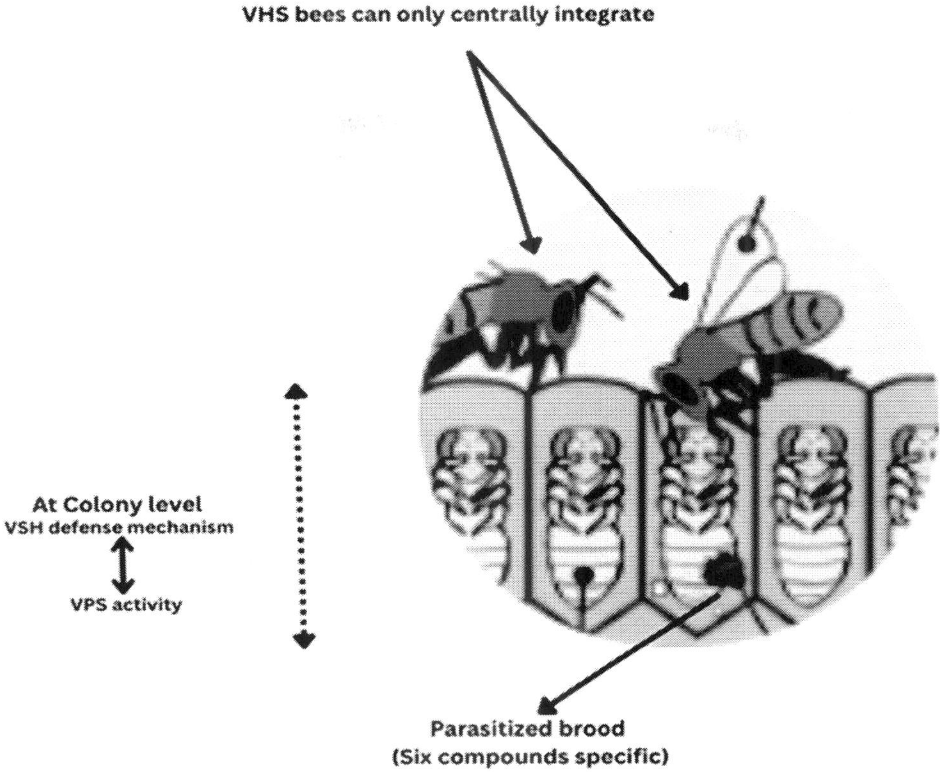

FIGURE 2.2 Chemical clues for the identification of parasitized broomechanism).

2.4 DAILY AND SEASONAL PATTERN OF ACTIVITY

Honey bees' ability to serve as pollinators depend on their ecology of feeding, which includes their foraging array, daily routines, and exploratory behavior (Young et al., 2021). Bees' pollen-foraging habits influence the fertility of the flowers they visit and the effectiveness of their colony's reproduction (Zych et al., 2013). Bees' foraging behavior varies depending on how much food is available at the flowers they visit (Lihoreau and Gómez-Moracho, 2022). Bees have a low frequency of foraging visits even in the best abiotic settings when there are not enough floral resources (Chen et al., 2022). A bee's choice of flower to visit may be significantly influenced by pollen supply (Su et al., 2022). Bumblebees are able to determine the quantity of pollen present in open flowers visually and regulate their feeding strategies accordingly (Xu et al., 2022). Pollinators can only collect a certain amount of pollen depending on an anther's dehiscence (Layek et al., 2022), while there are behavioral mechanisms by which some species can be active at night, for example, *Apis dorsata*.

According to behavioral data, *A. dorsata* foragers can only be active at night when there is some level of illumination. The moon must be at least half full, according to reports, for worker bees to continue foraging during the night (Dyer, 1985). Furthermore, even when the moon is below the horizon in cities, *A. dorsata* may be able to fly using the artificial sky glow produced by city lights (Young et al., 2021). In light of this, *A. dorsata*'s nocturnal foraging habits may serve as an example of how light can spur activity during a time when the species wouldn't ordinarily be active (Wcislo and Tierney, 2009).

The anticipated temperature changes over the following few decades will have an impact on the regional dissemination and action patterns of *Apis cerana* bees (Cui and Corlett, 2016). Low temperatures can stop bee activity for a portion of the year, especially close to the tropics' edge

(Corlett, 2011). According to Corlett et al. (2013), honey bees will adapt to climate variation more quickly than plants and may employ various climate factors as phenological signals, which could lead to future distribution and phenology mismatches between bees and plants (Burkle et al., 2013). On the other hand, it is anticipated that coexisting honey bee species, as well as other solitary and social bees in the Asian tropics, will protect the pollinator community from functional changes. It is necessary to understand how daily activity is influenced by temperature and other meteorological factors in order to forecast honey bee dispersion and activity in a warmer world. Studies on daytime and seasonal action patterns demonstrate that the weather and the availability of floral resources (nectar and pollen) impact the foraging behaviors of bees (Romiguier et al., 2016). It is challenging to distinguish between weather's direct and indirect effects on activity patterns because both impact flowering and nectar production.

A colony's activities change with the seasons. For a colony of honey bees, the months of September through December could be seen as the start of a new year. The colony's health at this point in the year greatly impacts how prosperous it will be the following year (Harris and Woodring, 1992). The amine levels peak between June and September, coinciding with a peak in colony foraging activities. These variations can be a result of seasonal variations in the diet, population, or brood-rearing activity of the colony (Harris and Woodring, 1992). In addition, the population of bees decreases and brood raising is lowered in the fall as a result of a decrease in nectar and pollen entering the hive (Flores et al., 2019). The proportion of elderly bees in the colony falls with the queen's age and egg-laying capacity. In contrast to the older bees, which slowly perish, the baby bees survive the winter (Sammataro and Avitabile, 1998). To prevent cold air from entering the hive, propolis harvested from tree buds is used to seal all openings and to constrict the entrance.

Significant correlations were found between the quantity of recruiting activity targeted at pollen resources and the number of brood combs, and all months other than May saw brood production. The seasonal feeding habits of *A. mellifera scutellata* in the Okavango were comparable to those of neotropical Africanized honey bees (Schneider and McNally, 1992). A significant amount of food must be gathered for winter survival during the relatively short, predictable foraging season that honey bees in temperate climates encounter (Switanek et al., 2017). Building up substantial food stores necessitates the development of substantial volumes of comb, the maintenance of substantial colony populations, gathering food over substantial areas of the environment, and focusing on nectar collection (Brodschneider and Crailsheim, 2010). However, tropical honey bees do not experience a lengthy cold period and can forage all year (Winston et al., 1983). As a result, numerous behavioral differences between temperate and tropical honey bees can be attributed to variations in the seasonal rhythms of colony foraging activity. Colony migration may also be linked to the prolonged foraging season. Although floral resources are available year-round in tropical ecosystems, their temporal and spatial incidence is frequently uncertain because of erratic rain patterns (Cui and Corlett, 2016). Therefore, factors affecting honey bees' brood-rearing habits might be seen as major initiators in the emergence of winter bees. In this approach, declining pollen supplies in the fall are primarily responsible for the arrival of winter bees. The effects of other triggers effects have been labeled as minor or indirect, or in some cases, it's unclear how they're involved (Knoll et al., 2020). In fact, the wild bee community in temperate regions consists of both sociable and solitary species (Gibbs et al., 2017). Solitary bees have brief flying intervals that are strongly related to the host plants they select (Wood et al., 2016). Colonies of sweat bees (Halictidae) or eusocial bumble bees (*Bombus* spp.) can persist for several months (Michener and Michener, 1974). Some wild bees, especially gregarious species, can forage more opportunistically and have broader diets (Williams et al., 2011). Therefore, it may be advantageous for tropical honey bee colonies to migrate at locally unfavorable times. Hence, recognizing seasonal honey bee data patterns can be helpful for a variety of endeavors, including swarming observations and colony absconding projections, particularly for those hives where resources are limited. In another study, six observation colonies in Botswana's Okavango Delta were used to examine seasonal feeding habits. Over the course of the study (11 months), pollen was collected, bees were released from the hive, and people were sought after as pollen and nectar suppliers. However, there were seasonal differences in the foraging activity's distribution throughout the day. For most of the year, colonies prioritized recruiting for pollen sites. With the exception of May, all months had brood production (Schneider and McNally, 2005). There is a great deal of variety in the activity patterns of different *A. mellifera* races around the world, according to investigations. For instance, in Ghana's rainy and dry seasons, respectively, Woyke (1992) noticed three peaks of foraging activity for *A. mellifera adansonii*: one (the highest) in the morning, one peak after dawn, and one peak before dusk. Similarly to this, Fletcher (1978) noted an activity peak for *A. mellifera scutellata* in South Africa just before sunrise. Silva and De Jong (1990), on the other hand, noted three activity maxima at 7:00, 14:00, and 16:00 hours for Africanized bees in Brazil. Gary (1967) attributed the U-shaped foraging activity pattern he observed to the desert environment's high temperatures and low humidity levels. Other activities can be changed according to the season, such as drone activity.

Drone activity starts sometime in the afternoon on a scale that is used worldwide, involving all species of *Apis* and *A. mellifera* races (Ayup et al., 2021). On a seasonal level, this activity is linked to the onset of mating behavior and the rise in ambient temperature during spring and summer in both hemispheres (Ruttner, 1966). *A. mellifera* drones undergo orientation flights immediately after emergence. These flights are brief (between 3 and 10 minutes), wherein the drones rise and fall in a spiral while figuring out where exactly their hive is in relation to its surroundings (Rangel and Fisher, 2019). Drones hunt for DCAs after discovering where their hive is (Ayup et al., 2021) and then either go back to their own hives or stray to neighboring hives to find food (Burgett, 1973).

2.5 CONCLUSIONS

In social insects, notably honey bee colonies, different castes cooperate and use social immunity to fight against parasites and build superorganisms. A honey bee colony consists of workers, drones, and a queen. The honey bee queen is fertile, and the production and delivery of semen to the queen during the mating flight is the responsibility of drones. The newest bees are involved in cleaning the nest, and nurse bees participate in brood care and other duties. Forager bees live for three to four weeks during spring and summer and exhibit significant behavioral and biochemical changes along with more activity during the day. Chemical communication is highly beneficial in maintaining colony integration, which is required to control the distribution of labor among workers based on age. More research is required to comprehend honey bees' physiology, behavior, and social interactions inside colonies.

REFERENCES

Akratanakul, P. (1986). *Beekeeping in Asia*. Food & Agriculture Org.

Al-Lawati, H., Kamp, G., and Bienefeld, K. (2009). Characteristics of the spermathecal contents of old and young honeybee queens. *Journal of Insect Physiology* 55(2), 117–122.

Amiri, E., Meixner, M.D., and Kryger, P. (2016). Deformed wing virus can be transmitted during natural mating in honey bees and infect the queens. *Scientific Reports* 6(1), 1–7.

Amiri, E., Strand, M.K., Rueppell, O., and Tarpy, D.R. (2017). Queen quality and the impact of honey bee diseases on queen health: Potential for interactions between two major threats to colony health. *Insects* 8(2), 48.

Anderson, C., and McShea, D.W. (2001). Individual versus social complexity, with particular reference to ant colonies. *Biological reviews* 76(2), 211–237.

Ayup, M.M., Gärtner, P., Agosto-Rivera, J.L., Marendy, P., de Souza, P., and Galindo-Cardona, A.J.I. (2021). Analysis of honeybee drone activity during the mating season in Northwestern Argentina. *Insects* 12(6), 566.

Baer, B., Collins, J., Maalaps, K., and den Boer, S.P. (2016). Sperm use economy of honeybee (Apis mellifera) queens. *Ecology and Evolution* 6(9), 2877–2885.

Beekman, M., Gilchrist, A.L., Duncan, M., Sumpter, D.J. (2007). What makes a honeybee scout? *Behavioral Ecology and Sociobiology* 61(7), 985–995.

Ben-Shahar, Y., Leung, H.-T., Pak, W., Sokolowski, M., and Robinson, G.E. (2003). cGMP-dependent changes in phototaxis: A possible role for the foraging gene in honey bee division of labor. *Journal of Experimental Biology* 206(14), 2507–2515.

Bernhard Kraus, F., Gerecke, E., and Moritz, R.F. (2011). Shift work has a genetic basis in honeybee pollen foragers (Apis mellifera L.). 41(2), 323–328.

Bishop, G.H. (1920). Fertilization in the honey-bee: The male sexual organs: Their historical structure and physiological functioning. *Journal of Experimental Zoology* 31(2), 225–286.

Bloch, G., and Grozinger, C.M. (2011). Social molecular pathways and the evolution of bee societies. *Philosophical Transactions of the Royal Society B: Biological Sciences* 366(1574), 2155–2170.

Boes, K. (2010). Honeybee colony drone production and maintenance in accordance with environmental factors: An interplay of queen and worker decisions. *Insectes Sociaux* 57(1), 1–9.

Bortolotti, L., and Costa, C. (2014). Chemical communication in the honey bee society. In *Neurobiology of Chemical Communication*, Ed. C. Mucignat-Caretta, CRC Press.

Brodschneider, R., and Crailsheim, K. (2010). Nutrition and health in honey bees. *Apidologie* 41(3), 278–294.

Burgett, D.M. (1973). Drone engorgement in honey bee swarms. *Annals of the Entomological Society of America* 66(5), 1005–1006.

Burkle, L.A., Marlin, J.C., and Knight, T.M. (2013). Plant-pollinator interactions over 120 years: Loss of species, co-occurrence, and function. *Science* 339(6127), 1611–1615.

Cassier, P., Tel-Zur, D., and Lensky, Y. (1994). The sting sheaths of honey bee workers (Apis mellifera L.): Structure and alarm pheromone secretion. *Journal of Insect Physiology* 40(1), 23–32.

Charbonneau, D., Blonder, B., and Dornhaus, A. (2013). Social insects: A model system for network dynamics. In Peter Holme and Jari Saramäki (Eds), *Temporal Networks*. Springer, 217–244.

Chen, S., Wang, Y., Li, Y., Zhang, X., and Wu, J. (2022). Effects of honeybee (Apis cerana) visiting behaviour on toxic plant (Tripterygium hypoglaucum) reproduction. *AoB PLANTS* 14(3), plac002.

Chittka, L., and Rossi, N. (2022). Social cognition in insects. *Trends in Cognitive Sciences* 26(7), 578–592. doi: 10.1016/j.tics.2022.04.001.

Collins, A.M., and Pettis, J.S. (2013). Correlation of queen size and spermathecal contents and effects of miticide exposure during development. *Apidologie* 44(3), 351–356.

Conte, Y.L., and Hefetz, A. (2008). Primer pheromones in social hymenoptera. *Annual Review Entomology* 53, 523–542.

Corlett, R.T. (2011). Honeybees in natural ecosystems. In *Honeybees of Asia*. Eds. H. Randall Hepburn and Sarah E. Radloff. Springer, 215–225.

Corlett, R.T., Westcott, D.A. (2013). Will plant movements keep up with climate change? 28(8), 482–488.

Costa, J.T., and Costa, J.T. (2006). *The Other Insect Societies*. Harvard University Press.

Couvillon, M.J., Hughes, W.O., Perez-Sato, J.A., Martin, S.J., Roy, G.G., and Ratnieks, F.L. (2010). Sexual selection in honey bees: Colony variation and the importance of size in male mating success. *Behavioral Ecology* 21(3), 520–525.

Cui, Q., and Corlett, R.T. (2016). Seasonal and diurnal patterns of activity in honeybees (Apis spp.) on the northern edge of the Asian tropics; their implications for the climate-change resilience of pollination. *Tropical Conservation Science* 9(3), 1940082916667142.

Czekońska, K., and Chuda-Mickiewicz, B. (2015). The ability of honey bee drones to ejaculate. *Journal of Apicultural Science* 59(2), 127–133.

Czekońska, K., Chuda-Mickiewicz, B., and Chorbiński, P. (2013). The effect of brood incubation temperature on the reproductive value of honey bee (Apis mellifera) drones. *Journal of Apicultural Research* 52(2), 96–105.

Czekońska, K., Chuda-Mickiewicz, B., and Samborski, J. (2015). Quality of honeybee drones reared in colonies with limited and unlimited access to pollen. *Apidologie* 46(1), 1–9.

Dyer, F.C. (1985). Nocturnal orientation by the Asian honey bee, Apis dorsata. *Animal Behaviour* 33(3), 769–774.

Ellis, J., Lawrence, J., Koeniger, N., and Koeniger, G. (2015). *Mating Biology of Honey Bees (Apis mellifera)*. Wicwas Press.

Fletcher, D.J. (1978). The African bee, Apis mellifera adansonii, in Africa. 23(1), 151–171.

Flores, J.M., Gil-Lebrero, S., Gámiz, V., Rodríguez, M.I., Ortiz, M.A., and Quiles, F.J. (2019). Effect of the climate change on honey bee colonies in a temperate Mediterranean zone assessed through remote hive weight monitoring system in conjunction with exhaustive colonies assessment. *Science of the Total Environment* 653, 1111–1119.

Friedman, D., Johnson, B., Linksvayer, T. A. (2020). Distributed physiology and the molecular basis of social life in eusocial insects. *Hormones and Behavior* 122, 104757.

Gary, N.E. (1967). Diurnal variations in the intensity of flight activity from honeybee colonies. *Journal of Apicultural Research* 6(2), 65–68.

Gibbs, J., Ascher, J.S., Rightmyer, M.G., and Isaacs, R. (2017). The bees of Michigan (Hymenoptera: Apoidea: Anthophila), with notes on distribution, taxonomy, pollination, and natural history. 4352(1), 1–160.

Gilley, D.C., Tarpy, D.R., and Land, B.B. (2003). Effect of queen quality on interactions between workers and dueling queens in honeybee (Apis mellifera L.) colonies. *Behavioral Ecology and Sociobiology* 55(2), 190–196.

Giraldo, Y.M., and Traniello, J.F. (2014). Worker senescence and the sociobiology of aging in ants. *Behavioral Ecology and Sociobiology* 68(12), 1901–1919. doi: 10.1007/s00265-014-1826-4.

Gregorc, A., and Škerl, M.I.S. (2015). Characteristics of honey bee (Apis mellifera Carnica, Pollman 1879) queens reared in Slovenian commercial breeding stations. *Journal of Apicultural Science* 59(2), 5–12.

Grozinger, C.M., Richards, J., and Mattila, H.R. (2014). From molecules to societies: Mechanisms regulating swarming behavior in honey bees (Apis spp.). *Apidologie* 45 (3), 327–346.

Harris, J.W., and Woodring, J. (1992). Effects of stress, age, season, and source colony on levels of octopamine, dopamine and serotonin in the honey bee (Apis mellifera L.) brain. *Journal of Insect Physiology* 38(1), 29–35. doi: 10.1016/0022-1910(92)90019-A.

Hatch, S., Tarpy, D., and Fletcher, D. (1999). Worker regulation of emergency queen rearing in honey bee colonies and the resultant variation in queen quality. *Insectes Sociaux* 46(4), 372–377.

Hoekstra, H.E., and Robinson, G.E. (2022). Behavioral genetics and genomics: Mendel's peas, mice, and bees. *Proceedings of the National Academy of Sciences* 119(30), e2122154119.

Hrassnigg, N., and Crailsheim, K. (2005). Differences in drone and worker physiology in honeybees (Apis mellifera). *Apidologie* 36(2), 255–277.

Huang, Z.-Y., and Robinson, G.E. (1992). Honeybee colony integration: Worker-worker interactions mediate hormonally regulated plasticity in division of labor. 89(24), 11726–11729.

Huang, Z.-Y., and Robinson, G.E. (1996). Regulation of honey bee division of labor by colony age demography. *Behavioral Ecology and Sociobiology* 39(3), 147–158.

Huang, Z.-Y., Robinson, G., and Borst, D.W. (1994). Physiological correlates of division of labor among similarly aged honey bees. 174(6), 731–739.

Hughes, T.P. (1984). Population dynamics based on individual size rather than age: A general model with a reef coral example. *The American Naturalist* 123(6), 778–795.

Johnson, B.R. (2005). Limited flexibility in the temporal caste system of the honey bee. *Behavioral Ecology and Sociobiology* 58(3), 219–226.

Johnson, B.R. (2010). Division of labor in honeybees: Form, function, and proximate mechanisms. *Behavioral Ecology and Sociobiology* 64(3), 305–316. doi: 10.1007/s00265-009-0874-7.

Keeling, C.I., Slessor, K.N., Higo, H.A., and Winston, M.L. (2003). New components of the honey bee (Apis mellifera L.) queen retinue pheromone. *Proceedings of the National Academy of Sciences* 100(8), 4486–4491.

Knoll, S., Pinna, W., Varcasia, A., Scala, A., and Cappai, M.G. (2020). The honey bee (Apis mellifera L., 1758) and the seasonal adaptation of productions. Highlights on summer to winter transition and back to summer metabolic activity. A review. *Livestock Science* 235, 104011. doi: 10.1016/j.livsci.2020.104011.

Kocher, S.D., Richard, F.-J., Tarpy, D.R., and Grozinger, C.M. (2008). Genomic analysis of post-mating changes in the honey bee queen (Apis mellifera). *BMC Genomics* 9(1), 1–15.

Kocher, S.D., Richard, F.-J., Tarpy, D.R., and Grozinger, C.M. (2009). Queen reproductive state modulates pheromone production and queen-worker interactions in honeybees. *Behavioral Ecology* 20(5), 1007–1014.

Koeniger, N., and Koeniger, G. (2000). Reproductive isolation among species of the genus Apis. *Apidologie* 31(2), 313–339.

Koeniger, G., Koeniger, N., Ellis, J., and Connor, L. (2014). *Mating Biology of Honey Bees (Apis mellifera).* Wicwas Press.

Kraus, F., Neumann, P., and Moritz, R. (2005). Genetic variance of mating frequency in the honeybee (Apis mellifera L.). *Insectes Sociaux* 52(1), 1–5.

Layek, U., Midday, M., Bisui, S., Kundu, A., and Karmakar, P. (2022). Floral biology, breeding system and pollination ecology of Justicia betonica L.(Acanthaceae): An assessment of its low reproductive success in West Bengal, India. *Plant Species Biology* 37(4), 278–293.

Lee, S.B., Chouvenc, T., Mizumoto, N., Mullins, A., and Su, N.Y. (2022). Age-based spatial distribution of workers is resilient to worker loss in a subterranean termite. *Scientific Reports* 12(1), 7837. doi: 10.1038/s41598-022-11512-1.

Leoncini, I., Le Conte, Y., Costagliola, G., Plettner, E., Toth, A.L., Wang, M., et al. (2004). Regulation of behavioral maturation by a primer pheromone produced by adult worker honey bees. *Proceedings of the National Academy of Sciences* 101(50), 17559–17564.

Lihoreau, M., and Gómez-Moracho, T. (2022). Where, what and with whom to eat: Towards an integrative study of foraging behaviour. In *The Routledge International Handbook of Comparative Psychology.* Eds. Todd M. Freeberg, Amanda R. Ridley, and Patrizia d'Ettore. Routledge, 124–135.

Lucas, C., and Ben-Shahar, Y. (2021). The foraging gene as a modulator of division of labour in social insects. 35(3), 168–178.

Marshman, J. (2022). *Bee Cities Andmore than Human Communities: Protecting Pollinators in the Anthropocene.* Theses and Dissertations (Comprehensive). 2437.

Mattila, H., Harris, J., and Otis, G.W. (2001). Timing of production of winter bees in honey bee (Apis mellifera) colonies. 48(2), 88–93.

Michener, C.D., and Michener, C.D. (1974). *The Social Behavior of the Bees: A Comparative Study.* Harvard University Press.

Moore, D., Angel, J.E., Cheeseman, I.M., Fahrbach, S.E., Robinson, G.E., and Sociobiology (1998). Timekeeping in the honey bee colony: Integration of circadian rhythms and division of labor. 43(3), 147–160.

Moritz, R.F. (1989). *The Instrumental Insemination of the Queen Bee.* Apimondia.

Moritz, R., and Southwick, E.E. (2012). *Bees as Superorganisms: An Evolutionary Reality.* Springer Science & Business Media.

Navas-Zuloaga, M.G., Pavlic, T.P., and Smith, B.H. (2022). Alternative model systems for cognitive variation: Eusocial-insect colonies. *Trends in Cognitive Sciences* 26(10), 836–848. doi: 10.1016/j.tics.2022.06.011.

Niño, E., Tarpy, D., and Grozinger, C. (2013). Differential effects of insemination volume and substance on reproductive changes in honey bee queens (A pis mellifera L.). *Insect Molecular Biology* 22(3), 233–244.

Niño, E.L., Malka, O., Hefetz, A., Teal, P., Hayes, J., and Grozinger, C.M. (2012). Effects of honey bee (Apis mellifera L.) queen insemination volume on worker behavior and physiology. *Journal of Insect Physiology* 58(8), 1082–1089.

Page Jr. R.E., and Peng, C.Y.-S. (2001). Aging and development in social insects with emphasis on the honey bee, Apis mellifera L. *Experimental Gerontology* 36(4–6), 695–711.

Pankiw, T. (2004). Cued in: Honey bee pheromones as information flow and collective decision-making. *Apidologie* 35(2), 217–226.

Peso, M., Niño, E., Grozinger, C., and Barron, A. (2013). Effect of honey bee queen mating condition on worker ovary activation. *Insectes Sociaux* 60(2), 123–133.

Rangel, J., Böröczky, K., Schal, C., and Tarpy, D.R. (2016). Honey bee (Apis mellifera) queen reproductive potential affects queen mandibular gland pheromone composition and worker retinue response. *PLOS One* 11(6), e0156027.

Rangel, J., and Fisher, A. (2019). Factors affecting the reproductive health of honey bee (Apis mellifera) drones—A review. *Apidologie* 50(6), 759–778.

Reim, T., and Scheiner, R. (2014). Division of labour in honey bees: Age- and task-related changes in the expression of octopamine receptor genes. *Insect Molecular Biology* 23(6), 833–841. doi: 10.1111/imb.12130.

Revely, L., Sumner, S., Eggleton, P., and Evolution (2021). The plasticity and developmental potential of termites. *Frontiers in Ecology and Evolution* 9, 552624.

Robinson, G.E. (1992a). Regulation of division of labor in insect societies. *Annual Review of Entomology* 37(1), 637–665.

Robinson, G.E., Page Jr. R.E., Strambi, C., and Strambi, A. (1989). Hormonal and genetic control of behavioral integration in honey bee colonies. 246(4926), 109–112.

Robinson, G.E. (1992b). Regulation of division of labor in insect societies. *Annual Review of Entomology* 37(1), 637–665.

Romiguier, J., Cameron, S.A., Woodard, S.H., Fischman, B.J., Keller, L., Praz, C.J., et al. (2016). Phylogenomics controlling for base compositional bias reveals a single origin of eusociality in corbiculate bees. *Molecular Biology and Evolution* 33(3), 670–678.

Rueppell, O., Page Jr. R.E., and Fondrk, M.K. (2006). Male behavioural maturation rate responds to selection on pollen hoarding in honeybees. *Animal Behaviour* 71(1), 227–234.

Ruttner, F. (1966). The life and flight activity of drones. *Bee World* 47(3), 93–100.

Sammataro, D., and Avitabile, A. (1998). *The Beekeeper's Handbook.* Cornell University Press.

Sandrock, C., Tanadini, M., Tanadini, L.G., Fauser-Misslin, A., Potts, S.G., and Neumann, P. (2014). Impact of chronic neonicotinoid exposure on honeybee colony performance and queen supersedure. *PLOS One* 9(8), e103592.

Schlüns, H., Moritz, R.F., Neumann, P., Kryger, P., and Koeniger, G. (2005). Multiple nuptial flights, sperm transfer and the evolution of extreme polyandry in honeybee queens. *Animal Behaviour* 70(1), 125–131.

Schmickl, T., and Crailsheim, K. (2004). Inner nest homeostasis in a changing environment with special emphasis on honey bee brood nursing and pollen supply. *Apidologie* 35(3), 249–263.

Schneider, S.S., and McNally, L.C. (1992). Seasonal patterns of foraging activity in colonies of the African honey bee, Apis mellifera scutellata, in Africa. *Insectes Sociaux* 39(2), 181–193. doi: 10.1007/BF01249293.

Schneider, S.S., and McNally, L.C. (2005). Seasonal patterns of foraging activity in colonies of the African honey bee, Apis mellifera scutellata, in Africa. 39(2), 181–193.

Seeley, T.D. (1989). Social foraging in honey bees: how nectar foragers assess their colony's nutritional status. *Behavioral Ecology and Sociobiology* 24(3), 181–199.

Silva, D., and De Jong, D. (1990). Diurnal flight activity of Africanized and European honey bees. *Apiacta* 25(3), 78–80.

Smith, C.R., Toth, A.L., Suarez, A.V., and Robinson, G.E. (2008a). Genetic and genomic analyses of the division of labour in insect societies. *Nature Reviews. Genetics* 9(10), 735–748.

Smith, C.R., Toth, A.L., Suarez, A.V., and Robinson, G.E. (2008b). Genetic and genomic analyses of the division of labour in insect societies. 9(10), 735–748.

Streinzer, M., Brockmann, A., Nagaraja, N., and Spaethe, J. (2013). Sex and caste-specific variation in compound eye morphology of five honeybee species. *PLOS One* 8(2), e57702.

Stürup, M., Baer-Imhoof, B., Nash, D.R., Boomsma, J.J., and Baer, B. (2013). When every sperm counts: Factors affecting male fertility in the honeybee Apis mellifera. *Behavioral Ecology* 24(5), 1192–1198.

Su, W., Ma, W., Zhang, Q., Hu, X., Ding, G., Jiang, Y., et al. (2022). Honey bee foraging decisions influenced by pear volatiles. *Agriculture* 12(8), 1074.

Switanek, M., Crailsheim, K., Truhetz, H., and Brodschneider, R. (2017). Modelling seasonal effects of temperature and precipitation on honey bee winter mortality in a temperate climate. *Science of The Total Environment* 579, 1581–1587.

Szabo, T. (1974). Behavioural studies on queen introduction in the honeybee. 2. Effect of age and storage conditions of virgin queens on their attractiveness to workers. *Journal of Apicultural Research* 13(2), 127–135.

Taha, E.-K.A., and Al-Kahtani, S. (2013). Relationship between population size and productivity of honey bee colonies. *Journal of Entomology* 10(3), 163–169.

Tanaka, E.D., and Hartfelder, K. (2004). The initial stages of oogenesis and their relation to differential fertility in the honey bee (Apis mellifera) castes. *Arthropod Structure and Development* 33(4), 431–442.

Tanaka, Y., Hojo, M.K., and Shimoji, H. (2022). Individual experience influences reconstruction of division of labour under colony disturbance in a queenless ant species. *Frontiers in Zoology* 19(1), 20. doi: 10.1186/s12983-022-00466-9.

Tarpy, D.R., Hatch, S., and Fletcher, D.J. (2000). The influence of queen age and quality during queen replacement in honeybee colonies. *Animal Behaviour* 59(1), 97–101.

Tarpy, D.R., Keller, J.J., Caren, J.R., and Delaney, D.A. (2012). Assessing the mating 'health'of commercial honey bee queens. *Journal of Economic Entomology* 105(1), 20–25.

Tibor Szabo, I., Mills, P.F., and Heikel, D.T. (1987). Effects of honeybee queen weight and air temperature on the initiation of oviposition. *Journal of Apicultural Research* 26(2), 73–78.

Trhlin, M., and Rajchard, J. (2011). Chemical communication in the honeybee (Apis mellifera L.): A review. *Veterinary Medicine* 56(6), 265–273.

Vergoz, V., McQuillan, H.J., Geddes, L.H., Pullar, K., Nicholson, B.J., Paulin, M.G., et al. (2009). Peripheral modulation of worker bee responses to queen mandibular pheromone. *Proceedings of the National Academy of Sciences of the United States of America* 106(49), 20930–20935.

Wcislo, W.T., and Tierney, S.M. (2009). Behavioural environments and niche construction: The evolution of dim-light foraging in bees. *Biological reviews of the Cambridge Philosophical Society* 84(1), 19–37.

Wharton, K.E., Dyer, F.C., Huang, Z.Y., and Getty, T. (2007). The honeybee queen influences the regulation of colony drone production. *Behavioral Ecology* 18(6), 1092–1099.

Williams, N.M., Cariveau, D., Winfree, R., Kremen, C., and Ecology (2011). Bees in disturbed habitats use, but do not prefer, alien plants. *Basic and Applied Ecology* 12(4), 332–341.

Winston, M. (1987). *The Biology of the Honey Bee*. Harvard University Press.

Winston, M.L., Taylor, O.R., and Otis, G.W. (1983). Some differences between temperate European and tropical African and South American honeybees. *Bee World* 64(1), 12–21.

Withrow, J.M., and Tarpy, D.R. (2018). Cryptic "royal" subfamilies in honey bee (Apis mellifera) colonies. *PLOS One* 13(7), e0199124.

Wood, T., Holland, J., Goulson, D. (2016). Diet characterisation of solitary bees on farmland: Dietary specialisation predicts rarity. *Biodiversity and Conservation* 25(13), 2655–2671.

Woyke, J. (1958). *Histologiczna budowa organów rozrodczych trutnia*. Panst. Wyd. naukowe.

Woyke, J. (1992). Diurnal flight activity of African bees Apis mellifera adansonii in different seasons and zones of Ghana. *Apidologie* 23(2), 107–117.

Woyke, J., and Ruttner, F. (1958). An anatomical study of the mating process in the honeybee. *Bee World* 39(1), 3–18.

Wu-Smart, J., and Spivak, M. (2016). Sub-lethal effects of dietary neonicotinoid insecticide exposure on honey bee queen fecundity and colony development. *Scientific Reports* 6(1), 1–11.

Xu, V., Plowright, C.M.S. (2022). Floral categorization by bumblebees: The perceptual and the functional. *Animal Behavior and Cognition* 9(3), 316–335.

Yerushalmi, S., Bodenhaimer, S., and Bloch, G. (2006). Developmentally determined attenuation in circadian rhythms links chronobiology to social organization in bees. *Journal of Experimental Biology* 209(6), 1044–1051.

Young, A.M., Kohl, P.L., Rutschmann, B., Steffan-Dewenter, I., Brockmann, A., and Dyer, F.C. (2021). Temporal and spatial foraging patterns of three Asian honey bee species in Bangalore, India. *Apidologie* 52(2), 503–523. doi: 10.1007/s13592-020-00839-1.

Yue, C., Schröder, M., Bienefeld, K., and Genersch, E. (2006). Detection of viral sequences in semen of honeybees (Apis mellifera): Evidence for vertical transmission of viruses through drones. *Journal of Invertebrate Pathology* 92(2), 105–108.

Zych, M., Goldstein, J., Roguz, K., and Stpiczyńska, M. (2013). The most effective pollinator revisited: Pollen dynamics in a spring-flowering herb. *Arthropod-Plant Interactions* 7(3), 315–322.

3 Sex and Caste Determination and Life Cycle of Honey Bees

Javaid Iqbal, Mujahid Manzoor, Tridip Boruah, Sajad H. Wani and Aijaz Ahmad Wachkoo

3.1 INTRODUCTION

Hymenoptera belong to a group of holometabolic insects that includes 200,000 species of bees, sawflies, wasps and ants. They have no sex chromosomes, and sex is usually determined by a single locus. Among the Hymenopteran insects, honey bees are social pollinators, and play a vital role in biodiversity and ecological balance (Pan et al., 2022). Queen bees and, sometimes, worker bees are the sole sources of male drones. A diploid queen has 32 chromosomes, compared with a haploid drone's 16 chromosomes. Sperm cells produced by drones are genetically identical and have entire genomes. Thus, the genetic composition of a female worker derives from both her mother and her father to equal degrees, whereas the genetic makeup of a drone derives exclusively from his mother (Choudhary et al., 2022). Honey bees are haploid organisms, and the mechanism of sex determination is dependent on the complementary sex determinants (CSD) locus (Netschitailo et al., 2021; Fridi et al., 2022), in which the sexes are determined by the haploid–diploid sex determination system (Pan et al., 2022). The CSD gene is located on chromosome 3; it consists of 9 exons and is 11,734 bp long (Fridi et al., 2022). It is developed by gene replication from the inherited copy of its paralogous gene feminizer (FEM) within the honey bee genus *Apis*. Heterozygous CSD derives from heterozigation of CSD through interaction with Transformer 2 (a protein containing an RNA binding domain), which leads to a female pathway leading to a spliced female FEM transcript. This protein product is determined and is responsible for female-specific splicing of the Am-DSX transcript. Homozygous CSD derives from the initial stop codon of male FEM mRNA, as it induces a male pathway that is interrupted by a truncated FEM protein. As a result, translation is stopped prematurely, and Am-DSX undergoes male-specific splicing to form a male-specific carboxy-terminal protein (Fridi et al., 2022).

3.2 SEX DETERMINATION IN QUEENS AND DRONES

The queen honey bee is a fertile female specialized for reproduction. The queen honey bee's main task is egg production (O'Brien et al., 2022). Queens and workers are heterozygous females, and drones are homozygous males (Paolillo et al., 2022). Females, developed from fertilized eggs, are diploid (having two sets of chromosomes) and produce genetically dissimilar egg cells (Pan et al., 2022), whereas drones are male bees produced from unfertilized eggs. There is a small number of drones (about 2–3%) in the hive, and their sole function is to fertilize and impregnate the queen during successful copulating flights and pass their genes on to the offspring (O'Brien et al., 2022). The overall health of the offspring and the survival of the colony are affected by the health of the drone (Phokasem et al., 2021). Males develop from unfertilized eggs, are haploid and produce genetically similar sperm cells in mitosis (Pan et al., 2022).

3.2.1 Role of Small RNA in the Reproduction of Honey Bees

Small RNAs play an important role in regulating honey bee gamete formation and reproduction and provide a robust but putative pathway for parental manipulation of offspring gene expression and

DOI: 10.1201/9781003354116-3

reproductive physiology. Small RNAs are present in the reproductive tissues of honey bees (ovaries, testes, sperm, and fertilized and unfertilized eggs) but frequently exist in extracellular vesicles (EVs), which are present in the sperm of numerous organisms. Sperm contains a smaller amount of Piwi-interacting RNA (piRNA) than eggs and ovaries, and piRNA and microRNA, which are involved in DNA regulation and development, are expressed differently in each tissue. The queen may utilize additional epigenetic mechanisms to determine the sex of the egg. The microRNAs are miniature non-coding molecules of RNA. Maternally inherited miRNAs play an important role in determining sex during embryonic development to adapt the genetic expression in offspring; miRNAs degrade the maternal mRNA stored in the nascent zygote. Again, piRNA plays a role in sex determination in honey bees because it is involved in the regulation of gene expression in sperm and somatic development, and numerous piRNAs target sex-determining loci, while short-interfering RNAs (siRNAs) play an important role in genetics, epigenesis and regulation of gene expression during development. The tRNA derived fragments (tRFs) are largely unclear, as some tRFs have the same role as miRNAs, while others are involved in the global translation of proteins in the ribosome. However, the reproductive tissue of honey bees has a characteristic small RNA profile. In oocytes derived from virgin and mating uterine oocytes, the profile of this expression is more ovarian than in other tissues. Ovaries and eggs contain more piRNA than other tissues, and the sperm cells are high in tRNA fragments. The sperm and testes contain more miRNAs. Lastly, the germ cells utilize small RNAs to regulate function during gamete formation and sex determination and development (Watson et al., 2022). Gene expression profiles differ between worker and queen bees but are similar between young and old bees. A high proportion of small genes is represented by the worker bee transcript, while a high proportion of intermediately expressed genes is expressed by the queen bee transcript. Genes are considerably over-expressed in queen bees in contrast to worker bees (Lu et al., 2021).

3.3 MATING PATTERNS

Emerging as an adult after approximately one week, the queen will mate more than once in the area where the drones are concentrated. Drone sperm are transferred to the Fallopian tubes of the uterus during mating, and only 3–5% of the sperm from each drone is stored in a special organ called the spermatheca. Queens undergo significant behavioural, physiological and biochemical changes during mating. Behaviorally, mated queens lose their affinity for light and stop making mating flights, showing that they are no longer sexually receptive. The newly mated queen's ovaries begin to become active, and the ovaries thicken and begin to produce eggs. Then, the queen honey bee begins to lay eggs, laying eggs for the rest of her life (Jasper et al., 2020). The storage of sperm in the spermatheca of the queen bee (*Apis mellifera*) is important for the reproduction of the bee (Park et al., 2021). For more than 8,000 years, bees and bee products have been part of human survival. Because they are social insects, honey bees dwell in dense communities known as colonies. There are several thousand individuals residing in the colony, and most of them are female workers. The only female who actively participates in reproduction is the queen bee. In the emergence of larval life, the only difference between worker ants and queen bees is the difference in food. During mating, at least once in her lifetime, the queen bee mates with several drones. During this flight, she collects sperm and stores it in a special organ for the rest of the queen's life. Honey bee reproduction can be divided into two categories. The first is breeding by colonies (colony-level reproduction) through swarming. The second is individual bee reproduction within a colony. A honey bee colony is always led by a queen bee. If a sufficient brood is available, the colony can produce another queen bee, but colonies led by such queens cannot be expected to show genetic improvements in economically important traits, namely, honey yield. As a result, raising and rearing a queen bee is very important for beekeeping (Moškrič et al., 2020).

Polyandry is the female practice of producing genetically diverse offspring in one or more arrangements by mating with multiple males and storing sperm. An increase in family diversity potential of more than 90% is achieved by the mother's tenth mating. In social honey bees, it plays a vital role in the biological variability and the labour efficiency of worker bees and leads to improved stamina in the colony (Delaplane et al., 2021). Polyandrous insects are multi-social insects, and these, such as honey bees, are prime candidates for parents to manipulate gene expression in the offspring (Watson et al., 2022). The frequency of queen bee mating is related to worker bee genetic diversity and colony fitness. This is achieved by mothers' effective mating, also called hyper polyandry (Delaplane et al., 2021). During their mating flight, the honey bee queen mates with as many as 40 drones; this is termed 'extreme polyandry'. Consequently, they will establish concomitant benefits for the colony and inherently elevate diversity (Bilodeau, 2022).

3.4 MORPHOLOGICAL PECULIARITIES AND REPRODUCTIVE BIOLOGY OF HONEY BEES

On the legs of the honey bee, a specialized structure is present, which assists the honey bee in collecting pollen and other materials and bringing them back to the hive. Although the honey bee has a simple nervous system, it is capable of carrying out complex behaviour and possesses an impressive memory (Faux, 2021; Iqbal & Mueller, 2007; Ali et al., 2021). The honey bee, *Apis mellifera ligustica* has a distinct tongue made up of about 120 segmental units, covered with 3,000 glossy hairs, which is used to maintain an upright position and collect nectar. These bristle hairs are supported by a hard, continuous, annular hair base embedded in a resilient elastic intersegmental membrane, making the entire tongue flexible and firm at the same time (Sun et al., 2021). Worker antennae express more genes involved in neural communication; on the other hand, drone antennae express more genes related to energy metabolism (Jain and Brockmann, 2020). The discrimination of lineage and subspecies of honey bees (*Apis mellifera*) is often determined by wing measurements. Subspecies are organized using geometric morphometry. It helps to determine and analyse both wing size and shape separately. The precision and repeatability of measurement depend on various factors, including seasonal variation and the difference between the left wing and the right wing. Wing sizes vary considerably depending on colony, month, and side of the body. The shape of the wings differs significantly only depending on the hive and the month, but not on the side of the body. The seasonal variation in wing shape is much less than that between colonies (Janczyk and Tofilski, 2021). A few of the commercially important types of honey bees are *Apis mellifera* (European bee), *Apis florea* (little bee), *Apis indica* (Indian bee), and *Apis dorsata* (rock bee).

3.4.1 Reproductive Biology of *Apis mellifera*

The reproductive biology of the genus *Apis* is full of oddities. The virgin queens mate successively with several drones in the air. The drones perish during copulation. The previous drone's mating sign is immediately covered by the next drone copulating. Each drone can inject over a million spermatozoa into a queen's reproductive tract. With typical mating rates ranging from 15 to 25, honey bee queens are hence extremely polyandrous. Additionally, the drone's testes are made up of 150 or more testicular tubules per testis, which is a significant number of serial units, similar to the quantity found in the queen's ovaries. They are distinct in this regard from the males of all other bee species, which typically only have four testicular tubules per testis. The entire genus *Apis* contrasts sharply with all other bee species in terms of these reproductive characteristics. *A. mellifera* drones can produce around 10 times as much sperm as stingless bee males and 100 times as much as bumble bee males due to their large number of testicular tubules (Lago et al., 2020).

3.4.2 Reproductive Biology of *Apis florea*

Apis florea, a miniature honey bee, displays a variety of unusual traits in its reproductive and mating behaviour. The drones fly during their mating rituals between 11:30 and 14:00 hours, peaking between 12:30 and 13:00. In locations where drone congregations have been detected, the virgin queens of *A. florea* engage in nuptial flights with numerous drones. Drones gather at a height of

2–4 m in these open areas around small, densely leaved trees with a diameter of 10–12 m. The drone ejaculates sperm directly into the queen during mating. Additionally, following copulation, the queen develops no mating indication. It is well known that *A. florea* uses mutually policing mechanisms to regulate worker reproduction (Nagaraja and Abrol, 2020).

3.4.3 Reproductive Biology of *Apis cerana*

Every region of Asia has a high population of *Apis cerana*. It can modify itself to harsh weather and environmental conditions through longer flight times, efficient grooming and hygiene behaviour, and cooperative defence at the group level. *A. cerana* is known to aggregate when colonies are exposed to dangers such as predators or invaders. Alarm pheromones are produced by guard worker bees and control group behaviour. The apicultural sector also gains a great deal from *A. cerana*' high-quality byproducts in highland regions (Chen et al., 2022).

3.4.4 Reproductive Biology of *Apis dorsata*

In the open, the Asian honey bee (*Apis dorsata*) builds its nest as a colony gathered around the honeycomb. In *A. dorsata*, the workers, drones and queens are of the same size. The *dorsata* genome contains a large number of genes that set it apart from other Hymenoptera as well as many transposable elements. *A. dorsata*, which is native to Southeast and East Asia, is a significant pollinator and producer of honey. *A. dorsata* is significantly larger and its workers are almost twice as long as those of *A. mellifera*. *A. dorsata* further differs from other honey bees in that there is no caste-based variation in body size. *A. dorsata* honey has a different flavour than *A. mellifera* honey. Biochemical tests on honey have revealed variations in its antibacterial effectiveness, protein concentration and glucose content. Most of this diversity can be explained by the nectar that bees use to produce nectar, but the quality of the nectar also varies by bee species. The behaviour of *A. dorsata*'s nesting is the most obvious distinction between it and other honey bee species. Instead of making open nests hanging from branches or rocks, *A. dorsata* nests in hollows. A layer of up to 100,000 worker bees is constantly covering the exposed comb. *A. dorsata* colonies are reluctant to live in conventional Langstroth hives used by commercial beekeepers, unlike species that nest in confined spaces such as tree trunks or crevices in rocks. The ability of humans to domesticate this species and produce honey commercially is limited by this unique lifestyle and exceptionally high levels of defensive hostility (Oppenheim et al., 2020).

3.5 LIFE CYCLE OF ECONOMICALLY IMPORTANT SPECIES

The life cycle of a bee is divided into four stages. Only the queen bee in the colony can lay 2,000 to 3,000 eggs per day. By the third day, the egg is standing straight and has started to lean to one side. The queen bee lays both fertilized and unfertilized eggs. A queen or female bee develops from a fertilized egg. Drone bees, or unfertilized eggs, hatch to produce male bees (Kushram et al., 2022). A worker bee can be distinguished from a queen bee 3 days after the egg develops into a larva and 6 days after the egg is laid in the hive. All larvae, including queens, workers and drone bees, are fed royal jelly during their first 3 days as larvae. During this period, the larva repeatedly moults. Only the female larvae that will develop into queen bees are fed royal jelly after this period. Finally, worker bees seal the tops of the cells with beeswax to protect the larvae and speed their development into pupae (Kushram et al., 2022). Pupae develop little body hairs, eyes, legs, and wings until they resemble adult honey bees (Kushram et al., 2022). When the pupa reaches adulthood, the new adult bee chews its way out of the narrow cell. From the egg stage until adulthood, the queen bee develops over the course of 16 days. Drone bees need 24 days to fully mature into adult bees, while worker bees take 18 to 22 days (Kushram et al., 2022).

3.5.1 Life Cycle of *Apis cerana*

Bees pollinate various crops and produce honey, propolis and wax. All of these contribute to income generation, so bees are important in maintaining the ecological balance of the ecosystem. The productive yield of agricultural crops is increased by the increasing variety and number of pollinators, such as bees (Ziegler et al., 2021). In eusocial insects, division of work during reproduction is a key characteristic. Within eusocial Hymenoptera, there are striking morphological differences between fertilized queens and sterile workers of honey bees and ants. Their queens have the highest fecundity, although the morphology and function of their reproductive organs severely limit worker reproduction (Gotoh and Sasaki, 2021). The Endopterygota, the most varied superorder of insects, have a different pattern with more distinct stages. A single-celled egg develops through the egg, larval, pupal and adult stages. The larval stage is growth-dominant, where development begins, and culminates in the pupal stage when the larva metamorphoses into an adult (Martin et al., 2021).

A dense outer shell (chorion), created by defunct follicular epithelial cells, surrounds the bilaterally symmetrical honey bee egg. The nucleus and cytoplasm stand in for the shell-restricted interior of the egg. The yolk occupies most of the cytoplasm and serves as a reservoir of nutrients for embryonic development. Bee eggs are curved and have an elongated cylindrical shape. As is common in Diptera and Hymenoptera, the dorsal and ventral surfaces of the egg are concave and convex, respectively. The anterior and posterior ends of the egg are separate. With its posterior end significantly shortened, the egg moves in the oviduct (Boguslavsky and Zakharova, 2021). Honey bee eggs are between 1 and 5 millimetres long and resemble small grains of rice. In the brood region of the comb, the queen deposits her eggs in discrete hexagonal wax cells. Eggs hatch and larvae appear 3 days later. Bee eggs are difficult to find, but their presence indicates the presence of adults in the colony (a very useful tool when inspecting managed bee colonies (Mortensen et al., 2013). *A. mellifera* queens have over 200 ovarioles and have the capacity to deposit up to 2,000 eggs each day up to a total of 1.5 million eggs before they pass away. Worker bees, on the other hand, have up to a few dozen ovaries and cannot reproduce due to pheromone control by the queen and broods. Despite being few in number, the ovaries of worker bees still have the capacity to lay eggs and can be plastically activated without the colony's queen and brood (Gotoh and Sasaki, 2021).

3.5.2 Life Cycle Ontogenesis of *Apis mellifera*

The ontogenesis of *A. mellifera*, which is one of the most well-known Hymenopteran insects, possesses a number of characteristics that set the honey bee apart as a special case study in developmental biology. The primary characteristic of the honey bee's biology is that it is a social insect living in a well-organized colony family. The bee colony's structure is regulated by social interactions (Gotoh and Sasaki, 2021). The queen bee, who is usually the sole egg-laying individual and lays up to 3,000 eggs daily from February to October, serves as the foundation of the *A. mellifera* colony. Between days 7 and 10 after leaving the queen cell, a uniquely large cell, the young queen makes one to several mating flights, during which she mates with numerous drones, whose sperm can survive in the seminal vesicles for up to eight years. The egg-laying queen bee cannot feed the larvae or manage the nest. All these tasks are performed by domestic worker bees (Gotoh and Sasaki, 2021).

A queen bee can live for up to eight years in the right conditions. All other bee family members have shorter lifespans: drones live for only a few months and die soon after mating or when the worker bees no longer feed them in late summer or early autumn. Worker bees that hatch in the summer usually live for about a month. Worker bees that hatch in the autumn live until spring. The queen bee may temporarily stop laying eggs if there isn't enough nectar available during the warm season, which could result in the colony's demise. Provided with sufficient nectar, a bee family may have multiple egg-laying queens, but only one remains at the end of the season (Boguslavsky and Zakharova, 2021).

If beekeepers supply pollen substitutes or nutritional supplements during seasons when food is scarce and bee colonies are shrinking due to a lack of pollen or nectar, this can help strengthen bee populations. Bee food is one of the key components in this process. A hive requires 19.9 kg of pollen and 6.8 kg of honey throughout the year, and 100 mg of pollen for development and growth from larvae to adults. This is in addition to any adverse weather conditions or a lack of food in the cell. Like any living thing, bees require nutrients such as carbohydrates, protein, water, minerals and lipids, which they receive through collecting (Mansor et al., 2021). Neem oil was applied in a single exposure to groups of *A. mellifera* larvae that were 1–2, 3–4 and 5–6 days old. This has been found to have a significant impact on survival at all developmental stages. Multiple exposures to neem oil did not significantly affect adult bee mortality, nor larval and pupae survival. However, larvae that are 1–2 days old are most vulnerable. Neem oil in sufficient amounts may be able to manage Varroa destructors without harming bee colonies. The most vulnerable honey bee life stages were eggs and first instar larvae, and brood mortality rose as neem oil content was increased. Adult bees derived from treated larvae that were between 1 and 2 days old showed morphological diversity (Kaur et al., 2022).

When opposed to larvae inside worker cells, those inside queen cells receive a distinct feeding schedule. Both larval diets are made up primarily of the mandibular and hypopharyngeal gland secretions, which are found in the worker bees' heads. Pollen lipids and amino acids, which worker bees eat as 'bee bread', are the source of these secretions. Therefore, larvae are never directly exposed to pollen and its metabolites from worker bee gland secretions. The ratio of mandibular to hypopharyngeal glandular secretions in a mandibular-destined larva is roughly 1:1, whereas in a worker-destined larva, the ratio is roughly 1:2. The larva transforms into a pupa inside a covered larval chamber during a period of rapid growth (Martin et al., 2021). Larvae are known as 'open brood' in the honey bee colony because the cells are uncapped. Each caste of the honey bee has a different average number of days as a larva (worker: 6 days, drone: 6.5 days, queen: 5.5 days). The white larvae are in the shape of a twisted 'C' and are deposited at the bottom of waxy cells. The adult caring for the larva covers the cell with a wax cap, and the mature larva expands in an upright position in the cell and prepares to pupate (Mortensen et al., 2013).

Bee larvae pupate behind a beeswax cover. The pupae are covered with a waxy coating until they mature into adults, chew and leave the cage. Because the cells are closed, the pupae are called 'closed offspring'. As with the larval stage, the caste influences the timing of pupal development (workers: 12 days, males: 14.5 days, queens: 8 days) (Mortensen et al., 2013). The spermathecal reservoir becomes larger, the tracheal network surrounding it grows, and the spermathecal gland grows longer in queens during the pupal stage. This developmental process has never been observed in workers in the pupal stage, unlike queens. In addition, in worker bees, the development of the valve leaf and semi-circular muscles is stopped and changed to a rudimentary form in the intermediate stage of the pupa. The bee in this instance has developed features similar to those of an adult bee, such as wings, eyes, legs, and little body hair (Gotoh and Sasaki, 2021).

The adult stage, which is characterised by reproduction, is said to be post-mitotic. The adult's job is to guarantee that the cycle continues by mating, producing eggs, and depositing those eggs on or near a food source. The development of social communities further complicates the adult stage of the honey bee, *Apis mellifera*, in which some female bees are 'workers' who feed their young, protect their nests, and store food within the protected environment of the honeycomb, while others are 'queens' whose mission is to produce eggs, some of which will evolve into the next generation bee queens (Martin et al., 2021). The head, thorax and abdomen of adult bees can be distinguished by branching hairs. Compound eyes and antennae are the main features of the head. The thorax has three pairs of legs and two pairs of wings. When the second abdominal part contracts, the thorax is tapered. Stinging is the most prominent external feature of the abdomen. Saliva is only present in female honey bees because saliva develops in the fertilized ovipositor (Mortensen et al., 2013).

3.5.3 The Indian Bee, *Apis cerana/Indica* Fabricius

This is the common Indian bee, which may be found all over our country in both forests and plains. It is smaller than a rock bee but larger than a small bee. In caves, tree hollows and other hidden places, it constructs several parallel combs that are parallel to the entrance in plains areas and at right angles to the entrance in colder areas. This bee is more likely to swarm and flee. It may be domesticated, is gentle and is frequently raised in South India. Each colony produces 2 to 5 kilograms of honey annually. A queen can lay 350–1000 eggs daily. This species has a race that can be found in 'hilly tracts' (Pradhan and Nayak, 2021). *Apis cerana* forages from early in the morning till late at night for around 10 hours each day. Pollen and nectar are collected year-round regardless of flowers and weather conditions, so studies of the gathering activity of *A. cerana* have shown that this species follows a typical collection cycle. How hot it is affects how bees forage in a big way. In warmer months, bee colonies usually start foraging earlier than they do in colder ones. According to our research, the best months for honey bee foraging were May, April and March, with March being the least practical. Bee activity peaked at 1:00 pm and peaked at 8:00 am in each of the three months (Poudel et al., 2022).

Apis cerana is holometabolic. This means that an individual goes through four distinct life stages (egg, larva, pupa and adult). The queen lays one egg in each breeding room. The eggs are small, white and oval. The larvae hatch from the eggs after 3 days. The eggs are located in the upper central cell, and first instar larvae can be seen in the surrounding cells (Egelie et al., 2015). At the bottom of the cell, newly born larvae are twisted into the shape of the letter C. The larvae have a wet gloss, are blind and are white in colour. The larvae receive baby food and/or royal jelly inside the cell until they are large enough to pupate. At this point, the adult worker bees close the larval cells (Egelie et al., 2015). Under the wax cover, the larvae moult into pupae, and the pupal cells remain intact until the pupation is complete. *Apis cerana* drone pupa cells have unique pores in their wax coating. The function of these holes is not yet known (Egelie et al., 2015).

The new adults emerge from their capped cells after emerging from their pupal stage. Honey bees are superorganisms, meaning that the entire colony is considered a biological unit, rather than an individual bee. Swarming or breeding at the colony level often occurs in spring and summer. However, swarms may be more common in tropical areas, where conditions are more favourable year-round. It is believed that nectar and pollen abundance and big colony sizes are the main causes of swarming. A large colony will begin to swarm, giving birth to 10–20 daughter queens. Up to two-thirds of queens and adult workers leave the colony in search of a new home when their daughter queen is at the end of the pupal stage (Egelie et al., 2015).

3.6 STRUCTURAL DISTINCTION BETWEEN QUEEN BEES AND OTHER CASTES OF BEES

The queen bee has a significantly more developed abdomen and lacks pollen-collecting structures. The shape of the mandible is generally bifurcated at the distal end with long hairs on the outer surface. The proboscis of the queen and drone is shorter than that of the worker bee because neither collects nectar. The queen can eat under normal conditions to obtain the energy required to fly (Floris et al., 2021). The ovaries, calyx, lateral oviducts, and enlarged right oviduct, or median oviduct, comprise the upper reproductive organs of the honey bee queen's reproductive system. The tracheal network covers the spermatheca. The lateral oviducts, median oviduct, spermatheca, spermathecal glands, common duct, spermathecal duct, and sperm pump are all located in the middle portion of the reproductive tract (Kozii et al., 2022). The drone's reproductive organs consist of paired testes, conducting tubules, ducts (vas efferens, vas deferens and ejaculatory ducts), and accessory sex glands (seminal vesicles and mucus glands). The endophallus unit of the insect testes is the testicular tubule, also called the testicular follicle or vas deferens (Klein et al., 2021). Drones use different mating strategies because of variations in their size. Larger drones have fewer mating flights and a higher chance of success. Small drones fly more but are less likely to succeed. Because they start their flight in the early stages of their life, they make more attempts, but their chances of a successful mating flight are lower (Metz and Tarpy, 2022). Honey bee sperm are long, filamentous cells with tapering ends. A sperm has a length of 250–270 μm and a width of 0.7 μm. Its head area is short and narrow (10 μm long and 0.4–0.5 μm wide), the transitional centriole appendage and the flagellum are all included (Peng et al., 1993). The linear nucleus and the acrosomal complex are the two sequential, equal-sized components of the sperm head. When the perforator is located in the deep fossa, the conical bi-lamellar acrosome vesicles that cover the perforator to the nuclear distal end form an acrosome complex. The chromatin is tightly packed, and the nucleus is lengthy and thick. An axon with a 9+9+2 microtubule pattern, two important mitochondrial derivatives, and two additional bodies combine to produce sperm flagella. The mitochondrial derivatives are asymmetric in length and diameter and parallel to the axons along the entire spermatogenic flagellum. The two auxiliary bodies are elongated objects that are situated between each mitochondrial derivative and the axoneme (Yániz et al., 2020).

3.7 NUPTIAL FLIGHT PATTERN AND SEASONS FOR MATING FLIGHTS

Queens and drones are the founders of the colony, but a drone's activity is less eminent than that of a queen. Although drones have a significant risk of getting lost, being preyed on, starving to death, or dying during mating, they still exit the hive because mating is their sole objective. During spring and summer, drone nuptial flights take place, usually in drone swarm areas, and their exit time is in the afternoons. Usually, drones remain active at dawn, morning and late morning (Ayup et al., 2021). After emerging, the virgin queens spend a limited time in the colony before mating. This time frame is determined by the subspecies, climatic factors like temperature and rainfall, and the accessibility of drones, particularly mating, which takes place in the air in a short period of time. Typically, a queen mates with 7 to 10 drones, but sometimes as many as 45. When compared to several inseminations, a single insemination produced fewer spermatozoa entering the spermatheca. It is desirable for queen bees to mate with as many males as possible because the productivity of colonies led by queen bees mated with 30 to 60 or more males is high. However, it is possible to say that seven or more mates are sufficient. While drones choose the fastest route to a mate, queens can mate up to 16.2 kilometres away. The frequency of a queen's nuptial flight depends on how successfully she mated during her prior nuptial flights. Sperm are stored in the spermatheca after mating. The queen's secretions have an impact on the viability of the sperm, however, they have no effect on the number of sperms deposited in the spermatheca prior to fertilization. Sperm can live for a very long time, and the spermathecal gland secretions of mated queens are necessary for sperm viability. When they get 50% or more of fresh semen, queens fertilize every egg. The mated queen is fed by workers – who are 1 to 11 days old – with a 47-second duration for each feed (Allen, 1955; Abou-Shaara et al., 2021).

Although spring has been found to be preferable to autumn in certain places, queens can naturally mate throughout the year, with the exception of winter in temperate regions. There is currently no proof that queen mating ever occurs within the hive. In open areas, the queen bee can mate with drones from the same colony or drones from nearby apiaries. A remote location such as an island or artificial insemination can be used to prevent the mating of the queen bee. Swarming and absconding are two of the most important features in this regard. Swarming is a common way for honey bee colonies to reproduce naturally. In this instance, a number of queen cells up to 20 or more are developed. Except in cases of swarming or absconding, queens rarely leave their nests. Swarming typically occurs during the active seasons. During absconding, every bee, including the queen, leaves the colony and builds a nest elsewhere. Beekeeping is severely harmed by this behaviour. Absconding occurs when the colony is not in a place where it can survive due to an intense attack by ants, hornets or other bee adversaries, when it is exposed to extreme cold or heat, or when there is constant disturbance. Wax moth colonies may flee as a result of severe infestation (Abou-Shaara et al., 2021).

3.8 CASTES OF HONEY BEE

A queen bee, hundreds of drones, and thousands of worker bees make up a bee colony. Worker ants are sterile female insects, drones are males, and queens are fertile and functional females. A colony has only one queen. In comparison to members of other castes, she is significantly larger. When compared with her torso, her wings are significantly smaller. She has a longer, more tapering abdomen than the other colony members, giving the impression that she is a wasp. The mother of all bees and the only person that lays eggs in a colony is the queen. In *Apis mellifera*, up to 2,000 eggs can be laid daily. She has one or more nuptial flights with drones 5 to 10 days after emergence. She will start laying eggs and stop mating once her spermatheca is full of sperm. She lives for 3 years. Queen's substance is the name given to the secretion produced by the queen's mandibular gland. The queen material, when present in sufficient amounts, keeps colonies from swarming and fleeing, stops worker ovary growth and preserves colony cohesion. Depending on the situation, the queen might lay fertile or sterile eggs. The amount and calibre of food provided to the larva determines how the worker and queen are differentiated. The hypopharyngeal gland secretion of worker bees, known as royal jelly, is fed to the larva that develops into the queen. The lower part of the comb contains huge finger-shaped cells where the queen is raised. A colony may only support one queen at a time, but during unfavourable seasons, two queens are sometimes seen. As soon as the new queen is impregnated, the old queen is killed. A queen bee can usually only be kept during crowded seasons, but if the queen accidentally dies, the bees can raise a new queen. Supersedure is the term for the occurrence of raising queens off-season. The colony has effective family planning. Pollen and nectar are resources that are found in nature that affect egg production and egg laying. The workers prevent the queen from laying eggs if food is in short supply. The queen is meticulously cared for by young workers known as housekeepers, who feed her, keep her clean and comb her hair. With the exception of a swarm, the queen never exits the hive.

The drone's body is shorter than a queen's, but it is still significantly stronger than that of a worker or queen bee. Drones lack a stinger and a proboscis that would be useful for obtaining nectar. As a result, they are physically unable to perform the basic duties of the hive. Their sole purpose is to impregnate a young queen, which they cannot do until they are about 10 days old. They also help keep the honeycomb temperature stable. Only during the warm midday hours do they leave the hive. A colony of drones typically contains hundreds or even thousands of them, making it a very huge population. During the breeding season, drones are raised and tolerated. Before the monsoon and the winter, they are driven from the hive and left to starve to death. Drones are created from the queen's unfertilized eggs or from those workers who take over reproduction in the absence of the queen. A drone typically lives for 57 days. Mating takes place openly when the queen is in flight. The drone passes away during or right after the incident. In order for the vaginal organ to function, its abdomen must rupture.

The workers are the smallest beehive residents. They form the majority of the bees. A colony might have anywhere between 1,500 and 50,000 workers. They are flawed females unable to produce eggs. Sometimes, when a colony needs a queen, some workers start laying eggs, producing only drones. Whenever a new queen appears or spawns in a colony, these workers, known as spawn workers, are immediately executed. During the first 2–18 days of life, honey bees perform internal tasks within the hive, including making combs when some young bees begin to secrete beeswax. Later, they became hunter-gatherers, collecting water, pollen, honey and propolis (non-blue). Pollen is a nitrogen-containing food, essential for rearing young bees. The beeswax that makes the hive is a secretion from the wax glands in the abdomen of a worker bee. To produce 1 kg of beeswax, bees consume 10 kg of honey. Thus, the lifespan of a worker can be divided into two phases: the first 3 weeks for domestic work and the rest of their life for outdoor work. Their household duties include making a wax comb from the wax glands, feeding the young larvae with royal jelly secreted from the hypopharyngeal gland, feeding older larvae with bee bread, a mixture of pollen and honey, feeding and attending the queen, feeding drones, cleaning, ventilating and cooling the hive, guarding the hive, and evaporating nectar and storing honey. Their outdoor duties include collecting nectar, pollen, propolis and water, ripening honey in the honey stomach, and caste determination.

3.9 PHYSIOLOGY ASSOCIATED WITH CASTE DEVELOPMENT

In social insects, the pheromones of the queen suppress the development of ovaries in other workers and prevent the worker's reproduction. In *Apis mellifera*, the queen's mandibular gland secretes a mixture of fatty acids and aromatic chemicals, which are necessary to maintain the reproductive structure of the colony by inhibiting the growth of a functioning ovary (Strauss et al., 2008). Adults can be influenced by pheromones to treat larvae in different ways, which leads to the development of either workers or queens. If a worker bee kills a larva that has turned into a queen bee, the adult is unable to produce a new queen because the queen pheromone controls what the larvae eat. In addition, the queen releases two pheromones, a brood pheromone (BP), which consists mostly of esters, and the highly volatile E-β-ocimene, which slows the growth of worker ovaries. The queen-rearing process appears to be entirely regulated by worker actions in honey bees, which is an excellent example of how pheromone effects on worker activity lead to varying feeding for larvae. Caste in honey bees is decided by the production of queen cells by the workers in situations where the queen fails or dies, or when the colony is preparing to reproduce through fission. The amount of diet has a big impact on the caste and fitness of offsprings. The queen and worker castes in honey bee can be regulated by the total diet quantity offered to the larvae (Slater et al., 2020). Regardless of nutritional quality, larger meal portions improved adults' queenliness. Despite being raised on a middling-quality diet, all of the ad libitum individuals (20 out of 20) grouped with the commercially raised queens. In contrast to the lowest-quantity treatment (160 μl), which generated no queens regardless of diet quality, the majority of the 370 μl treatment (19 out of 33) clustered with commercially bred queens. During the measurement of queenliness, the dietary intake revealed a considerable impact on PC1 in the generalized linear mixed models (GLMM's). This study, which is validated by the observation of nursing behavior, investigated how diet size affects honey bee queen growth. Throughout growth, nurse bees provide queen larvae with more food than workers. When urgent queen rearing is required, nurses enlarge the cell size of worker-destined larvae to rear queens. Increased cell size likely enables nurse bees to boost queen-rearing provisioning. Because this quantity creates workers, the smallest diet quantity (160 μl) was used in earlier in vitro procedures. To create the additional treatments, the quantity was increased from 160 μl to 370 μl in 30-l increments. Larvae were fed more than they could eat as part of an additional ad libitum treatment. A total of 110 μl of feed was given to all larvae over the first 5 days of development: 10 μl on days 1 and 2, 20 μl on days 3 and 4, 30 μl on days 4 and 5, and 40 μl on days 5 and 6. The overall feed quantity ranged from 160 μl to 370 μl during the larvae's sixth day of development, depending on the diet quantity treatment. Larvae in the ad libitum treatment received 200 μl of food daily until gut purging. Individuals were placed into pupation plates after a gut purge (Slater et al., 2020). Since honey bees have a social structure, they are frequently referred to as superorganisms. A healthy colony must be able to perform its duties and resist various abiotic and biotic stressors, as well as support disease-free and functioning members (workers and reproductive organs). Adequate intake of high-quality pollen has been shown to reduce susceptibility to *Nosema ceranae,* reduce pathogen load, increase bee immunity and hibernation success, improve drone sperm quality, and produce healthier bees that are more resistant to pesticide stress and disease. The behaviour of bees and their diet have been linked in other research. Nutrition can affect the physiology of emerging spring workers, which in turn, can change their behaviour. When larvae did not have access to the best nourishment, they grew up to be ineffective foragers and waggle dancers. The feeding choices of pollinators like bees may also be influenced by the pollen protein:lipid ratio. Therefore, proper nutrition can be considered the first line of defence for bee colonies to better withstand both biotic and abiotic stress (Tsuruda et al., 2020). Caste determination is a physiological phenomenon. Pheromones and hormones have a big effect. Juvenile hormone (JH) stands out in the first category. It is obvious that the quantity and components of royal jelly play crucial roles in the caste division of social Hymenopteran insects. The effect of higher JH titres is only seen in termites, possibly after numerous moults and metamorphosis in other social logging. The division of labour and the worker's seasonal status are included in this. The use of juvenoid insect growth regulators is significantly impacted by the endocrine foundation of caste distinction. Entire generations of insect physiologists have been interested in the exciting field of programming the development and function of hormonally modified castes and the underlying triggering mechanisms of endocrine activity (Hossain et al., 2022).

3.10 CONCLUSION

The caste and sex determination process of the honey bee is a complex phenomenon. Various factors, including morphological, ecological, physiological, genetic and epigenetic factors, take part in the process to make it a well-structured web of biological complexities. The immense commercial value of honey always attracts researchers to look beyond the traditional understanding of the physiological processes associated with honey bees. The conventional 'queen and worker' concept is good enough for preliminary investigations, but future scope always lies in the complexities of the molecular and ecological behaviours of the honey bee. A few researchers have reported on the transition or shift of sex in the honey bee populations in their recent works. A few reports are emerging in regard to chromosomal aberrations in honey bees, which may lead to the formation of a new kind of sex as well. Although a huge amount of work has been conducted in the past, looking at its immense market value, we can conclude that this fascinating field of honey is only getting started, and future researchers need to dig deep to reveal all the mysteries and myths associated with the sex and caste determination of honey bees.

REFERENCES

Abou-Shaara, H. F., Adgaba, N. & Al-Ghamdi, A. A. 2021. Current knowledge about behaviors of honey bee queens with highlighting of the importance future studies. *The Journal of Basic and Applied Zoology*, 82, 1–7.

Ali, H., Iqbal, J., Raweh, H. S. & Alqarni, A. S. 2021. Proboscis behavioral response of four honey bee Apis species towards different concentrations of sucrose, glucose, and fructose. *Saudi Journal of Biological Sciences*, 28, 3275–3283. doi:10.1016/j.sjbs.2021.02.069

Allen, M. D. 1955. Observations on honeybees attending their queen. *British Journal of Animal Behaviour*, 3, 66–69.

Ayup, M. M., Gartner, P., Agosto-Rivera, J. L., Marendy, P., Souza, P. D. & Galindo-Cardona, A. 2021. Analysis of honeybee drone activity during the mating season in Northwestern Argentina. *Insects*, 12, 566.

Bilodeau, L. 2022. Genetic diversity and structure in a closed breeding. *Journal of Economic Entomology*, 115, 682–687.

Boguslavsky, D. V. & Zakharova, I. S. 2021. Role of external factors in embryogenesis of *Apis mellifera*. *Russian Journal of Developmental Biology*, 52, 422–429.

Chen, S., Wang, Y., Li, Y., Zhang, X. & Wu, J. 2022. Effects of honeybee (*Apis cerana*) visiting behaviour on toxic plant (*Tripterygium hypoglaucum*) reproduction. *AoB Plants*, 14, plac002.

Choudhary, R., Chand, S., Singh, T., Singhal, R. K., Chourasiya, V. K. & Indu 2022. Sex determination. In: Çalişkan, M. (ed.) *Genetic Polymorphisms-New Insights*. IntechOpen.

Delaplane, K. S., Given, J. K., Menz, J. & Delaney, D. A. 2021. Colony fitness increases in the honey bee at queen mating frequencies. *Behavioral Ecology and Sociobiology*, 75, 1–12.

Egelie, A. A., Mortensen, A. N., Gillett-Kaufman, J. L. & Ellis, J. D. 2015. Asian honey bee - *Apis cerana*. Entomology and Nematology Department, EENY-616.

Faux, C. M. 2021. Honey bee anatomy. In: Kane, T. R. & Faux, C. M. (eds.) *Honey Bee Medicine for the Veterinary Practitioner*. Wiley Publisher.

Floris, I., Pusceddu, M., Niolu, P. & Satta, A. 2021. Where is the honey bee queen flying? The original case of a foraging queen. *Insects*, 12, 1035.

Fridi, R., Aoul, N. T., Catays, G., Basso, B., Bienefeld, K., Gregorc, A., Vignal, A. & Kamila, C.-T. 2022. Genetic diversity and population genetic structure analysis of *Apis mellifera* subspecies in Algeria and Europe based on complementary sex determiner (CSD) gene. *Apidologie*, 53, 1–14.

Gotoh, A. & Sasaki, K. 2021. Caste differentiation of spermatheca and organs related to sperm use and oviposition in the honeybee, *Apis mellifera*. *Apidologie*, 52, 262–271.

Hossain, S., Rahman, M. M., Riaz, G. M. & Rahman, H. 2022. Insights of caste determinations in social insects. *Acta Scientifica Malaysia*, 6, 48–54.

Iqbal, J. & Mueller, U. 2007. Virus infection causes specific learning deficits in honeybee foragers. *Proceedings of the Royal Society B-Biological Sciences*, 274, 1517–1521. doi:10.1098/rspb.2007.0022

Jain, R. & Brockmann, A. 2020. Sex-specific molecular specialization and activity rhythm-dependent gene expression in honey bee antennae. *Journal of Experimental Biology*, 223, jeb217406.

Janczyk, A. & Tofilski, A. 2021. Monthly changes in honey bee forewings estimated using geometric morphometrics. *Journal of Apicultural Science*, 65, 139–146.

Jasper, W. C., Brutscher, L. M., Grozinger, C. M. & Niño, E. L. 2020. Injection of seminal fluid into the hemocoel of honey bee queens (*Apis mellifera*) can stimulate post-mating changes. *Scientific Reports*, 10, 1–18.

Kaur, G., Singh, R., & Singh, A. 2022. Impact of neem oil on developmental stages of honey bee *Apis mellifera* L. *Indian Journal of Entomology,* 84**,** 783–787.

Klein, C. D., Kozii, I. V., Wood, S. C., Koziy, R. V., Zabrodski, M. W., Duylyuk, I., de Mattos, I. M. D., Moshynskyy, I., Honaramooz, A. & Simko, E. 2021. Testicular changes of honey bee drones, *Apis mellifera* (Hymenoptera: Apidae), during sexual maturation. *Journal of Insect Science*, 21, 1–12.

Kozii, I. V., Sarah, C. W., Roman, V. K. & Elemir, S. 2022. Histomorphological description of the reproductive system in mated honey bee queens. *Journal of Apicultural Research*, 61, 114–126.

Kushram, T., Sahu, M. K. & Bairwa, P. L. 2022. Package and practices of apiculture. *The Pharma Innovation Journal,* SP-11, 719–724.

Lago, D. C., Martins, J. R., Dallacqua, R. P., Santos, D. E., Bitondi, M. M. G. & Hartfelder, K. 2020. Testis development and spermatogenesis in drones of the honey bee, *Apis mellifera* L. *Apidologie*, 51, 935–955.

Lu, C.-Y., Weng, Y.-T., Tan, B. & Hsu, C.-Y. 2021. The trophocytes and oenocytes of worker and queen honey bees (*Apis mellifera*) exhibit distinct age-associated transcriptome profiles. *GeroScience*, 43, 1863–1875.

Mansor, M. S., Al-Bajari, H. A. J. & Ibrahim, M. K. 2021. Effect of feeding with vitamins, sugars and pollen on the biology of honey bee *Apis Mellifera* L. (Hymenoptera : Apidae) during the autumn season and the use of various wintering methods. *IOP Conference Series: Earth and Environmental Science*, 923, 012018.

Martin, N., Hulbert, A. J., Bicudo, J. E. P. W., Mitchell, T. W. & Else, P. L. 2021. The adult lifespan of the female honey bee (*Apis mellifera*): Metabolic rate, AGE pigment and the effect of dietary fatty acids. *Mechanisms of Ageing and Development*, 199, 111562.

Metz, B. N. & Tarpy, D. R. 2022. Variation in the reproductive quality of honey bee males affects their age of flight attempt. *PeerJ*, 10, e13859.

Mortensen, A. N., Schmehl, D. R. & Ellis, J. 2013. *European Honeybee-Apis mellifera*. Entomology and Nematology Department, University of Florida.

Moškrič, A., Formato, G., Smodiš Škerl, M. I. & Prešern, J. 2020. Reproductive technologies in the honeybee (*Apis mellifera*). *Reproductive Technologies in Animals*. Academic Press.

Nagaraja, N. & Abrol, D. P. 2020. Reproductive biology of *Apis florea*. In: Abrol, D. P. (ed.) *The Future Role of Dwarf Honeybees in Natural and Agricultural Systems*. CRC Press.

Netschitailo, O., Raub, S., Kaftanoglu, O. & Beye, M. 2021. Sexual diversification of splicing regulation duringembryonic development in honeybees (*Apis mellifera*),A haplodiploid system. *Insect Molecular Biology*, 31, 170–176.

O'Brien, W., Tashakkori, R., Parry, R. M., Hamza, A. & Graber, J. 2022. Estimating the number of drones at the entrance of a honey bee hive using machine learning tools. *SoutheastCon 2022*. IEEE.

Oppenheim, S., Cao, X., Rueppel, O., Krongdang, S., Phokasem, P., Desalle, R., Goodwin, S., Xing, J., Chantawannakul, P. & Rosenfeld, J. A. 2020. Whole genome sequencing and assembly of the Asian honey bee Apis dorsata. *Genome Biology and Evolution*, 12, 3677–3683.

Pan, L. X., Hu, W. W., Cheng, F. P., Hu, X. F. & Wang, Z. L. 2022. Transcriptome analysis reveals differentially expressed genes between the ovary and testis of the honey bee *Apis mellifera*. *Apidologie*, 53, 1–12.

Paolillo, G., De Lorio, M. G., Filipe, J. F. S., Riva, F., Stella, A., Gandini, G., Pagnacco, G., Lazzari, B. & Minozzi, G. 2022. Analysis of Complementary Sex-Determiner (CSD) allele diversity in different honeybee subspecies from Italy based on NGS data. *Genes*, 13, 991.

Park, H.-G., Kim, B.-Y., Kim, J.-M., Choi, Y.-S., Yoon, H.-J., Lee, K.-S. & Jin, B.-R. 2021. Upregulation of transferrin and major royal jelly proteins in the spermathecal fluid of mated honeybee (*Apis mellifera*) queens. *Insects*, 12, 690.

Peng, C.Y.S., Yin, C.M. & Yin, L.R.S. 1993. Ultrastructure of honey-Bee, *Apis mellifera*, sperm with special emphasis on the acrosomal complex following high-pressure freezing fixation. *Physiological Entomology*, 18, 93–101. doi: 10.1111/j.1365-3032.1993.tb00454.x

Phokasem, P., Liuhao, W., Panjad, P., Yujie, T., Li, J. & Chantawannakul, P. 2021. Differential viral distribution patterns in reproductive tissues of *Apis mellifera* and *Apis cerana* drones. *Frontiers in Veterinary Science*, 8, 1–8.

Poudel, P. K., Baral, B., Shah, R., Cb, R. K. & Singha, N. B. 2022. Study of foraging behavior of *Apis cerana* during summer in Sundarbazar, Lamjung, Nepal. *Big Data in Agriculture (BDA)*, 4, 43–46.

Pradhan, K. & Nayak, P. 2021. Apiculture. In: Ghoneim, K. (ed.) *Advances in Agricultural Entomology*. AkiNik Publications.

Slater, G. P., Yocum, G. D., & Bowsher, J. H. 2020. Diet quantity influences caste determination in honeybees. *Proceedings. Biological Sciences*, 287, 20200614.

Strauss, K., Scharpenberg, H., Crewe, R., Glahn, F., Foth, H. & Moritz, R. 2008. The role of the queen mandibular gland pheromone in honeybees (*Apis mellifera*): Honest signal or suppressive agent?. *Behavioral Ecology and Sociobiology*, 62, 1523–1531. 10.1007/s00265-008-0581-9

Sun, Y., Zhang, J., Tang, X., Wu, Z., Gorb, S. N. & Wu, J. 2021. Specialized morphology and material properties make a honey bee tongue both extendible and structurally stable. *Acta Biomaterialia*, 136, 412–419.

Tsuruda, J.M., Chakrabarti, P. & Sagili, R.R. 2020. Honey bee nutrition. *Veterinary Clinics of North America: Food Animal Practice*, 37(2021), 505–519. doi: 10.1016/j.cvfa.2021.06.006

Watson, T. O., Buchmann, G., Young, P., Lo, K., J.Remnant, E., Yagound, B., Shambrook, M., F. Hill, A., P. Oldroyd, B. & Ashe, A. 2022. Abundant small RNAs in the reproductive tissues and eggs of the honey bee, *Apis mellifera*. *BMC Genomics*, 23, 1–21.

Yániz, J. L., Silvestre, M. A. & Santolaria, P. 2020. Sperm quality assessment in honey bee drones. *Biology*, 9, 174.

Ziegler, C., Sinigaglia, T., Martins, M. E. S. & Souza, A. M. 2021. Technological advances to reduce Apis mellifera mortality:A bibliometric analysis. *Sustainability*, 13(15), 8305.

4 Honey Bee Behavior

Muhammad Adnan Bodlah, Alishbah Mohsin, Ayesha Younas, Shabbir Hussain, Aneela Ashiq, Shumaila Khan, Imran Bodlah, Ahmad Bilal Arif, Ammara Gull-E-Fareen, Muhammad Tariq Rasheed and Muhammad Amjad Bashir

4.1 INTRODUCTION

As social insects, honey bees are special and crucial for acquiring honey, getting food, eliminating poverty, finding a job, being paid, preventing pollution, and taking care of human health (AU-IBAR, 2019). The purpose of raising honey bees is to produce honey, preserve biodiversity, pollinate crops, and provide beeswax, propolis, royal jelly, venom, apitherapy, and financial gain. According to one report (FAO, 2019), humans first started beekeeping and searching for honey 4500 years ago. About three-quarters of the angiosperm plants on the planet are pollinated by honey bees. During ancient times, honey bee farming was done to produce honey for domestic use and for sale only (Aizen and Harder, 2009). Natural honey harvesting generated US$ 3.3 billion in revenue in 2011 alone (FAO, 2015), and the nations that produce it observed significant increases in their economies (Thompson, 2012). But occasionally, the number of honey bee colonies and valued goods declined (Aizen and Harder, 2009; Chantawannakul and Ramsey, 2018).

Honey bees are social creatures by nature, and one bee cannot live on its own for very long without the assistance of its nest-mates. A bee leaves the hive, travels over distances of many kilometers, and quickly and efficiently visits hundreds of flowers to collect food (nectar and pollen). Furthermore, it roams to obtain information and collects water and resin. Both motor skills and sensory abilities are highly developed. Bees sense color, forms, and patterns, as well as motions with great temporal resolution (Giurfa, 2003). Bees also see the world in color (Srinivasan et al., 1998; Srinivasan 2010). A wide variety of smells may be distinguished by their olfactory sense (Laska et al., 1999), and their mechanosensory perception is also very strong due to the millions of hairs that cover their bodies and the sensory organs that provide information on position and movement. Building a wax comb is an example of a complex controlling behavior where innate mechanisms or learning are dominant (e.g. effectively handling complicated flower structures for extracting food) (Free et al., 1992). In general, the hive and food supplies are the locations of interest, and natural selection has developed the learning ability for these local cues (Menzel, 1985). The ability to learn landmark constellations and celestial cues for navigation (such as the azimuthal position of the sun and the ray pattern of the blue sky) ensures a secure return to the nest and enables maximum foraging effectiveness (Collett and Zeil, 1998).

Crop pollination is the most significant bee activity in terms of increase in economic importance for humans that is obtained from bees' pollination services. In comparison to nectar-gathering worker activities, pollination is more effectively accomplished when bee workers start gathering pollen from flowers to utilize it as a protein-rich source for the hive (Cane and Schiffhauer, 2001). The overall quantity of the pollen collected to store in hives was subjected to bi-directional selection, which demonstrated that the trait for more pollen collection is genetically regulated, and the quantity depends on the number of specialized pollen collection foragers (Page et al., 2000). This chapter reviews the overall activity of bee behavior inside honey bees' nests.

 DOI: 10.1201/9781003354116-4

4.2 HONEY BEE BEHAVIOR WITHIN THE HIVE

4.2.1 Larval Emergence and Egg Migration

The egg is immobile after oviposition until the larva emerges. It is forced down inside the cell when the workers move inside the cell base as deeply as is feasible. Workers may enter the cells for the purpose of creating or preserving heat within the honeycomb (direct incubation). During this procedure, the worker and its antennae are immobile. This finding supports earlier claims that an embryonic "tipping over" is not a natural element of the development phase (Dupraw, 1961; Siefert P. et al., 2020). As a result, a decreasing number of eggs represents the likelihood of how frequently workers entered the cells for temperature regulation. In certain cases, thorough checks may cause the egg to shift slightly. Flexing and bending motions gradually get harder and stronger until the wax surface is hit by the larva's anterior end, which is ultimately useful to start the process of the larva hatching from its upright position. After that, the hatched larvae don't drag themselves to the upper side of the cell, but slowly move down toward the cell's base and the side walls in some situations. The egg membranes completely disintegrate during the hatching process after the larva hatches, when it receives its first feeding (Dupraw, 1961).

4.2.2 Larval Examination and Eating

Examinations involve the input and processing of sensory data to ascertain things like the age, placement, and condition of the brood as well as the composition of the cells. Fast antennal movement is the key inspectional trait. It was observed that the workers showed no antennal moves during resting or heating behavior, which is differentiated by the frequency of pumping movements of their abdominal body parts (Kleinhenz et al., 2003). When there are no other behaviors present after the examination, it lasts either for a very short time, after which the worker does not try to enter the cell, or for a very lengthy time, which is typical inside a hive containing very young larvae.

The examination that occurs before the meal is given out takes from almost 11 to 63 percent of a total feeding visit's time. The worker inspects before feeding, during which it will flex its antennae vigorously and point its mouth and antennae at the larva. Following the observation, the worker begins to slowly approach the larva while vibrating its mandibles. All tagmata typically remain immobile while food is being provided, though the antennae may occasionally twitch slightly. For immature larvae, food is placed carefully, but workers can place food anywhere near an adult larva, which could be almost three days old or older. The overall larval moves occur after feeding events, suggesting that the larva's goal remains focused on getting to the just-delivered food. For instance, after receiving a bite from a worker's mouth, larvae do not start the migration. However, this was contrary to what was previously believed (Nelson et al., 1924), that the workers cannot only provide food close to the larva's mouth, but mouth-to-mouth feeding was also observed. According to the observation in a study, after the larva has finished eating, the process of cocooning starts, followed by different tapping moves of the larval anterior ends where its silk glands are present. These moves trigger the cell's shift, twisting from a transversal to a longitudinal direction (Jay, 1964).

4.2.3 Building Combs and Capping Cells

The wax material used inside the comb has been observed inside the workers' mandibles in two different configurations: a non-transparent string-like structure and transparent wax scales. The latter has mostly been observed in our observations during emergencies, such as when the combs are immediately fixed to the adjacent glass in the aperture just after the colonies are set up. But later in the colony's growth, the utilization of wax strings is frequently seen. Bees use these wax strings in colony growth; they allow for quick alterations in the division of labor, because comb remodeling involves labor workers with underdeveloped wax glands. The worker extends the thread between the mandibles and rotates its head fast back and forward like a pecking bird to produce a string of wax.

The wax strings extend just below the caput and the thorax, and can be several millimeters long. To move lengthy strings, workers use their mandibles and prothoracic legs. Moreover, workers use the basitarsal brushes of their legs to remove waxy scale from intersternal pockets (Ro sch et al., 1927). It needs roughly 5 seconds to retrieve the scale from the pocket, and the lateral movements of the scale to the mouthparts with the current busy leg take just 400 milliseconds.

4.2.4 Preservation of Pollen and Nectar

Workers enter the cell ventrally and climb upward to store nectar and honey (Park, 1925). A recurrent half-circular movement spreads the meal from the worker's stomach to the upper cell wall. The mandibles dive into any liquid food that may already be present in the cell. The mandibles are held open and the proboscis is held folded during the entire procedure. Without a focus on the lower half of the cell, the worker can fill the cell uniformly; even if the food sticks on the upper surface of the cell wall, it is drawn downward by gravitational force. The worker's proboscis absorbs liquid nutrition from the filled cells, an action that is achievable whether the workers are aligned toward the cell or not. The foraging individuals use their prothoracic legs to cling on to the lower cell wall, close to the examined cell, after determining that it's a cell that is suitable for storing pollen. The forager places its bent abdominal part against the lower portion of the appropriate cell wall, and its meta-thoracic legs grip the top wall. Workers were never seen to enter the cubicle with their meta-thoracic legs extended and hanging loosely within as claimed in other publications (Casteel, 1912).

4.2.5 Thermoregulation

Honey bees use a variety of behaviors to regulate the temperature in order to prevent deformities observed in the brood or in developing adults when this range is surpassed (Tautz et al., 2003). Honey bees keep the temperature of the brood region surrounding a colony at 33–36°C as long as there is a brood present (Kleinhenz et al., 2003). Individual workers congregate in the brood area, metabolic heat is produced, and direct incubation in which workers warm their thorax by muscular contractions is used to heat the region, while individual workers disperse, fans are used, and water is evaporated during the cooling process (reviewed in Jones and Oldroyd, 2006). During a study, up to six observation hives were placed in trials between 3 and 5 m from heat-emitting radiators (Siefert et al., 2020). One further hive, however, was situated directly above the heat source and was used for macro recordings. Because of this, the majority of hives maintained a lower temperature of 28°C; for more information, see the discussion in Siefert P. et al. (2020) that the hive above the radiator temperature was not determined. Workers in the cell were found to be motionless in hives far from the heat source for up to 90 minutes. It was noticed that a discharge of transparent fluid dripped into the cell in hives that were directly above the radiator. As a result, the workers' inactivity probably served to actively produce heat through the vibration of their thoracic muscles (Esch et al., 1991) or to passively preserve heat by the clustering (Free and Simpson, 1963) of sleeping bees throughout the combs. During this time, no antennal movements were displayed by bees. By observing their abdominal movements, heat-generating bees can be distinguished from those that are resting. The abdominal pumping actions of workers who are at rest are irregular and interrupted, whereas those of individuals who are producing heat are swift and constant (Kleinhenz et al., 2003). Intriguingly, both short- and long-term cell occupancy was noticed in cells at distinct phases of growing brood without observing antennal movements. The workers frequently went increasingly deep inside the cell throughout the procedure, suggesting either active or passive thermoregulation.

Honey bees scatter throughout the comb and abruptly depart the hive if the colony's temperature rises above acceptable levels. Workers begin fanning the hive entryway as foragers gather water to evaporate inside the hive. The brood nest has a relative humidity of about 30–50 percent, which is typically attained by the evaporation of stored nectar.

4.2.6 Sanitation and Cannibalism

The prevention and elimination of fungi, parasites, and mold that threaten the colony's survival are examples of hygienic behavior. Workers engage in a variety of hygienic behaviors, including cannibalism and cleaning of surfaces, cells, larvae, other nest-mates, or themselves (auto-grooming and allo-grooming). The colony uses cannibalistic behavior to efficiently recycle the protein and to stop fungi and mold from developing on decomposing individuals. Workers can consume progeny at any stage of development, except the final 72 hours when workers' cuticula hardens, in cases of male development or diploid drones (Woyke, 1963). According to observations during a research study, cannibalization frequently happened without a clear reason (such as when offspring turned dark), which suggested that the workers employ chemical information to recognize dead, sick, parasitized, or malformed offspring (McAfee et al., 2018).

A worker who is grooming can still chop off the legs or damage the shell with its mandibles to reduce *Varroa* mite infestation even when the mite's cuticula has already hardened (Moosbeckhofer, 1992). Nest-mates for grooming get invitations from the worker bees to engage in a specific behavior. The "grooming invitation dance" entails fast leg motions for self-cleansing as well as body wriggling and bending (Bozic and Valentincic, 1995). Different *Apis* species engage in different amounts of auto- or allo-grooming with *Tropilaelaps clareae* and *Varroa destructor*, and social grooming is positively connected with the severity of the infectious tracheal mite (*Acarapis woodi*) (Pettis and Pankiw, 1998). A "rocking movement" of the honey bees, in which the tarsi of worker bees' front legs and mandibles are utilized as scrapers, is another name for the mechanical cleaning process of surfaces inside a hive (Alfonsus, 1932).

4.3 FORAGING BEHAVIOR

When a worker switches to foraging, this begins a new phase of her existence, for the first time in the environment outside the colony, which is combined with the multiple foraging duties and environmental interactions with the environment, as presented in the upcoming paragraphs.

4.3.1 Foraging Tasks

There are mostly two types of foraging bees: the scout bees, with the task of looking after the reticent bees and the best sources of food, and reticent bees. Generally, about 40 to 90 percent of the entire foraging population are reticent bees (Nest and Moore, 2012). Forager bees need to sleep at night, and lack of sleep may affect their ability to remember where they've been (Beyaert et al., 2012). Water, pollen, nectar, and resin foraging are the four categories that foraging bees' activities fall under. Occasionally, foraging bees gather wax from the scales of insect species like *Ceroplastes* sp. (Dimou and Thrasyvoulou, 2007).

Whether for pollen or nectar, this kind of foraging is included in a colony-based character with a genetic component that is controlled by the genotypic characters of the bee species (Hunt et al., 1995). The mentioned duties also depend on the decisions made by foraging bees, both collectively and individually. Sucrose response thresholds are yet another crucial element for this kind of foraging task. The susceptibility to sucrose of different honey bee workers varied. The bees with water foraging characters had the lowest susceptibility, followed by pollen to nectar foragers, and lastly, foragers of both nectar and pollen. Researchers made a connection between these variations and how foraging bees divided their labor. According to Simone-Finstrom et al. (2010), resin foragers had lower sucrose response thresholds than pollen foragers. Based on recent data, bee workers with both high and low sucrose threshold responses have widely been observed while foraging for nectar and pollens respectively (Pankiw, 2005). Honey bee nests increase the number of pollen-foraging individuals without increasing foraging places when pollen is insufficient or of poor quality (Pernal and Currie, 2010).

4.3.2 Foraging Time

The bees' foraging activity starts in the morning and mostly stops in the evening. According to various studies, honey bee workers start their foraging at 6:15 am (Joshi and Joshi, 2010); however, the region can have a massive effect on this start time. Alqarni (2006) discovered that under desert conditions, more foragers left the colony earlier at 8:00 am than at 10:00 am. The foragers' activity generally varies all day, starting from the morning to the evening. According to the study by Reyes-Carrillo et al. (2007), the pollen collection was highest early in the morning and lowest in the afternoon. According to Pernal and Currie (2010), the mean foraging rate was higher in the afternoon than in the morning. According to Yucel and Duman (2005), onion flowers were visited by honey bee workers from 8:15 to 16:30 hours, with a peak foraging hour falling from 11:00 to 12:00. Similarly, in another research study on the *Sysirinchium palmifolium* plant (Silva et al., 2013), the foragers were able to recall the time of day when they find the larger food supplies in their area. The visit time for Chinese cabbage, broccoli, and kohlrabi blossoms was 6.92, 6.50, and 5.54 s, respectively (Sushil et al., 2013).

4.3.3 Foraging Distance

The energy hypothesis suggests that foragers determine the feeder travel distance toward food resources based on the energy used during foragers' flight and on the energy found to be expended during speed. However, the retina plays a key role in distance calculation and estimation during the flight activity of these worker honey bees. *Apis mellifera* bees had a mean of 1526 meters' foraging distance, followed by the pollen-collecting bees, which had a mean of 1743 (simple landscape) and 1543 (complex landscape) meters' foraging distance (Steffan-Dewenter and Kuhn, 2003). During the month of July, *Apis mellifera*'s average foraging distance was about 670 meters. Similarly, during August environmental conditions, it was about 1430 meters for small and 2850 for larger colonies (Beekman et al., 2004).

4.3.4 Foraging Preference

Foraging bees choose to gather pollen, water, resin, and nectar from specific sources. Larger water containers and continuous water sources were preferred by water foragers over stable ones and smaller ones, respectively (Abou-Shaara, 2012). Additionally, foraging bees prefer to obtain water from different peculiar sources, like cow manure, and also from stagnant clean water (Butler, 1940). Sometimes, nectar foragers favor one food source over another, as well as one flower's location over another. According to Sushil et al. (2013), broccoli, kohlrabi, and Chinese cabbage gathered the most honey bee foragers, with 6.05, 5.35, and 5.05 plants, respectively. During a research study, Mayer and Lunden (1988) observed more nectar foragers on the apex of Manchurian crabapple flowers than on Red Delicious apple flowers. Similarly, on hairy vetch (*Vicia vitlosa* Roth) flowers, Weaver (1965) observed variations in honey bee foraging behavior; some bees utilized the flower base, while others were found to be using the flower mouth. The apple tree twigs or branches in the middle of trees are preferred by honey bees over those that are higher up or lower down in the tree (Mattu et al., 2012). Similar preferences can be found in pollen and resin foragers.

4.3.5 Monitoring of Foraging Activity

The total number of bees leaving or returning, or both behaviors, are important factors that can be utilized to measure the foraging activity of the bees (Beekman et al., 2004; Pernal and Currie, 2010; Abou-Shaara et al., 2013). Similarly, other factors include fluctuations and peaks of foraging time, speed, distance, foraging commencement, and rest or quiet time (Malerbo-Souza, 2011). Moreover, there are foraging factors, such as foraging individuals per flower, the number of flowers visited, time spent, the method used to gather pollen or nectar, and foraging positions (Mayer and Lunden, 1988; Mattu et al., 2012; Sushil et al., 2013; Brittain et al., 2013). The bee workers' foraging flight paths can be monitored using harmonic radar (Smith, 2002). The important factor that can be focused on while studying bee foraging activity is the equal strength of the studied bee colonies, particularly the quantity of pollen and brood frames; the presence of any diseases in the studied colonies; the time of day and year; temperature and relative humidity; and the presence of bee competitors or predators. Using forceps in front of the colonies and an aspirator, forager bees can be gathered from the hive entrance (Yucel and Duman 2005). Additionally, specialized tools (like bee scans) can be used to count foraging bees.

4.3.6 Foraging Behavior of Honey Bee Subspecies

The Carniolan, Italian and Yemeni bee species were found to differ in their foraging behavior. Yemeni bees had the highest foraging activity in dry desert conditions, followed by Carniolan and Italian bees (Alqarni, 2006). Additionally, Ali (2011) discovered that Yemeni honey bee foraging rates were higher than those of Carniolan bee species throughout the months of June and August and at various monitoring times, including 6 to 7 am, 11 to 12 am, and 4 to 5 pm. The same pattern was discovered by Abou-Shaara et al. (2013), who discovered that under desert conditions, Yemeni honey bees were more active foragers than Carniolan honey bees. The variations in the morphological characters of honey bee subspecies can help to explain some of the variances in their foraging behavior. According to reports, bees with large wings can fly farther than those with short wings (Mostajeran et al., 2006). Bees with broken wings flew closer to the hive and made fewer foraging trips, according to research by Higginson et al. (2011). The foragers' activity, sealed brood area, and bee population all showed positive associations (Abou-Shaara et al., 2013). Similarly, the foraging behavior of honey bee subspecies may vary depending on how well they have adapted to their environment (Alqarni, 2006).

4.3.7 The Value of Foraging Activities

Plant pollination depends on the foragers' activities, like the collection of nectar, pollen, water, and resin. Various plant species depend on pollination by foragers; apples, blueberries, pear varieties, almonds, cantaloupes, and varieties of rape have been discovered to be honey bee–pollinated plants (Mayer and Lunden, 1988; Reyes-Carrillo et al., 2007; Blazyte-Cereskiene et al., 2010). Bees were found to play a significant role in boosting the seed production of three crops: Chinese cabbage, broccoli, and kohlrabi (Sushil et al., 2013). Additionally, it was discovered that the Valencia cultivar of the onion *Allium cepa* produced more and higher-quality seeds (Yucel and Duman, 2005). According to Mishra et al. (2013), foragers also provide other benefits besides pollination, such as the application of nitrogen (in the form of feces) to plants while they are present. They discovered that a 5,000-bee colony produces bee frass at a mean rate of 2.27 to 2.69 g nitrogen per month. Forager bees can also spread some biocontrol agents, such as *Erwinia herbicola* Eh252 of fire blight, into apple blooms and nashi flowers (Cornish et al., 1998). A new, high-performance "Triwaks" dispenser was created to optimize the contribution of foraging bees to the dispersal of biocontrol chemicals (Bilu et al., 2004). According to Balayiannis and Balayiannis (2008), honey bee foraging behavior is a crucial bio-indicator for research that indirectly examines pesticide contamination of the environment. Proboscis extension reflex conditioning can even be used to educate foraging bees to recognize TNT. Honey bee foraging behavior has also been used to keep track of the species of flowering plants in a region. Bees employed for foraging can also recognize a pest infestation (such as fruit flies; Chamberlain et al., 2012). Beekeepers can take advantage of their colonies' foraging habits by installing pollen traps or venom collection boards in front of the hives to gather pollen or bee venom, respectively. Computer science also has a role to play in foraging behavior. Although there may be a wide variety of accessible food resources, forager bees are known to be able to choose their food sources in the best way possible (Thuijsman et al., 1995). As a result, computer scientists employed honey bee foraging behavior and related abilities in food scouting and gathering (swarm intelligence) to tackle several optimization challenges. Currently, swarm intelligence is a crucial area in artificial intelligence (Kumar and Govindaraj, 2013). Honey bee foraging (HBF), an algorithm developed by Baig and Rashid, was modeled after honey bee swarming and was intended to be used for multimodal and dynamic optimization challenges (Baig and Rashid, 2007).

4.3.8 Disadvantages of Foraging Activity

Even though foraging behavior is extremely important, there are some drawbacks to it as well. The bacterium that causes the fire blight of apples and pears, *Erwinia amylovora*, can be spread by honey bee foragers (Keitt, 1941). Moreover, Boylan-Pett et al. (1991) found that forager bees play a significant part in the dissemination and transmission of pollen-borne blueberry leaf mottle virus (BBLMV). For at least 10 days, this virus can continue to spread inside honey bee colonies. Some plants cannot be successfully pollinated by honey bees; bumble bees, for instance, outperformed honey bees in the pollination of cranberries (Mackenzie, 1994). Additionally, the exposure of bee foragers to other bees can cause the spread of bee parasites from one forager to another. Also, honey bees can spread various mite types among plants or even inside their colonies. Some plant species have deadly pollen that foragers can collect and store in their colonies, with detrimental effects on the colony's health.

4.4 SCOUTING

The topmost and most spectacular example of decentralized decision-making in animal groups is a swarm of honey bees selecting a new home among a variety of potential nest sites. Swarms of honey bees, which typically occur in the spring as part of the colony's reproductive cycle, are made up of one queen and thousands of workers. The former nest site has been given to one of the daughter queens, leaving the honey bee swarm momentarily homeless (the biology of swarming is reviewed in Winston, 1987). So, immediately after the swarm has formed, a large number of scout bees take off to search the area for a suitable new home.

4.4.1 Dance Following

The number of dances that are currently accessible determines how likely it is that you will discover one: P_{find} 1/4 K 0.005D (D = the total number of dances that are now available and K = constant). A searching bee randomly chooses one dance from the various dances. Depending on how many other bees are participating in the dance, the selected dance may or may not be followed by the selected bee. This is calculated using the formula 0.2^min{f,2} (f = the number of additional bees already following the dance) up to a maximum of seven followers. The probability of discovering and joining a dance activity was chosen so that the ensuing behavior resembled the number of dance followers for foraging discovered by Tautz and Rohrseitz (1998). A dance follower bee waits until the dancer is finished before flying off to find the designated location. The number of waggle runs a bee has made for a certain spot determines the likelihood that it will locate there. If the bee has been to the location before, it will undoubtedly discover it again. Otherwise, the possibility is predicted using observations of the foraging success rate for the number of waggle runs that were followed (Mautz, 1971). To determine whether a bee was successful in finding the advertised place, Mautz (1971) kept track of how many waggle runs each bee had previously followed. With the use of these data, the researchers were able to calculate the cumulative distributions of the number of waggle runs followed by successful and failed followers, respectively. Based on the quantity of waggle runs we performed, we replicated these two distributions to determine the likelihood of discovering a location.

4.4.2 Resting

A seeking bee has a P_{rest} = 1/4 0.002/s chance of beginning to rest. This is equivalent to the possibility that a bee will decide to start looking for a dance after waking up from a nap. As a result, $1/P_{rest}$ = 500 s is the anticipated time it will take a bee to switch between these two states. This probability is predicated on Visscher and Camazine (1999).

4.4.3 Search Strategy

When there are sites available at various distances, how should the swarm examine its surroundings to guarantee that the best site is selected? The swarm should select the site that is nearest to it when given the option between two locations of equal quality but different distances because of the reduced trip time (unless there are other reasons why a more distant site should be preferred, but the researchers have no reason to believe that is the case). However, regardless of distance, a better-quality location should be chosen when one is available (presumably up to a certain maximum distance). Which location is selected relies on how scouts search their surroundings or the possibility that a site at a specific distance is found. The scouts should, ideally, prioritize close areas while maintaining the ability to select a site farther away if it is of higher quality. How does the swarm's search approach take these requirements into account? Using a specific probability distribution for finding a nest location at a specific distance, we discuss a swarm's search method.

4.4 DEFENSE OF THE HIVE ENTRANCE

4.4.1 Intra-specific Defense

Guards fly above and antennate any other bees that land at the hive entrance to see whether they are nest-mates. The observation of chemically formed cues supplied by the flying bee is the premise for nest-mate recognition (Pradella et al., 2015).

The guard bee inspects typically in the last 1–5 seconds, and most of the bees being examined do not even stop when being antennated by guards (Butler and Free, 1951). Similarly, inspections can rarely last about 30 seconds or longer. When this happens, the examinee assumes a passive role and warms its thorax, rendering it likely that chemical evaporation is increased and making identification easier (Butler and Free, 1951). If an approaching bee is identified as an intruder, the guards maul it and drag it away while it is still in the submissive position. Intruding bees may still be investigated when present on the combs and carried back to the entrance of the hive even if they manage to enter the colony, indicating the presence of guard bees inside the hive (Butler and Free, 1951).

The behavioral services presented in previous paragraphs are mostly targeted at foragers who unintentionally land at the wrong hive or returning nest-mates. However, bees may steal honey from the stores of other nearby colonies. Because of their distinctive swaying flight, which involves them going back and forth in front of the hive "as if anticipating for a chance to exit unobserved through defender bees, these mentioned robber's bees were recognized before they emerge" (Butler and Free, 1951). As soon as the would-be thief lands, guard bees race toward it and begin mauling it without appearing to need olfactory inspections, despite the fact that they will produce another bee which will carry the scent of its own hive. When captured in this manner, a robber will attempt to flee right away and if successful, restart its wavering flight. If the guards remain successful in keeping hold of the thief, a one-on-one struggle occurs in which the parties attempt to sting one another (Free, 1954). Since both the guard and the thief have an equal chance of winning, such conflicts finish with the death of the opponent who is successfully stung.

The defensive behavior of honey bees toward non-nest-mates is influenced by several variables. The colony's resource availability significantly affects the behavior of the guardians. When the colony has enough resources, they rarely attack non-nest-mates who arrive at the hive entrance;

nevertheless, when there is food scarcity, they reject or even kill non-nest-mates (Butler and Free, 1951). The nest's empty combs, which have been connected to a considerable rise in colony defensiveness, may act as a mediator for this impact (Collins and Rinderer, 1985). In contrast, under intense predation pressure, guarding and foraging diminish. Moreover, for instance, if a colony has been without a queen for a long period, honey bee workers begin to produce eggs (creating haploid males), and non-nest-mates with active ovaries will be rejected very easily by the guards. A queen's presence or absence significantly affects a honey bee's defensive behavior. All bees become generalists without a queen and take part in colony defense (Naeger et al., 2013). Similarly, they'll refuse most non-nest-mates to save the nest from social infestation (Chapman et al., 2009). Extended absence of the queen, however, results in hives being submissive, indicating that the queen directly affects hive defense to preserve her survival (Delaplane and Harbo, 1987).

4.4.2 Defense Against Other Insects

Other insects, including ants, are attracted to honey storage as well. Bees at the entrance to the hive display a stereotypical response to these attackers: they begin by moving away from the ants before fanning their wings at a very high frequency (275 Hz on average, higher than the wing-beat frequency during flight) to drive the tiny insects off the landing board (Spangler and Taber, 1970). This method effectively exterminates the ants without coming into contact with them, because ants move more quickly than bees and are not afraid to bite them. *Apis mellifera capensis* performs alternate circles in both clockwise and anticlockwise directions to cover a large region, while *Apis mellifera ligustica* uses its hind legs to attack ants (Spangler and Taber, 1970) as well as beetles (Atkinson and Ellis, 2011).

While most insects, besides ants, are opportunistic, other insects have established strong parasite interactions with honey bee colonies. The pests well-known to beekeepers include the pathogen destructor mites, *Galleria mellonella*, a greater wax moth, and *Aethina tumida*, a small hive beetle (Rosenkranz et al., 2010). Unscreened top entrances allow wax moths to enter the hive, where they lay their eggs in crevices where bees cannot reach them. The emerging larvae devour the combs by digging just beneath the cell tops, feasting on abandoned nests, and consuming wax and hive debris. Honey bees extract wax moth larvae from nests by biting and tugging them out (Papachristoforou et al., 2012). Several guard bees typically maul a group of trespassing beetles. The little hive beetle, on the other hand, has developed a turtle-like ability to retract its appendages under its body and a protective exoskeleton. This beetle frequently seeks a smaller, inaccessible location in the colony to hide, allowing it to prevent guards from grasping or stinging due to its body structure and behavior. The beetles will be confined to this region by guard bees that will surround them. Through trophallaxis, a mouth-to-mouth food exchange, the beetles may trick their hosts into feeding them; therefore, they are still able to thrive under these circumstances (Atkinson and Ellis, 2011; Ellis and Hepburn, 2006). Honey bees (*Apis mellifera capensis*) will also encase the beetles in propolis within the beetles' ecological habitat (Neumann et al., 2001). The bees may leave, which is a sort of swarming in which they quit their hives all at the same time if the beetle infestation becomes too severe. It's interesting to note that honey bees react more defensively to the special small hive beetle as compared with other beetle species that may unintentionally inhabit hives, indicating that they have evolved a defense mechanism against this particular invader (Atkinson and Ellis, 2011).

Finally, hornets that hunt honey bees are another threat. These enormous insects swoop on foraging adult honey bees as they return to the hive, usually hovering close to the entrance. One huge honey bee colony may be eradicated in a single day by a handful of Japanese giant hornet (*Vespa mandarinia*) workers, who then feast on the larvae and pupae (Matsuura and Sagakami, 1973). Hornets' tough cuticles make it very hard for honey bees to sting them. To defend themselves against such attacks, the bees first gather in large groups at the entrance to the hive. The bees use their front legs and mandibles to grab the hornet while adhering to one another to form a "carpet." When they are successful, they will immediately enclose a hornet in a huge bee ball (Baracchi et al., 2010). It's interesting to note that while this behavior is common across the *Apis* genus, it has been developed to address the different relationships between each honey bee subspecies or species as well as the local hornet species. There are six kinds of hornets in Asia: *Apis cerana* honey bees are especially effective at enlisting around 30 workers to form a "living ball" where the hornet will be trapped or destroyed (at a high temperature of about 45°C). By tightening their thoracic muscles, bees raise their body temperature. Although the temperature at the center of the ball is higher as compared with the hornet's limit, which is less than 50°C, it is safe for the bees themselves (Ken et al., 2005).

When hornets are around, honey bees have been seen to make humming sounds or "hisses," also referred to as "shimmering" (Baracchi et al., 2010). Given that this behavior also occurs in response to electric shocks, hissing appears to be an intrinsic reaction to noxious stimuli (Wehmann et al., 2015). It is unknown whether such sounds are utilized as a distress call to the hives, as a warning to hornets, or as a threat.

4.4.3 Defense Against Large Predators

In addition, guards serve as the first line of defense from large predators such as birds, rodents, wolves, and people. They can go to investigate disturbances that are taking place close to the colony (Moore et al., 1987). Their main triggers include black hues, quick activities, the smell of mammals, and rough textures (Free, 1961). Some guards flee toward a large predator as soon as they see it, but other guards eject their claws, lift their abdomens, and fly inside the colony while flapping their wings (Collins et al., 1980), releasing the alarm pheromone secreted by the sting function and warning their nest-mates of the imminent threat. Only guardians can battle with a powerful predator. More bee recruitment is thus required for the colony's defense.

The honey bee will begin seeking out the potential target once it has been attracted. The animal's mobility is what initially draws them in (Wager and Breed, 2000). The warning chemicals left by earlier guards are potent and attractive, leading the honey bees to swiftly concentrate on the solitary most-stung goals, according to an examination of the number and distribution of stings left in moving targets shown concurrently (Millor et al., 1999). The majority of bees harass the localized opponent in what is perceived as a dangerous manner by flying quickly about it, frequently colliding with it while emitting an audible increased buzzing (Cunard and Breed, 1998).

4.4.4 Alarm Pheromones: Defensive Communication

Alarm pheromones are molecules that let members of the same species communicate with one another (Karlson and Luscher, 1959). Primer pheromones produce long-term physical changes and in the short-term releaser pheromones play the main role in behavioral adjustments (Wyatt, 2003). These chemicals play a crucial role in social insects' ability to maintain colony cohesion, and honey bees' protective behavior is no exception. In the following, we look at two pheromones that are crucial to honey bee defense mechanisms.

4.4.5 The Sting Alarm Pheromone

4.4.5.1 Production and Dispersal

A pheromonal blend that alerts honey bees to hazards to the entire colony is one of the important components in their protective behavior, as already mentioned. Beekeepers know that every time the bees are disturbed, a distinctive perfume that smells like bananas is released from the hive. Early studies showed that the sting apparatus emits alarm pheromones, which can attract honey bees and incite them to attack (Free, 1961). According to anatomical studies, the proximal portion and the Koschewnikow glands of the sting sheaths both produce the sting alarm material (Cassier et al., 1994; Grandperrin and Cassier, 1983). The secretory mixture that was secreted travels into the sting chamber and builds up on the setaceous membrane (Mauchamp and Grandperrin, 1982).

When the string is extruded, the structure's numerous setae create a wide surface area that allows a rapid release of pheromones (Lensky and Cassier, 1995). The primary constituent of this pheromonal blend is isoamyl acetate (IAA). It is not produced by newly emerging bees when they are 3 days old, and levels remain very low for a week, although they can detect it (Allan et al., 1987)

However, as the honey bee ages, the amount of alarm pheromones it produces each sting quickly rises to roughly 4–5 g before leveling off at around 2 g. It's interesting to note that regardless of the bee's age, the peak of production coincides with the beginning of foraging and guarding behaviors (Allan et al., 1987).

4.4.5.2 Composition

IAA (also known as isopentyl acetate), the first recognized and primary element of the alarm pheromone, was previously mentioned. The guards are drawn to and alerted by an IAA-labeled stationary item near the hive's main entrance, yet only a moving item causes them to sting (Free, 1961). While honey bees respond highly to IAA, this fragrance alone cannot fully explain the response seen with sting extracts (Free and Simpson, 1968). Later, the presence of a second substance, (Z)-11-eicosen-1-ol, in comparable concentrations was found. In contrast to IAA, this chemical draws honey bees toward small targets but is not static. IAA activity on stationary objects is prolonged, and a full response can be elicited by a combination of these two chemicals (Pickett et al., 1982). But, more than 40 other substances are included in the combination of the pheromones (Collins and Blum, 1983). Even though the cause of this complexity is unknown, it might be used to give this pheromone a special signature.

4.4.5.3 Alert and Orient Functions

Pheromone releasers, which cause rapid reactions, are used in all alarm chemicals. lists some of the chemicals found, along with how well they agitate young bees raised in cages (Collins and Blum, 1983; Collins and Blum, 1982). By putting some of these compounds on targets that move or remain stationary in front of the hive entrance, Wager and Breed (2000) were able to further examine the functionality of some of these chemicals. They discovered that most were specifically concerned with attracting more guards, such as 1-octanol and 1- butanol, while others simply directed the bees toward the target (such as octyl acetate), and some possessed both qualities: IAA, 1-butyl acetate, and hexanol. The only studied substance that increased flying capacity was IAA (Wager and Breed, 2000). Interestingly, guard bees do not reject non-nest-mates more quickly if IAA is blasted at the entrance of the hive, contrary to what might be expected, suggesting that the sting alarm pheromone's alerting function may only be used in interactions with other species (Couvillon et al., 2010). Therefore, another mechanism that has not yet been identified must be responsible for the sudden drop in the acceptance requirement of the soldiers to occur after a large number of non-nest-mate honey bees have attempted to access the colony.

4.4.5.4 Primer Effects and Long-Term Exposure

The long-term effects of IAA exposure on physiology and behavior have been the subject of some research. The adaptation of bees to their alarm pheromones was first shown. Bees become less prone to sting and fail to distinguish between scented and non-scented targets within an hour of the placement of a dispenser carrying artificial alarm chemicals in a hive (Al-Sa'ad et al., 1985; Free, 1988). Adaptation is extremely rare under natural circumstances due to the significant volatility of IAA. Furthermore, disturbed colonies are known to remain aroused for a considerable amount of time.

A colony's recruitment of bees to the entrance increased during the course of trials until reaching a plateau when IAA was applied frequently (Alaux and Robinson, 2007). IAA may function as a pheromone primer, causing permanent alterations in brain gene regulation by inducing the expression of the immediate early gene and transcription factor c-Jun in the antennal lobes (Alaux and Robinson, 2007).

Finally, it has been suggested that IAA can elicit analgesia by activating the narcotic system (Nez et al., 1998), a result that prevents recruits from quitting the fight. This could be a key factor in social organization, making enrollment easier after demanding and protracted offensive events. Additionally, IAA can impair appetitive learning after prolonged exposure, and this effect can last up to 24 hours (Urlacher et al., 2010). This may be a component of a larger system that primes honey bees for defense by focusing their attention on events important for colony defense rather than appetizing stimuli. It's interesting to note that after being exposed to IAA while feeding in a feed patch, honey bees prevent other honey bees from sending out recruits to the specific spot after their return to the colony (Nieh, 2010). Studies recently found that appetizing flowery scents can stop honey bees from attacking in response to IAA (Nouvian et al., 2015). A long-term adaptive strategy for avoiding predator-infested areas may therefore include the blocking effect of floral odors. By keeping bees from defending themselves and possibly perishing, this system ensures that they'll return to the hives, and bees are alerted to the risk at the forage place.

4.4.6 The Mandibular Alarm Pheromone

The glands of the worker mandibles store another substance with an alerting function (Shearer and Boch, 1965). This chemical, known as 2-heptanone, is created in relatively substantial concentrations (15–23 g per bee). This secretion is discharged from the mandibular glands through a pore on their internal faces, and a groove on the mandibles guides it in the direction of the spatula's pointed tip (Papachristoforou et al., 2012). A small amount of 2-heptanone is produced by young bees, but as they get older, this amount gradually increases. Mandibular alarm pheromone is produced at the lowest level in bees working inside, at an intermediate level in guards, and at its highest level in hunters (Vallet et al., 1991), casting doubt on the substance's purported defensive function. According to Kerr et al. (1974), there was an association between stinging behavior and high levels of 2-heptanone; however, further research disproved this claim (Lensky and Cassier, 1995). It could have been a result of the initial study's use of a similar colony (Kerr et al., 1974), in which case the result might just point to a biological relationship between these two elements. The efficiency of 2-heptanone as an alerting pheromone was already widely questioned. It causes guard bees to act defensively when used on screw caps at the entrance of the hive (Shearer and Boch, 1965). As in other investigations, Balderrama et al. (2002) discovered that it enhances sensitivity to electric shocks (measured by the sting extension). indicating that bees prefer to attack a 2-heptanone-treated ball over a standard one. Collins and Blum (1982) observed that it agitates immature cage-reared bees (Free and Simpson, 1968).

Two moving balls were simultaneously presented with 2-heptanone and IAA in one study, and the bees did not attack either one more frequently than the other (Free and Simpson, 1968). In some real-world tests, guard bees either don't react to 2-heptanone or repel it (Butler, 1966; Papachristoforou et al., 2012). In terms of colonial defense, 2-heptanone has recently been found to serve a distinct purpose: it causes local anesthesia and paralysis when administered to parasites by biting, making it easier to remove them from the hive. Small parasites like Varroa mites can even be killed by the injection of 2-heptanone (Papachristoforou et al., 2012). This discovery shows that this chemical may be more crucial in the context of defense against other insects, along with the limited efficiency of 2-heptanone as an attracting pheromone in modeled mammal attacks. For instance, it could entice nest-mates to remove parasites rather than sting, or it could serve as a deterrent to outsiders attempting to join the colony. The utilization of 2-heptanone in foraging is also distinct. It serves as a foraging pheromone, keeping foragers away from freshly visited and nectar-emptied flowers and saving them both effort and time. It enables honey bees to operate in concert with the other workers to efficiently graze in a patch of flowers (Giurfa, 1993).

4.5 CONCLUSION

This important chapter has presented a comprehensive review of knowledge related to bee behaviors, including behaviors within the hive, foraging, scouting, and defense behaviors, which ultimately meets the needs of beekeepers, scientists, taxonomists, and all other individuals interested in studying the nature, behavior, ecology, diversity, and natural history of insects, mostly focused on social insects.

REFERENCES

Abou-Shaara HF (2012). Notes on water collection by honey bees. *Bee World* 89, 50–51.

Abou-Shaara HF, Al-Ghamdi AA, Mohamed AA (2013). Honey bee colonies performance enhance by newly modified beehives. *J. Apic. Sci* 57, 45–57.

Aizen MA, Harder LD (2009). The global stock of domesticated honeybees is growing slower than the agricultural demand for pollination. *Curr. Biol.* 19(11), 915–918.

Alaux C, Robinson GE (2007). Alarm pheromone induces immediate–early gene expression and slow behavioral response in honey bees. *J. Chem. Ecol.* 33, 1346–1350.

Alfonsus EC (1932). The rocking movements of bees. *J. Econ. Entomol.* 25(4), 815–820.

Ali MAM (2011). Comparative study for evaluating two honey bee races, Apis mellifera jementica (indigenous race) and Apis mellifera carnica (carniolan race) in brood production, population development and foraging activity under the environmental conditions of the central region of the Kingdom of Saudi Arabia. *Ann. Agric. Sci.* 56(2), 127–134.

Allan SA, Slessor KN, Winston ML, King GGS (1987). The influence of age and task specialization on the production and perception of honey bee pheromones. *J. Insect Physiol.* 33(12), 917–922.

Alqarni AS (2006). Tolerance of summer temperature in imported and indigenous honeybee Apis mellifera L. Races in central Saudi Arabia. *Saudi J. Biol. Sci.* 13, 123–127.

Al-Sa'ad BN, Free JB, Howse PE (1985). Adaptation of worker honeybees (Apis mellifera) to their alarm pheromones. *Physiol. Entomol.* 10(1), 1–14.

Atkinson EB, Ellis JD (2011). Adaptive behaviour of honeybees (Apis mellifera) toward beetle invaders exhibiting various levels of colony integration. *Physiol. Entomol.* 36(3), 282–289.

AU-IBAR (Africa Union Inter African Bureau for Animal Resources) (2019). The inaugural report on the status of apiculture in Africa 2019: Opportunities and strategies for development of a sector.

Baig AR, Rashid M (2007). Honey bee foraging algorithm for multimodal and dynamic optimization problems. *Proceedings of the 9th Annual Conference on Genetic and Evolutionary Computation*, 7–11 July.

Balayiannis G, Balayiannis P (2008). Bee honey as an environmental bioindicator of pesticides' occurrence in six agricultural areas of Greece. *Arch. Environ. Contam. Toxicol.* 55(3), 462–470. 10.1007/s00244-007-9126-x.

Balderrama N, Núñez J, Guerrieri F, Giurfa M (2002). Different functions of two alarm substances in the honeybee. *J. Comp. Physiol. A* 188, 485–491.

Baracchi D, Cusseau G, Pradella D, Turillazzi S (2010). Defence reactions of Apis mellifera ligustica against attacks from the European hornet Vespa crabro. *Ethol. Ecol. Evol.* 22(3), 281–294.

Beekman M, Sumpter DJT, Seraphides N, Ratnieks FLW (2004). Comparing foraging behaviour of small and large honey-bee colonies by decoding waggle dances made by foragers. *Funct. Ecol.* 18(6), 829–835.

Beyaert L, Greggers U, Menzel R (2012). Honeybees consolidate navigation memory during sleep. *J. Exp. Biol.* 215(22), 3981–3988. 10.1242/ jeb.075499.

Bilu A, Dag A, Elad Y, Shafir S (2004). Honey bee dispersal of biocontrol agents: An evaluation of dispensing devices. *Biocontrol Sci. Technol.* 14(6), 607–617.

Blazyte-Cereskiene L, Vaitkeviciene G, Venskutonyte S, Buda V (2010). Honey bee foraging in spring oilseed rape crops under high ambient temperature conditions. *Zemdirbyste Agric.* 97, 61–70.

Boylan-Pett W, Ramsdell DC, Hoopingarner RA, Hancock JF (1991). Honeybee foraging behavior, in-hive survival of infectious, pollen-borne blueberry leaf mottle virus and transmission of the virus in highbush blueberry. *Phytopathology* 81(11), 1407–1412.

Boz˘ič J, Valentinčič T (1995). Quantitative analysis of social grooming behavior of the honey bee Apis mellifera carnica. *Apidologie* 26(2), 141–147.

Brittain C, Williams N, Kremen C, Klein A-M (2013). Synergistic effects of non-Apis bees and honey bees for pollination services. *Proceedings of the Roal Society B 280*, 20122767. 10.1098/rspb.2012.2767.

Brouwers EVM, Ebert R, Beetsma J (1987). Behavioural and physiological aspects of nurse bees in relation to the composition of larval food during caste differentiation in the honeybee. *J. Apic. Res.* 26(1), 11–23.

Butler CG (1940). The choice of drinking water by the honeybee. *J. Exp. Biol.* 17(3), 253–261.

Butler CG (1966). Mandibular gland pheromone of worker honeybees. *Nature* 212(5061), 530.

Butler CG, Free JB (1951). The behaviour of worker honeybees at the hive entrance. *Behaviour* 4(1), 262–291.

Cane JH, Schiffhauer D (2001). Pollinator genetics and pollination: Do honey bee colonies selected for pollen-hoarding field better pollinators of cranberry Vaccinium macrocar pon? *Ecol. Entomol.* 26(2), 117–123.

Cassier P, Tel-Zur D, Lensky Y (1994). The sting sheaths honey bee workers (Apis mellifera L.): Structure and alarm pheromone secretion. *J. Insect Physiol.* 40, 23–32.

Casteel DB (1912). The behavior of the honey bee in pollen collecting. *USDA Bur Entomol. Bull.* 121, 1–36.

Chamberlain K, Briens M, Jacobs JH, Clark SJ, Pickett JA (2012). Use of honey bees (Apis mellifera L.) to detect the presence of Mediterranean fruit fly (Ceratitis capitata Wiedemann) larvae in Valencia oranges. *J. Sci. Food Agric.* 92(10), 2050–2054. 10.1002/jsfa.5742.

Chantawannakul, P., Ramsey, S. (2018). The overview of honey bee diversity and health status in Asia. In: Chantawannakul P., Williams G., Neumann, P. (eds) Asian Beekeeping in the 21st Century. Singapore: Springer. https://doi.org/10.1007/978-981-10-8222-1_1.

Chapman NC, Makinson J, Beekman M, Oldroyd BP (2009). Honeybee, Apis mellifera, guards use adaptive acceptance thresholds to limit worker reproductive parasitism. *Anim. Behav.* 78(5), 1205–1211.

Collett TS, Zeil J (1998). Places and landmarks: An arthropod perspective. In: Healy S (ed.) *Spatial Representation in Animals.* Oxford: Oxford University Press; pp. 18–53.

Collins AM, Blum SM (1982). Bioassay of compounds derived from the honeybee sting. *J. Chem. Ecol.* 8(2), 463–469.

Collins AM, Blum MS (1983). Alarm responses caused by newly identified compounds derived from the honeybee sting. *J. Chem. Ecol.* 9, 57–65.

Collins AM, Rinderer TE (1985). Effect of empty comb on defensive behavior of honeybees. *J. Chem. Ecol.* 11(3), 333–338.

Collins AM, Rinderer TE, Tucker KW, Sylvester HA, Lackett JJ (1980). A model of honeybee defensive behavior. *J. Apic. Res.* 19(4), 224–231.

Cornish DA, Voyle MD, Haine HM, Goodwin RM, Vanneste JL (1998). Distribution of beneficial bacteria on nashi and apple flowers using honey bees. *Proceedings of 51st New Zealand Plant Protection Conference*, 107–111.

Couvillon MJ, Barton SN, Cohen JA, Fabricius OK, Karcher MH, Cooper LS, Silk MJ, Helantera H, Ratnieks FL (2010). Alarm pheromones do not mediate rapid shifts in honey bee guard acceptance threshold. *J. Chem. Ecol.* 36(12), 1306–1308.

Cunard SJ, Breed MD (1998). Post-stinging behaviour of worker honey bees (Hymenoptera: Apidae). *Ann. Entomol. Soc. Am.* 91(5), 754–757.

Delaplane KS, Harbo JR (1987). Effect of queenlessness on worker survival, honey gain and defence behaviour in honeybees. *J. Apic. Res.* 26(1), 37–42.

Dimou M, Thrasyvoulou A (2007). Collection of wax scale (Ceroplastes sp.) by the honey bee Apis mellifera. *J. Apic. Res.* 46(2), 129.

DuPraw EJ (1961). A unique hatching process in the honeybee. *Trans. Am. Microsc. Soc.* 80(2), 185.

Ellis JD, Hepburn HR (2006). An ecological digest of the small hive beetle (Aethina tumida), a symbiont in honey bee colonies (Apis mellifera). *Insectes Soc.* 53(1), 8–19.

Esch H, Goller F, Heinrich B (1991). How do bees shiver? *Sci. Nat.* 78(7), 325–328. 10.1007/ BF01221422 PMID: 26969012.

FAO (Food and Agriculture Organization of the United Nations). (2015). Retrieved February 14, 2020. www.fao.org.

FAO (2019). The state of food and agriculture 2019. Moving forward on food loss and waste reduction. *Rome.* Licence: CC BY-NC-SA 3.0 IGO.

Free JB (1954). The behaviour of robber honeybees. *Behaviour* 7(1), 233–240.

Free JB (1961). The stimuli releasing the stinging response of honeybees. *Anim. Behav.* 9(3–4), 193–196.

Free JB, Ferguson AW, Simpkins JR (1992). The behaviour of queen honeybees and their attendants. *Physiol. Entomol.* 17(1), 43–55.

Free JB, Simpson SJ (1963 September). The respiratory metabolism of honey-bee colonies at low temperatures. *Entomol. Exp. Appl.* 6(3), 234–238.

Free JB, Simpson J (1968). The alerting pheromones of the honeybee. *Z. vergl. Physiol.* 61(3), 361–365.

Giurfa M (1993). The repellent scent-mark of the honeybee Apis mellifera ligustica and its role as communication cue during foraging. *Insectes Soc.* 40(1), 59–67.

Giurfa M (2003). The amazing mini-brain: Lessons from a honey bee. *Bee world*, 84(1), 5–18.

Grandperrin D, Cassier P (1983). Anatomy and ultrastructure of the Koschewnikow's gland of the honey bee, Apis mellifera L.(Hymenoptera: Apidae). *Int. J. Insect. Morphol. Embryol.* 12(1), 25–42.

Higginson AD, Barnard CJ, Tofilski A, Medina L, Ratnieks F (2011). Experimental wing damage affects foraging effort and foraging distance in honeybees Apis mellifera. *Psyche*. 10.1155/2011/419793.

Hunt GJ, Page RE, Fondrk MK, Dullum CJ (1995). Major quantitative trait loci affecting honey bee foraging behavior. *Genetics* 141(4), 153.

Jay SC (1964). The cocoon of the honey bee, Apis mellifera L. *Can. Entomol.* 96(5), 784–792.

Jones AK, Sattelle DB (2006). The cys-loop ligand-gated ion channel superfamily of the honeybee, Apis mellifera. *Invert. Neurosci.* 6, 123–132.

Jones JC, Oldroyd BP (2006). Nest thermoregulation in social insects. *Advances in Insect Physiology*. 33, 153–191.

Joshi NC, Joshi PC (2010). Foraging behaviour of Apis spp. on apple flowers in a subtropical environment. *N. Y. Sci. J.* 3, 71–76.

Karlson P, Luscher M (1959). Pheromones-new term for a class of biologically active substances. *Nature* 183(4653), 55–56.

Keitt GW (1941). Transmission of fire blight by bees and its relation to nectar concentration of apple and pear blossoms. *J. Agric. Res.* 62, 745–775.

Ken T, Hepburn HR, Radloff SE, Yusheng Y, Yiqiu L, Danyin Z, Neumann P (2005). Heat-balling wasps by honeybees. *Naturwissenschaften* 92(10), 492–495.

Kerr WE, Blum MS, Pisani JF, Stort AC (1974). Correlation between amounts of 2-heptanone and iso-amyl acetate in honeybees and their aggressive behaviour. *J. Apic. Res.* 13(3), 173–176.

Kleinhenz M, Bujok B, Fuchs S, Tautz J (2003). Hot bees in empty broodnest cells: Heating from within. *J. Exp. Biol.* 206(23), 4217–4231.

Kumar SS, Govindaraj M (2013). A detailed study about foraging behavior of artificial bee colony (ABC) and its extensions. *Int. J. Eng. Technol.* 5, 992–997.

Laska M, Galizia CG, Giurfa M, Menzel R (1999). Olfactory discrimination ability and odor structure-activity relationships in honeybees. *Chem. Senses* 24(4), 429–438.

Lensky Y, Cassier P (1995). The alarm pheromones of queen and worker honey bees. *Bee World* 76(3), 119–129.

Mackenzie KE (1994). The foraging behavior of honey bees (Apis mellifera L) and bumble bees (Bombus spp.) on cranberry (Vaccinium macrocarpon Ait.). *Apidologie* 25(4), 375–383.

Malerbo-Souza DT (2011). The corn pollen as a food source for honeybees. *Acta Sci. Agron.* 33(4), 701–704.

Matsuura M, Sagakami SF (1973). A bionomic sketch of the Giant Hornet, Vespa Mandarinia, a Serious Pest for Japanese Apiculture. *J. Fac. Sci. Hokkaido Univ. Ser. VI Zool.* 19, 125–162.

Mattu VK, Raj H, Thakur ML (2012). Foraging behavior of honeybees on apple crop and its variation with altitude in Shimla hills of western Himalaya. *Int. J. Sci. Nat.* 3, 296–301.

Mauchamp B, Grandperrin D (1982). Chromatographie en phase gazeuse des composés volatils des glandes a phéromones des abeilles: Méthodes d'analyse directe. *Apidologie* 13(1), 29–37.

Mautz D (1971). Der Kommunikationseffekt der Schwa¨nzelta¨nze bei Apis mellifica carnica (Pollm.). *Z. vgl. Physiol.* 72, 197–220.

Mayer DF, Lunden JD (1988). Foraging behavior of honey bees on Manchurian crabapple and red delicious apple. *J. Entomol. Soc. BC* 83, 67–71.

McAfee A, Chapman A, Iovinella I, Gallagher-Kurtzke Y, Collins TF, Higo H, et al. (2018). A death pheromone, oleic acid, triggers hygienic behavior in honey bees (Apis mellifera L.). *Sci. Rep.* 8(1), 1–13. 10.1038/s41598-017-17765-5 PMID: 29311619.

Menzel R (1985). Learning in honey bees in an ecological and behavioral context. In: Hölldobler B, Lindauer M (ed.) *Experimental Behavioral Ecology* Stuttgart: Gustav Fischer; pp. 55–74.

Michener CD, Winston ML, Jander R (1978). Pollen manipulation and related activities and structures in bees of the family Apidae. *The University of Kansas Science Bulletin* 51(19), 575.

Millor J, Pham-Delegue M, Deneubourg JL, Camazine S (1999). Selforganized defensive behavior in honeybees. *Proc. Natl Acad. Sci. U. S. A.* 96(22), 12611–12615.

Mishra A, Afik O, Cabrera ML, Delaplane KS, Mowrer JE (2013). Inorganic nitrogen derived from foraging honey bees could have adaptive benefits for the plants they visit. *PLOS One* 8(7), e70591. 10.1371/journal.pone.0070591.

Moore AJ, Breed MD, Moor MJ (1987). The guard honey bee: Ontogeny and behavioral variability of workers performing a specialized task. *Anim. Behav.* 35(4), 1159–1167.

Moosbeckhofer R (1992). Beobachtungen zum Auftreten bescha¨digter Varroamilben im natu¨rlichen Totenfall bei Vo¨lkern von Apis mellifera carnica. *Apidologie* 23(6), 523–531.

Mostajeran MA, Edriss MA, Basiri MR (2006). Analysis of colony and morphological characteristics in honey bees (Apis mellifera meda). *Pak. J. Biol. Sci.* 9, 2685–2688.

Naeger NL, Peso M, Even N, Barron AB, Robinson GE (2013). Altruistic behavior by egg-laying worker honeybees. *Curr. Biol.* 23(16), 1574–1578.

Nelson JA, Lineburg B, Sturtevant AP (1924). *Growth and Feeding of Honeybee Larvae*. Washington, DC: U. S. Dept. of Agriculture.

Nest BNV, Moore D (2012). Energetically optimal foraging strategy is emergent property of time-keeping behavior in honey bees. *Behav. Ecol.* 23(3), 649–658.

Neumann P, Pirk C, Hepburn H, Solbrig A, Ratnieks F, Elzen P, Baxter J (2001). Social encapsulation of beetle parasites by Cape honeybee colonies (Apis mellifera capensis Esch.). *Naturwissenschaften* 88(5), 214–216.

Nieh JC (2010). A negative feedback signal that is triggered by peril curbs honey bee recruitment. *Curr. Biol.* 20(4), 310–315.

Nouvian M, Hotier L, Claudianos C, Giurfa M, Reinhard J (2015). Appetitive floral odours prevent aggression in honeybees. *Nat. Commun.* 6, 10247.

Page Jr RE, Fondrk MK, Hunt GJ, Guzman-Novoa E, Humphries MA, Nguyen K, Greene AS (2000). Genetic dissection of honeybee (Apis mellifera L.) foraging behavior. *J. Hered.* 91(6), 474–479.

Pankiw T (2005). The honey bee foraging behavior syndrome: Quantifying the response threshold model of division of labor. *Proceedings of Swarm Intelligence Symposium*, 8–10 June; 1–6.

Papachristoforou A, Kagiava A, Papaefthimiou C, Termentzi A, Fokialakis N, Skaltsounis AL, Watkins M, Arnold G, Theophilidis G (2012). The bite of the honeybee: 2-heptanone secreted from honeybee mandibles during a bite acts as a local anaesthetic in insects and mammals. *PLOS One* 7(10), e47432.

Park W (1925). The storing and ripening of honey by honeybees. *J. Econ. Entomol.* 18(2), 405–410.

Pernal SF, Currie RW (2010). The influence of pollen quality on foraging behavior in honeybees (Apis mellifera L.). *Behav. Ecol. Sociobiol.* 51(1), 53–68.

Pettis JS, Pankiw T (1998). Grooming behavior by Apis mellifera L. in the presence of Acarapis woodi (Rennie) (Acari: Tarsonemidae). *Apidologie* 29(3), 241–253.

Pickett JA, Williams IH Martin AP (1982). (Z)-I 1-eicosen-1-ol, an important new pheromonal component from the sting of the honey bee, Apis mellifera L. (Hymenoptera, Apidae.). *J. Chem. Ecol.* 8, 163–175.

Pradella D, Martin SJ, Dani FR (2015). Using errors by guard honeybees (Apis mellifera) to gain new insights into nestmate recognition signals. *Chem. Senses* 40(9), 649–653.

Reyes-Carrillo JL, Eischen FA, Cano-Rios P, Rodriguez Martinez R, Camberos UN (2007). Pollen collection and honey bee forage distribution in Cantaloupe. *Acta Zool. Mex.* 23, 29–36.

Rosenkranz P, Aumeier P, Ziegelmann B (2010). Biology and control of Varroa destructor. *J. Invertebr. Pathol.* 103(suppl. 1), S96–S119.

Shearer DA, Boch R (1965). 2-heptanone in the mandibular gland secretion of the honey-bee. *Nature* 206, 530.

Siefert P, Hota R, Ramesh V, Gru¨newald B (2020). Chronic within-hive video recordings detect altered nursing behaviour and retarded larval development of neonicotinoid treated honey bees. *Sci. Rep.* 10(1), 8727. 10.1038/s41598-020-65425-y PMID: 32457387.

Silva DP, Moisan-De-Serres J, Souza DC, Hilgert-Moreira SB, Fernandes MZ, Kevan PG, Freitas BM (2013). Efficiency in pollen foraging by honey bees: Time, motion, and pollen depletion on flowers of Sysirinchium palmifolium (Asparagales: Iridaceae). *J. Pollination Ecol.* 11, 27–32.

Simone-Finstrom M, Gardner J, Spivak M (2010). Tactile learning in resin foraging honeybees. *Behav. Ecol. Sociobiol.* 64(10), 1609–1617.

Smith DR (2002). Genetic diversity in Turkish honey bees. *Uludağ Arıcılık Dergisi* 2(3), 10–17.

Spangler HG, Taber S III (1970). Defensive behavior of honey bees toward ants. *Psyche* 77(2), 184–189.

Srinivasan MV, Zhang SW, Zhu H (1998). Honeybees link sights to smells. *Nature* 396(6712), 637–638.

Srinivasan MV (2010). Honey bees as a model for vision, perception, and cognition. *Annual Review of Entomology* 55, 267–284.

Steffan-Dewenter I, Kuhn A (2003). Honeybee foraging in differentially structured landscapes. *Proc. R. Soc B* 270(1515), 569–575.

Sushil SN, Stanley J, Hedau NK, Bhatt JC (2013). Enhancing seed production of three Brassica vegetables by honey bee pollination in north-western Himalayas of India. *Univers. J. Agric. Res.* 1(3), 49–53.

Tautz J, Maier S, Groh C, Rossler W, Brockmann A (2003). Behavioral performance in adult honey bees is influenced by the temperature experienced during their pupal development. *Proc. Natl Acad. Sci.* 100(12), 7343–7347. 10.1073/pnas.1232346100 PMID: 12764227.

Tautz J, Rohrseitz K (1998). What attracts honeybees to a waggle dancer? *J. Comp. Physiol. A* 183(5), 661–667.

Thompson HM (2012). *Interaction between Pesticides and Other Factors in Effects on Bees Food and Environment Research Agency*. Sand Hutton, York YO41 1LZ. www.efsa.europa.eu/publications.

Thuijsman F, Peleg B, Amitai M, Shmida A (1995). Automata, matching and foraging behavior of bees. *J. Theor. Biol.* 175(3), 305–316.

Urlacher E, Francés B, Giurfa M, Devaud J-M (2010). An alarm pheromone modulates appetitive olfactory learning in the honeybee (Apis mellifera). *Front. Behav. Neurosci.* 4, 157.

Vallet A, Cassier P, Lensky Y (1991). Ontogeny of the fine-structure of the mandibular glands of the honeybee (Apis mellifera L.) workers and the pheromonal activity of 2-heptanone. *J. Insect Physiol.* 37(11), 789–804.

Visscher PK, Camazine S (1999). The mystery of swarming honey bees: from individual behaviors to collective decisions. In: Detrain C, Deneubourg JL, Pasteels JM (ed.) *Information Processing in Social Insects*. Basel: Birkhaser; pp. 355–378.

Wager BR, Breed MD (2000). Does honey bee sting alarm pheromone give orientation information to defensive bees? *Ann. Entomol. Soc. Am.* 93(6), 1329–1332.

Weaver N (1965). The foraging behavior of honeybees on hairy vetch. *Insectes Soc.* 12(4), 321–326.

Wehmann H-N, Gustav D, Kirkerud NH, Galizia CG (2015). The sound and the fury – Bees hiss when expecting danger. *PLOS One* 10(3), e0118708.

Winston ML (1987). *The Biology of the Honey Bee.* Cambridge: Harvard University Press.

Woyke J (1963). What happens to diploid drone larvae in a honeybee colony. *J. Apic. Res.* 2(2), 73–75.

Wyatt TD (2003). *Pheromones and Animal Behaviour: Communication by Smell and Taste.* Cambridge: Cambridge University Press.

Yucel B, Duman I (2005). Effects of foraging activity of honeybees (Apis mellifera L.) on onion (Allium cepa) seed production and quality. *Pak. J. Biol. Sci.* 8, 123–126.

5 Pests and Diseases of Honey Bees

Uzma Aarifie, Umer Bin Farook,
Shahnawaz Ahmad Dar, A.R. Malik, Ruqqaya Javid,
Nadeya Khaliq and Aijaz Ahmad Wachkoo

5.1 INTRODUCTION

The honey bee is a true social insect, and it is important to examine the colony frequently as a whole in order to spot any variation from the colony's typical state of health. Like all other living things, honey bees are plagued by illnesses and pest infestation. Some of these are more harmful to honey bee colonies than others, but it's crucial for apiarists to be able to identify symptoms that could be caused by a disease or pest and take appropriate action (Murilhas, 2002). Losses of honey bee colonies have been reported during summer as well as winter by countries across the globe (Seitz et al., 2015). Countries having different climatic conditions are reported to have varied diseases and colony losses (Meixner et al., 2014). Moreover, not all of the reasons causing population loss or decreased honey production are related to illness or pests. Pesticides and/or nutritional deficits, for instance, can also harm colonies. Therefore, it is crucial for apiarists to be as knowledgeable as possible and to get as much help as possible regarding the identification of honey bee diseases/pest issues (Sanford, 1987).

Honey bees are social animals and cannot survive as individuals. A colony has been created as their social environment, and three castes inhabit it: worker, drone and queen. The term "diploid" refers to females with 32 chromosomes, who develop from fertilized eggs to become the queen and workers. The male drone, which is haploid and develops from an unfertilized egg, contains only half as many chromosomes as females. The distinction in caste between the queen and the worker bees is difficult to explain by genetic variation; instead, it results mostly from the way larvae are fed during the growth process (Coffey, 2007).

A normal honey bee colony has one queen and about 10,000 mature workers in the early spring, and that number rises to about 50,000 to 60,000 in the summer. Between May and August, the colony can bear the living expenses of 250–1000 male drones, but after autumn, the worker bees eliminate them to minimize food losses. The colony contains varying numbers of broods at various developmental stages for most of the year in addition to the adult bee population. Most illnesses are unique to either the adult stage or the brood. The infections that affect the brood are the most dangerous. Even though the cuticle, midgut and tracheal system serve as various anatomical and physiological barriers preventing the entry of microbes into the haemolymph, bees are still susceptible to infection from a variety of viruses, bacteria, protozoa and spiroplasmas. The food transfer within the colony is done by trophallaxis, a mouth-to-mouth transfer that exposes the colony to the quick spread of disease-causing agents within the colony. The effective management of honey bee colonies against pests and diseases calls for proper knowledge and skill for identification and keeping the colony healthy.

The biotic stress honey bee colonies undergo can be broadly divided into different categories such as diseases, insect pests and mites. Diseases and mites directly affect the adult bees as well as the brood. Insect pests affect the honey bee colony directly, by affecting the bee population, as well as indirectly, by affecting the physical contents of the hive. (Chantawannakul et al., 2016).

DOI: 10.1201/9781003354116-5

5.2 DISEASES OF HONEY BEES AND THEIR MANAGEMENT

5.2.1 Brood Diseases

5.2.1.1 American Foulbrood

Wherever honey bee colonies are kept, American foulbrood disease is a common problem. It is unquestionably the most dangerous brood disease for honey bees (Brodsgard et al., 1998). The spore-forming bacterium *Paenibacillus larvae* is the causal organism and generates more than a billion spores in each infected larva. The spores are heat and drought-resistant and can germinate at any moment in a favourable environment. Worker, drone and queen larvae are all susceptible to the illness. Infected queen and drone larvae are rarely observed in the wild. The bacterium targets the worker brood specifically, typically in the pre-pupal stage. The older larvae are attacked mostly, literally being digested by the bacterium's enzymes (Genersch, 2008).

An American foulbrood comb with severe infection has a mix of healthy capped brood, cells carrying the remnants of diseased larvae, and empty cells, giving the comb a mottled look. The disease-carrying cells' cappings seem damp and discoloured. As the condition worsens, the normally convex cell caps of infected larvae become concave. Punctured capping is another sign of the disease that is frequently present. A healthy larva seems to be pearly white and shimmering. Normally, it starts out its development curled up at the cell's base. As it develops, it grows all the way along the cell's length. The larva or pupa passes away while it is stretched. The larva or pupa first turns creamy brown as the infection worsens before finally turning deep brown. The leftovers turn ropy and are thus extracted as threads that are an inch long or longer. At this point, a vile odour starts to develop. The smell is reminiscent of infrequently used animal glues. Finally, the infected brood's remnants dry down to produce scales, which firmly attach to the cells' lower surfaces. The mouth parts may be attached as a fine thread to the upper edge of the cell if death happens during the pupal stage. This is an indication that American foulbrood disease is present (Riessberger-Galle et al., 2001).

Nurse bees/spores left in the brood cell might spread the virus to a larva. The primary mode of spreading the disease from colony to colony is via exchanging combs that contain infected larvae remains or honey, or both, that are contaminated with *P. larvae* spores. In order to stop the disease from spreading further, early detection is important. An American foulbrood–weakened colony may be plundered, and the neighbouring bees unintentionally transport honey containing *P. larvae* spores to healthy colonies.

It can be treated by giving infected hives sulphathiazole or terramycin (oxytetracycline) in sugar syrup, although this treatment has significant downsides. The vegetative stage of the disease is treatable by the medications, which means that careful and time-consuming equipment adjustments are required to stop the sickness from spreading while receiving treatment. Second, there is a chance that the organism could develop drug-resistant variants. Finally, the medicine may contaminate extracted honey, in violation of the Food and Drug Regulations. These factors make it unwise to treat the condition chemically. Another strategy for controlling the disease involves burning diseased honey, equipment and bees from affected hives after blasting them with a cyanide chemical (Genersch et al., 2006). American foulbrood diagnosis takes expertise, which is regularly available from a licensed bee inspector. The use of the "ropy" test to generate a working diagnosis is handy. The infected larva "sticks" firmly, and the contents can be dragged out into a long string or "rope" once a stick or twig is introduced and removed from the cell. Another alternative test is the Holst Milk Test, which is based on the fact that high levels of proteolytic enzymes are produced by sporulating *Paenibacillus larvae*. A smear of a diseased larva is placed in a test tube or glass vial containing 4 ml of 1% powdered skim milk in water. After that, the tube is incubated at 37°C. If American foulbrood is present, the suspension should clear in 10 to 20 minutes. There are multiple modes of transmission of these diseases, including feeding contaminated honey and pollen, using contaminated tools. In almost all cases, using sugar syrup instead of honey to feed bees is preferred, because honey from unidentified sources is inherently questionable. To prevent potential colony contamination, it is strongly advised that tools and hand instruments are as neat and clean as possible at all times (Ratnieks, 1992).

5.2.1.2 European Foulbrood Disease

The disease typically strikes when colonies are reaching their peak numbers. *Melissococcus plutonius*, a non-spore-forming bacterium, is the main causal organism of this illness. *Apolactobacillus kunkeei*, *Paenibacillus alvei*, *Enterococcus faecalis* and *Brevibacillus laterosporus* are also frequently linked to the illness (Bailey, 1983).

A superficial inspection of infected combs reveals the same puncturing and mottled appearance as in American foulbrood. European foulbrood can affect larvae of all types, including worker, drone and queen, and death typically happens in the larval stage.

European foulbrood larvae occur in a variety of shapes when they pass away. Some are extended, while others are coiled. A healthy larva initially appears pearly white, then turns dull-white, then yellow, and eventually brown. The disease's secondary organisms create ropiness and bad odour. The ropy material's flexibility is lower than that of American foulbrood. When the larvae turn brown, the tracheae can be seen as tiny, silvery tubes just below the skin. European foulbrood is characterized by this symptom in particular. The disease can also be distinguished from American foulbrood by its loosely adhering scales.

Equipment and food storage facilities that have been contaminated can spread European foulbrood. The spring is typically when the sickness is at its worst, and in summer, with an abundance of nectar and pollen, it gets better. However, late summer outbreaks of European foulbrood are not unusual.

The symptoms of American foulbrood and European foulbrood are very similar. *Melissococcus plutonius*, the bacterium that causes the disease, does not create spores, making it less dangerous than American foulbrood. Only younger larvae are typically attacked by the bacterium. As a result, perforated cappings might not be present, and the infected individual might still be coiling, giving the scale a "twisted" appearance. Similarly to American foulbrood, European foulbrood–infected bees eventually turn into a dark scale. The diagnosis for European foulbrood and American foulbrood is very similar.

European foulbrood is transmitted similarly to American foulbrood. Again, it is advised that when operating several colonies, hands and hive instruments should be kept free of detritus.

In comparison to American foulbrood, European foulbrood typically receives milder treatment. Burning is only used in really dire circumstances. Re-queening with more disease-resistant stock typically cures the condition. This interrupts the cycle of brood production and introduces fresh genetic material – possibly more resilient genetic material – to the colony. Additionally, regular use of the antibiotic terramycin as a preventative measure is successful in preventing both European and American foulbrood.

The occurrence of European foulbrood is frequently connected with the amount of stress that a colony is under. When raising a lot of brood or when nectar is scarce, symptoms may show up. Therefore, the importance of effective management practices that lessen colony stress to help prevent European foulbrood and the majority of other bee diseases cannot be overstated.

5.2.1.3 Chalkbrood Disease

Chalkbrood is caused by fungus and affects the brood. When removed from the comb, the disease's afflicted brood, known as "mummies," appear as solid clusters resembling chunks of chalk. The hue of the mummies might range from white to dark grey or black (when fruiting bodies are present). Outbreaks of chalkbrood can occur in the summer and autumn, but they are more common in the spring (Flores et al., 1996).

The fungus *Ascosphaera apis* is the main pathogen responsible for the disease. The fungus is most contagious in larvae that are 3 to 4 days old and under stress due to cool, wet weather or poor nutrition. Chalkbrood disease affects worker, drone and queen larvae equally (Spiltoir and Olive, 1955). The mycelia of the fungus infiltrate infected larvae, which causes them to die. The mycelia-filled larvae eventually dry out to form the classic chalkbrood sickness rigid white mummies. Larvae with disease can be speckled, entirely grey or black, and they can also be stretched out in their cells. The black fruiting structures that form on the outside of the larvae cause the colour variance. By looking for mummies in the combs, entrance and bottom boards of the hives, the disease can be identified. Nurse bees can easily remove the mummies, since they do not adhere to the cells. The mummified larvae smell faintly of yeast (Borum and Ulgen, 2008).

Adult bees and equipment tainted with *Apis* spores can spread the chalkbrood sickness. The illness seems to go away on its own but could come back eventually during the season or the next year. Rarely does the disease completely wipe out a colony, but it lowers bee populations, which can impact honey production (Aronstein and Murray, 2010). It has been speculated that the rise in cases of the fungus illness chalkbrood is related to the importation of pollen from other countries. High moisture, chilly temperatures and colony stress all tend to promote the growth of the pathogen *Ascosphaera apis*. A colony will typically be weakened rather than killed, which will result in less honey. Certain factors, including moisture, bee stock sensitivity, poor nutrition and other illnesses, also encourage the disease (queenlessness, laying workers, chilled brood). The appropriate technique to alleviate the symptoms is to re-queen with resistant bee stock. There is no advised chemical therapy for the infection; re-queening a colony frequently is an efficient remedy. The colony's good hygiene practices, such as workers' prompt evacuation of the corpses, seem to help alleviate the symptoms (Zaghloul et al., 2005).

5.2.1.4 Sacbrood Disease

Sacbrood virus is the causative agent and rarely causes a colony to die. The beekeeper should be able to differentiate between sacbrood and other similar diseases. The pre-pupal skin separates from the larval skin, creating a sac as the illness worsens. There is an accumulation of fluid between these two layers of skin. The outer skin becomes tougher, making it possible to pick up the whole larva without causing the fluid to leak.

The larva's colour gradually shifts from pearly white to off-white, then to brown, and lastly to blackish. Usually, the larva's head coils upward from the cell floor. The larval remnants are transformed into a loosely adhering scale. It resembles a foulbrood scale but has no odour and is bacteria-free (White, 1917).

Similarly to European foulbrood, sacbrood occurs most frequently during springtime. There is no chemotherapeutic treatment for sacbrood. Although re-queening might be beneficial, most colonies seem to recover naturally from the illness (Bailey et al., 1964).

5.2.1.5 Stonebrood Disease

The fungi *Aspergillus flavus* or less frequently, *Aspergillus fumigatus*, widely found in soil, are the main causal organisms of the disease stonebrood. Stonebrood is also caused by *Aspergillus niger* and in rare cases, by other *Aspergillus* species (Foley et al., 2014). It can only grow at temperatures between 7°C and 40°C, with a preferred range of 33–37°C. Both adult bees and larvae can be impacted by stonebrood. Oral transmission of the illness occurs during feeding (by sharing nectar from one bee to another) or hive cleaning. Adults, larvae and pupae can all contract the infection because it spreads also via the cuticle and through the consumption of conidia (Lopes et al., 2015). Although uncommon in adults, overt stonebrood symptoms are rather prevalent in larvae (Batra et al., 1973). Additionally, the fungus has the capacity to grow on the bees' body surface, doing harm from the outside. Both capped and uncapped broods are affected. The larvae start looking white and fluffy before turning yellow (for *A. flavus)* or greenish brown (for *A. fumigatus*), mummified, with firmness. The appearance is very reminiscent of larvae infected by chalkbrood. Adult bees show behavioural changes, such as agitation, weakness, paralysis and the inability to fly out from the hive, as well as morphological abnormalities such as an enlarged abdomen and subsequent mummification.

When bee larvae eat *Aspergillus* spp. conidia, these remain in the gut until first defecation, at which point they hatch, quickly develop into hyphae and form a ring-like collar close to the head (Seyedmousavi et. al., 2015). The larvae change colour and turn black. They are then tough to break, like little stones, and are covered in powdery fungal spores; hence the disease's original name, stonebrood (Sarwar, 2016). Stonebrood mummies are stuck in the cells and cannot be retrieved by worker bees.

The condition cannot be cured with current therapies. It is transitory and vanishes on its own. However, proper apiary location (avoiding wetlands and having good exposure), adequate management (preventing water penetration inside the hives), and routine cleaning of beekeeping equipment (such as sterilizing by torch flames) are effective preventative strategies. Even though the fungus can cause entire colonies to die, the disease is typically transitory and goes away on its own, especially if the prior steps are taken.

5.2.1.6 Other Honey Bee Brood Diseases

The nutritional ailment known as purple brood affects larvae and pupae. The nectar or pollen from the plant *Cyrilla racemiflora*, also known as southern leatherwood, black titi or red titi, is thought to be the most likely cause of the illness. Purple is the colour of sick larvae and pupae (Katie, 2012).

Cold and starving brood are additional situations that resemble infectious illnesses. The lack of enough bees to maintain a warm environment for the brood area results in chilled brood. As a result, chilled brood is typically seen around the periphery of the brood nest. As the brood pattern intensifies, brood may resemble European foulbrood, but is easily removed by nurse bees. Usually, a lack of nectar or honey results in starving brood. Wax moth damage to developing bees might occasionally make them appear sick (Rhodes, 2005).

5.3 ADULT BEE DISEASES

5.3.1 Nosemosis

Nosema is the most pervasive and dreadful disease. The fungus *Nosema apis* is the agent behind this sickness. Adult bees have a shorter life span as a result of *Nosema* sickness. It can lead to queen supersedure and lower the honey production of afflicted colonies.

N. apis spores develop and grow in the epithelial cells of the ventriculus after swallowing and start the illness cycle. The hypopharyngeal glands of infected worker bees and the ovaries of infected queens deteriorate in addition to having an impact on the digestive system. Workers, queens and drones are all affected by the illness. Infected colonies of bees or individual bees may not exhibit any external symptoms. The ventriculi of some diseased organs may swell and turn white. In order to identify *Nosema* disease, spores of *N. apis* must be present. However, since additional life stages can be present, the absence of spores does not guarantee the absence of illness. The disease has a yearly cycle, with most spores produced in the spring months. The quantity of spores decreases in the summer, and occasionally, a minor increase can be seen in the autumn.

5.3.2 Bee Paralysis Virus

A number of viruses can cause bee paralysis, but certain nectars and pollens can also cause similar symptoms. Both hairless black syndrome and chronic bee paralysis are caused by the same virus. Bees are more swiftly damaged by acute bee paralysis when infection is transmitted in larval jelly by asymptomatic infected adult bees to developing larvae. When transmitted orally from adult to developing bee, it causes covert infections (no visible symptoms).

Affected bees tremble and are unable to fly. They frequently have no hair on their thorax and look oily. When healthy bees fight sick bees, or when food is shared between healthy and sick bees, the disease is spread to healthy bees.

Queens, drones and workers are all vulnerable to long-term bee paralysis. It appears that the queen is the source of susceptibility to the disease. Re-queening diseased colonies may therefore completely cure the colony of all symptoms.

5.3.3 Septicaemia

Pseudomonas aeruginosa bacteria are a rare cause of septicaemia in adult honey bees. The bacteria enter the haemolymph through an unidentified route, quickly multiply and ultimately result in the host's demise.

Bees that succumb to septicaemia appear to lack connective structures and are easily dismembered. Even with light handling, body parts, such as head, wings, thorax, etc., separate. The fluid of sick bees may also appear milky to creamy white. For verification, it might be necessary to isolate and identify *P. aeruginosa* bacteria from haemolymph.

5.3.4 Amoebiasis

Malpighamoeba mellificae is the causative agent of amoeba illness. This disease and *Nosema* disease are sometimes encountered in combination, and the two together may be more dangerous than either disease by itself. The attack is restricted to the Malpighian tubules and spread by tiny amoeba cysts found in bee excretions. The amoeba is thought to interfere with the Malpighian tubules, which finally causes the bee to die. The Malpighian tubules are microscopically examined for the presence of amoeba cysts to make the disease diagnosis.

Malpighamoeba mellificae is a single-celled organism that affects adult bees' excretory organs, the Malpighian tubules. All castes are vulnerable, but only rarely do drones and queens contract the disease. Infection results from the consumption of latent forms of the parasite. The parasite then develops, infiltrates the Malpighian tubules and multiplies there at the expense of the bee's excretory cells. Cysts enter the rectum and are released along with the faeces (Schäfer et al, 2022).

Only a microscopic examination allowing the identification of the amoebic cysts can provide a conclusive diagnosis of *M. mellificae*. Dysentery, spring decreasing and a reduction in the lifetime of infected bees have all been linked to *M. mellificae* infections. Since Nosemosis and *M. mellificae* infections are frequently seen together, a honey bee's health is likely to suffer more from a dual infection.

As with *Nosema* and other illnesses, maintaining excellent hygiene and effective treatment is essential to prevent the spread of the causal organism. The following may help to lower the risk of occurrence, even if there is no certified proprietary material/product registered for the checking of *M. mellificae* at this time:

1. Autumn feed intake needs to be finished before it gets cold to give the colony time to consume the sugar syrup and bring the water content down to a safe level. Avoid feeding fermented syrup and honey.
2. Refined sucrose or white table sugar should be used for generating one's own sugar syrup. Use white sugar only. Feed alternatively with a prepared syrup.
3. Change the comb in a hive frequently, and try to avoid using ones older than 3 years. New combs from diseased colonies can be sterilized with acetic acid.

5.4 PESTS OF HONEY BEES

5.4.1 Mites Infesting Honey Bees

5.4.1.1 Acarine Disease

Acarine disease is caused by the tracheal mite *Acarapis woodi*. The adult female enters the spiracle of the host by clinging to the thoracic hair of a newly emerged bee (about 5 days old). It begins to lay eggs once it penetrates the trachea. Mites are typically present in the prothoracic trachea, while they can occasionally be found in the air sacs of the thorax, abdomen and head. The host's haemolymph is consumed by the larvae and adults once they penetrate the breathing tubes. When more than 30 per cent of honey bees in a colony are afflicted, these mites are often linked to the demise of a whole bee colony in the autumn/early spring.

The tracheal mite is extremely tiny; males are 125–136 micrometres in length, while females range from 143 to 174 micrometres in length. The body is oval and white with a smooth cuticle. It has several long hairs sticking out from its torso and legs as well as a long mouthpiece that resembles a beak. Except for a brief period of migration, the tracheal mite spends its entire life cycle inside the honey bee's respiratory system. The female mite produces five to seven eggs while she is in the reproductive stage. The eggs hatch after 3 to 4 days and grow into larvae that feed, moult and finally become non-feeding nymphs. After one final moult, the adult mite emerges. Males take 11–12 days from egg to adult development, compared with 14–15 days for females. Gravid females exit the trachea through the spiracle in order to spread the disease again. They then connect to the bees' hair tips, making it simple for the disease to spread from one bee to another.

The easiest way to positively identify tracheal mites is by dissecting deceased honey bees; however, external examination of dead bees in colonies for the occurrence of tracheal mites is also possible. A normal healthy bee's trachea is consistent, smooth, and white or fleshy in appearance. On the other hand, in infected colonies, the trachea turns black as a result of mite activity in both the left and right tracheal trunks. The discolouration (bronzing) of the trachea along some of its length is followed by black specks or streaks that eventually combine, as signs of the early and middle phases of infestation.

Infestations with tracheal mites can persist for many years without showing any symptoms. The lifespan of adult bees and their capacity for thermoregulation are, however, both shortened by mite infestation. Therefore, a colony's near-death symptoms include a significant number of bees crawling out of the hive, clinging to grass stems and occasionally vibrating their wings.

Mite transport is facilitated by close bee contact. Only bees less than 5 days old are vulnerable, and dispersion occurs mostly at night. Only one generation can occur per host in the short-lived summer bees, but throughout the winter, each bee can support numerous generations. Drifting, swarming, crawling and robbing bees are the main causes of infestation across colonies. The rate of tracheal mite infestation in young bees is increased in bee colonies where feeding is suppressed. The mites are lost in the field after becoming dislodged from foragers during foraging trips; as a result, there is a greater probability that newly emerging bees will avoid infestation during their brief window of vulnerability. The ability of honey bees to maintain brood nest temperatures at 30–35°C is also diminished by moderate infestation. Reduced brood temperature makes newly emerging bees more susceptible to tracheal infection. The colony won't be able to effectively replace aged winter bees in the spring because nest growth will be constrained. As the sick colony disappears, the nest temperatures will drop even further, making the tracheal mite infections more expansive. It has been proposed that auto-grooming helps bees defend against parasitic tracheal mite infection. Genetically resistant honey bee strains groom more persistently and have lower response thresholds to migratory tracheal mite stimulus (such as observed motions, etc).

In the UK and Ireland, there is currently no approved proprietary product registered for the control of acarine. "Apiguard," despite being predominantly advertised as a varroacide, works well against tracheal mite infestations. Re-queening with a nearby colony has proven tracheal mite resistance and is also advised because bees' grooming prowess can vary greatly among populations. Treatment of the hive with the miticide chlorobenzilate is necessary for the disease to be controlled as effectively as possible.

5.4.1.2 Varroa Mite (Varroosis)

Varroa destructor, sometimes known as "varroa," is a highly destructive bug that can significantly lower honey output. The tiny mite impairs foraging, increases drifting, and is associated with a high winter mortality rate. The Asian bee *Apis cerana* is the original host of the mite, with a history of coevolution. *Varroa* has, however, been transmitted to *Apis mellifera*, a host with far less resilience to the parasite. The demise of untreated colonies of *Apis mellifera* within 2 years of infection indicates that the host–parasite relation is still evolving.

The mature female mite measures 1.1 mm in length and 1.5–1.6 mm in width. It is reddish brown in colour with oval, flattened shape, and its dorsal shell completely encloses the idosoma. Its ellipsoid shell is punctured on one side by four pairs of legs and an unclear head. Due to its flattened shape, suckers at the base of each tarsus, and the stiff hairs on the ventral side of the body, the mite is challenging for the bee to remove through grooming. The mites are typically found on first abdominal segments; however, they can also be seen between head and thorax and between thorax and abdomen. The intersegmental barrier can more easily be pierced in certain areas, allowing the mite to gain access to the infected bee's haemolymph. The male is yellow-white in hue and smaller (0.7 mm × 0.7 mm). Males cannot feed and are never discovered outside the brood cells, with their chelicerae modified for sperm transmission.

The phoretic stage and the larval stage comprise the varroa mite's life cycle:

The Phoretic Phase: The adult bee serves only as an intermediary host and a vehicle for the mite. The female mite consumes small amounts of adult haemolymph, but also requires larval haemolymph to promote egg laying. On adult bees, mites can survive for several months. The brood ratio and the mites' age both affect the rate of infestation. However, a few days after being released into the colony, the majority of mites enter brood cells to begin reproducing.

The Larval Phase: Five to six hours prior to capping, the adult female enters the brood cell. Drone larvae are favoured over worker larvae, and only heavily infested areas allow queen cells to be invaded. Until the cell is sealed and the larva has finished spinning the cocoon, the mite submerges itself in the liquid brood food at the bottom of the cell. The developing larva is subsequently fed on by the mite, which then begins to lay eggs 3 days later. The female will lay 25 eggs at various periods. Within the egg, a six-legged larva becomes an eight-legged protonymph within 24 hours. The protonymph feeds on the pupa's haemolymph for 1–2 days after hatching and before moulting into a deutonymph. Before moulting to the adult stage, feeding lasts for a further 3–4 days. The first egg turns into a male, while the others become females. Male mites mature in 6.8–6.9 days, whereas females do so in 6.0–6.2 days. Mating occurs between males and females within the cell. Haplo-diploid breeding is exhibited by the varroa mite, allowing very low levels of harmful recessive genes, a pre-adaptation to inbreeding.

Individual bees infected with one or more varroa mites during development can survive and seem normal. Deformed wings, inflated abdomens and discoloured legs, however, are typical signs of sickness in bees that have been more severely infected. The body weight of infected bees is also significantly reduced, and the haemolymph proteins are diminished. Additionally, the lifespan of the bee is decreased, and the bees are generally more susceptible to other bacterial and viral diseases.

Due to their ineffective metabolic activity, mites have relatively high food requirements. Up to 25% of the nutritional reserves that the pupae built up in the tissue during the larval stage are consumed by the mites. Together with viral particle transmission, this causes deformity and weakness in individual bees, which eventually results in colony collapse. Both worker and drone brood can be parasitized; however, the drone brood is preferred. This preference may result from a variety of factors, including: (1) chemical stimuli released by the growing larva; (2) larger drone cells; (3) cells that are open for longer; and (4) cells that are capped for longer, allowing more daughter mites to mature sexually. Reduced body weight, life expectancy and survival of the emerging bee are the results of parasitism during the pupal stage. Additionally, infected drones generate less sperm and have shorter flight times, both of which are crucial during the numerous mating rituals. Furthermore, the infected drones might be a source of viral infection and a potential sexual channel for viral particles to be transmitted.

Depending on the severity of the infestation, the honey bee colony may experience varroa mite symptoms. Low infestation levels have no noticeable effects; however, medium to high infestation levels are indicated by a patchy brood pattern and deformed drones and workers on the comb.

The colony eventually loses its social structure and disbands when the mite population rises to a level it can no longer support. This phenomenon is known as colony collapse. A dramatic drop in the number of adult bees, dead bees in the hive, different deformities of the brood (such as bald and cold brood), and a large number of varroa mites on the remaining bees and in the brood cells are typical signs of colony collapse. Control of varroa requires early detection thorough observation, rapid treatment and sound beekeeping techniques. By keeping an eye on natural mite fall, sampling drone brood or doing the ether roll test, one can determine the mite levels in the hive.

It is difficult to prevent this pest; however, regular mite population monitoring and subsequent management/treatment may lessen the negative impact that the feeding mite has on both growing larvae and adult bees. Coordinated treatment over a region is crucial, since it not only lowers the danger of re-infestation but also prevents the spread of virus infections between hives.

5.4.1.3 *Tropilaelaps* Mite

The *Tropilaelaps* mite is a quarantine pest with two main species, *Tropilaelaps clareae* and *Tropilaelaps koenigerum*. *T. koenigerum* is only known from Sri Lanka and Nepal, but *T. clareae* is widespread across the globe. Although the giant honey bee *Apis dorsata* serves as the native host and the source of *T. clareae*, these parasites have also been found in *A. florea*, *A. mellifera* and *A. laboriosa*. *A. laboriosa* is the sole additional host discovered for *T. koenigerum*. Female *Tropilaelaps clareae* measure roughly 1.0 mm long and 0.6 mm wide with a light brown tint. Although they are less sclerotized, the males are smaller than females. *T. koenigerum* is 0.7 mm × 0.5 mm in size, oval-shaped and brown in colour. Using a magnifying glass, both species can be distinguished from varroa. *Varroa* glides slowly on the comb and is wider than long. *Tropilaelaps*, on the other hand, is a fast-moving, elongated mite with a highly sclerotized holoventral shell.

On adult larvae, right before they are capped, the colonizing female lays one to four eggs. Although both worker and drone brood can become infected, mites favour drone brood. The larvae go through several nymphal transformations before becoming adults as the eggs hatch after 12 hours. The haemolymph of the developing bee is consumed by females of all stages, which results in deformity of the brood. Approximately 1 week is needed for development from egg laying to adulthood. The juvenile males and females, as well as the initial invading mother mite, leave the cell as the bee emerges. Adult bees' integument cannot be penetrated by *Tropilaelaps* mites; hence, phoretic survival on bees is quite brief (1-2 days). These mites are supposedly incapable of surviving in broodless colonies, and gravid female mites will pass away in 2 days if they don't lay any eggs. *T. clareae* populations grow more quickly than varroa mite populations, which can be attributed to their short lifespan and brief stay on adult bees. *T. clareae* outcompetes the varroa mite in colonies where both mites are present.

Numerous larvae die as a result of a *Tropilaelaps* mite infestation, and newly emerged juvenile bees are malformed and exhibit symptoms similar to those of bees infested with deformed wing virus . Most infected colonies have an uneven brood pattern, and worker bee sanitation activities can lead to many pierced cappings. Fluvalinate (Apistan) and formic acid are used successfully as a control against *Tropilaelaps*. Since the mite cannot feed on adult bees and can only survive for a short time outside enclosed brood, infestations may also be eliminated using non-chemical techniques like queen caging. The agile *Tropilaelaps* mites can easily travel from one bee to another within a colony. However, the beekeeper usually relocates infected colonies to new locations, or movements between colonies are brought on by organic processes including drifting, robbing and swarming.

5.5 INSECT PESTS OF HONEY BEES

5.5.1 Bee Louse (*Braula coeca*)

The braula fly is a wingless dipteran and doesn't harm or parasitize honey bees at any point in their life cycle. However, it feeds by stealing food directly from bees during trophallaxis, and when present in high numbers on a queen, it may make it difficult for her to lay eggs.

The adult braula are sturdy, reddish-brown, tiny (1.0–1.5 mm in length) flies. They lack halteres and wings. To firmly attach to the bees, their tarsi are modified with comb-like features. *Braula* can move fast over a bee's body surface but usually hangs around on the dorsal surface between abdomen and thorax. When starving, they climb up to the bee's head and consume regurgitated nectar straight from its mouth (Smith and Caron, 1984).

The braula eggs have a diameter of 0.42–0.84 mm. They can be deposited throughout the hive in empty cells, the caps of brood cells, wax remnants on the bottom board, and capped honey comb. Only eggs laid on honey combs with caps will hatch in 2–7 days, depending on the temperature. The larvae emerge from their eggs and burrow beneath the cappings, leaving the honey comb's surface with a pattern of 1-mm-wide, slender tracks. Before pupating, the larvae go through three instars. Larval stage duration can range from 7 to 11 days. The pupa is 1.4–1.7 mm in length and 0.5–0.75 mm wide, and has a creamy-white tint. This phase lasts only 1–3 days before the adult hatches. Depending on the temperature and the presence of capped comb honey in the hive, the time it takes for an egg to develop into an adult can be anywhere from 10 to 23 days. Braula overwinter as adults on adult bees. The larval tunnels give honey cappings the fissured appearance.

5.5.2 Wax Moths (*Achroia grisella* and *Galleria mellonella*)

The normal associations between wax moth species and bee colonies include two species. Greater wax moth adults (*Galleria mellonella*) have wings that wrap over their bodies like a roof and are greyish-brown in colour. They are active from June to October. *Achroia grisella*, a smaller wax moth, is seen more frequently. Its sheened brown-grey wings, with a wingspan of 16–24 mm, appear silvery on newly emerging adults. Although lesser wax moth larvae are known to also feed on dried fruit and dead insects and are crucial in the breakdown and decomposition of old, abandoned comb in the wild, wax moths typically require comb to thrive. The wax moths are not major pests but occasionally pose issues in a colony that is already struggling and cause issues when the comb is not being used (Ritter and Akratanakul, 2006).

Adults enter hives at night, sneaking past guard bees, or look for dead or abandoned colonies. The lesser wax moth may deposit between 300 and 600 eggs, while the greater wax moth may lay as many as 1800 eggs with ~60 eggs per day. They lay their eggs in batches in dark crevices. A week after the eggs are laid, they moult, and the larvae dig into the wax to begin their voracious feeding. Wax moth larvae's colour varies depending on their environment and food. Greater wax moth larvae are typically described as white, whereas lesser wax moth larvae are typically described as pinkish. The larvae have a series of pseudolegs and three pairs of legs behind the head. Contrary to their name, wax moths do not consume beeswax; instead, they feed on comb imperfections. They rarely harm new or untried comb, preferring comb that has been used for brood rearing. The wax moths often lay down silken webbing and make clear pathways inside the comb. Fully developed larvae construct a silk cocoon in the comb, among the debris on the hive floor, or affixed to a frame or the hive body after several months of eating (temperature dependent). The life cycle will restart a few weeks later when young adults break free of their cocoons. At the moment of cocoon creation, increased numbers of wax moth larvae can harm wooden hives (Ramachandaran and Mahadevan, 1951).

Low numbers of wax moths are common in many hives, but the guard bees are constantly on the lookout for them. Wax moths only become an issue in colonies that are weakened by illness or a lack of food. However, over the summer, they may increase to comparatively high densities in dead or abandoned hives. Moth infestations can be detected by galleries and webbing on the frames and supers. Cocoons are frequently covered in silk webbing and fastened to the frame and body of the hive. Before landing, the adult moth can be observed flying swiftly over the frames and comb. Combs may collapse and become stained with moth faeces in cases of severe infestation (Viraktamath, 1989).

5.5.3 Small Hive Beetle (SHB)

The Nitidulidae family of coleopterans includes the little hive beetle, *Aethina tumida*, a quarantine pest of honey bees. They are indigenous to sub-Saharan Africa, where they can be found both as scavengers and symbiotic organisms. The beetle can complete its life cycle on a fruit, but in its natural habitat, it feeds on pollen, honey and bee brood. It is typically regarded as a minor nuisance in African honey bees, despite the fact that it may operate as a parasite that destroys sick and weak colonies. In contrast, SHB infestation has a more negative impact on European honey bee colonies (Gulati and Kaushik, 2004). Within an infected colony, the beetle and its larvae can grow into enormous populations and devour honey, pollen, bee eggs and bee brood. Beetles can breed, consume honey bees' food and decimate existing colonies without necessarily waiting for a recently abandoned nest. In 2 weeks, weak and stressed colonies can possibly die from the beetle infestation (Caron, 1990).

The adult beetle is light yellowish brown when it is emerges. It then turns brown, dark brown and eventually black. The insect measures 5–7 mm in length and has an ovoid form. Young beetles are agile, fly easily and start flying after the first day or two after emergence. Later, they exhibit less activity and remain in the bee colony's darker areas. Adult beetles can occasionally infest a honey bee colony because they are aggressive flyers and can fly singly or in swarms. The SHB can identify stressed colonies at a distance of 13–16 kilometres; however, the exact method by which this is done is yet unknown.

During their life period, SHBs go through four unique stages: egg, larva, pupa and adult. All stages can be found inside the nest, with the exception of the pupal stage. The majority of the time, eggs are found in clusters in grooves and crevices. They are small, about two-thirds the size of bee eggs, and white in colour. They are occasionally discovered in brood cells or pollen cells during peak periods of reproduction. SHB larvae start to feed by digging into honeycombs and devouring bee larvae after 2–3 days post-emergence. Similar to wax moth larvae in appearance, SHB larvae are cream in colour and 10–11 mm long. Both species have three pairs of legs close below the head, but the SHB lacks the wax moth larvae's unique series of paired pro-legs. The larvae leave the hive after 10 to 14 days, when they have finished growing, and bury themselves in the ground beneath or in front of the hive to pupate. The pupae are initially pearly white in colour, but as they develop into adults, colouration starts to take over, first in the eyes, then at the base of the wings, and finally covering the entire body. The duration of time in the soil ranges from 15 to 60 days. After 3–4 weeks in the ground, the bulk of beetles begin to emerge (Taber, 1999).

Honey Bee Behavioural Adaptations to the SHB: A variety of strategies are used by European and Africanized honey bees to thwart or delay effective beetle reproduction. Aggression, social isolation, patrolling, worker grouping, espionage and migration are some examples of these behavioural adaptations. While breeding efforts in managed European populations have selected against undesirable traits in commercial bees, such as evasion, aggression and excessive propolis use, these traits are more visible in Africanized subspecies.

Combative Means: The SHB typically assumes a turtle-like protective position or flees the assaulting bees by rushing, falling and hiding. Africanized subspecies defend themselves by actively fighting both adults and larvae.

Social Encapsulation: SHB reproduction can be decreased through social encapsulation which restrict beetles from freely roaming over the comb if they manage to get into the hive. A few workers propolize around captured beetles, while others watch over the enclosures non-stop for up to 57 days. Even so, sometimes beetles succeed in eluding encapsulation, particularly at night when bees are less active. They can ensure their survival inside the cage by breeding, cannibalism and behavioural mimicry. The latter occurs when the beetles go toward the jail guard bee, extend their heads forward and make contact with the guard bee's antennae (mimicking normal bee trophallaxis). Although the guard bee may at first react violently, persistence on the side of the beetle typically persuades the bee to vomit a drop of honey, which it appears to swallow directly from the bee's mouth.

Patrolling and Worker Aggregation: Strong colonies rarely have SHB on the comb because the guard bees patrolling help to limit beetle entry to the brood regions. Beetles still occasionally lay eggs on the outside frames even in strong colonies with more bees per square inch of brood space. When insects trespass, Africanized workers congregate and clear the area by removing neighbouring pollen, honey and brood to expose the hiding beetles, which are subsequently killed. European bees also remove beetle eggs and larvae.

Absconding and Migration: Africanized bees are significantly more mobile than European bees, and any type of colony movement will lower SHB colony infestation levels because the non-phoretic beetles are left behind. Absconding is abandoning a colony, leaving brood and honey behind. Migration is a phenomenon that can be predicted by the season, and bees typically get ready for it by reducing egg laying, waiting for the brood to hatch, and eating a lot of stores. Although both movements disrupt the beetle's life cycle, migration has a more severe population-limiting impact than absconding because fewer food reserves are left behind for the beetles to use.

A quick way to search for bugs is to remove the lid and set it upside down on the ground. Stack the brood box atop the tipped-up lid. If beetles are present, they will leave the brood box and crawl under the lid away from the light. If the frame is in the sun, beetles may also be seen "surfing" across the comb. There are several different types of controls that have been created in areas where SHB is regarded as a nuisance. These include chemical control, soil insecticide treatment, colony management for prevention, and cleanliness.

5.5.4 Wasps and Hornets

Numerous wasp and hornet species prey on honey bees, causing serious injury to bee colonies and ultimately the destruction of entire apiaries (Sharma and Deshraj, 1985). The wasp and hornet invasion in commercial apiaries begins after spring and continues throughout the monsoon season, with colonies suffering the most harm from July to September. Due to wasp attacks, about 20–25% of bee colonies abandon their nests each year (Adlakha and Sharma, 1975). *Vespa* spp., the largest species of social wasp, are capable of consuming honey bees easily. Some of the species that decimate weak and queenless colonies for honey and the brood in apiaries and foragers in the field include *Vespa orientalis* (yellow-banded brown wasp), *V. magnifica* (large black wasp), *V. ducalis* and *V. auraria* (golden wasps) (Cappa, 2021).

A new wasp colony is started by a single mated queen (foundress queen) in the spring; the population grows during the monsoon and reaches its peak in the autumn. They fly to look for tree sap after emerging from their hibernation locations (dark, wet areas, such as cavities in trees, under roofs, etc.). They relocate to new nesting locations and begin building "embryo nests" after collecting tree sap (15–60 cells till the first emergence of the workers). After the initial group of workers appears, the division of labour occurs, and the colony then grows quickly. There is an emergence of new males and queens. For 1–2 months, males and queens are produced continuously. Until new queens are crowned, the foundress queen reigns. The males fly and wait for the virgin queens at the entrances to the nests, mate outside of the nest, and leave afterwards. Only 30–50% of queens mate successfully in spite of the high male density. It is seen that the queen only mates once, after which she begins looking for a place to hibernate. Queens occasionally hibernate in groups.

Wasps typically attack colonies within a flight radius of 800 metres to 1.5 kilometres. They do not sting the bees to kill them during the attack. It's likely that all *Vespa* species attack in a similar manner. In the beginning, a hornet is seen to catch and kill bees one at a time during a "hunting phase." Later, a "slaughtering phase" occurs, in which the wasps gather together at the hive's entrance and massacre the bees in large numbers. The wasps infiltrate the hive at last, during what is known as the "occupation phase," after this phase has lasted long enough and the colony under attack has lost the majority of its protectors. They rob nectar and brood in their nests. However, during the "hunting phase," they use various tactics to capture the prey at the entrance or other holes in the hive.

Wasps ignore the adult bees' heads and abdomens, favouring just the thorax part. Some species attack frequently at hive entrances to reduce the field force of the colony, while others wait on the back of the hive and capture bees emerging through fissures. Some species feed on the brood and young bees inside the hive after killing the guard bees. Bees that approach cracks, landing boards or hive entrances are captured by *Vespa orientalis*, but *V. magnifica* uses a collective predation method (Morse and Nowogrodzki, 1990).

Bees of the *A. mellifera* and *A. cerana indica* species deter/execute wasps by shimmering defence behaviour. The intruder dies either from being stung or from asphyxia and the high temperature at the shimmering ball's centre (43–60°C) (Abrol and Kakroo, 1998). One of the control methods is to destroy wasp nests. Another method of reducing wasp damage is to burn nests with kerosene or spray insecticide at night. Nest fumigation with calcium cyanide (0.5 g) at the entrance to the nest resulted in 100% mortality in a hornet nest. Protective screens, such as queen gates or guard boards installed at hive entrances or wire screens or cages, offered a physical barrier against wasp attack (Rana et al., 2000). However, *V. auraria* was able to quickly adapt to this alteration and began pulling bees just outside the cage, making the technique ineffective. The usage of flap in the apiary is another typical control strategy. Continuous flapping for 30 minutes has been found to keep wasps and hornets away from the apiary for up to 3 hours. By using various baits, such as lures or poison baits, and by combining fermented honey and water, wasp attacks are reduced (Shah and Shah, 1991).

5.6 GOOD BEEKEEPING PRACTICES

Good beekeeping practices are the holistic operations used by apiarists to ensure optimum well-being for human people, honey bees and the ecosystem. As a result, the implementation of "good beekeeping practices" would benefit colony health and society while also endorsing better production expectations. Such strategies are overall assessments that are applicable to beekeeping tasks and are widely accepted around the world. They are not disease-specific and are intended to be used by apiarists in the production of colony products. They are a resource that beekeepers can use to skilfully tackle the difficulties of day-to-day apiary monitoring.

1. Ensure adequate site selection for apiaries with an ample supply of bee flora available year-round at suitable settings that are far from pollution sources like intensive agriculture and industrial estates.
2. Verify the health state of newly purchased swarms, colonies and queen bees, and be sure to choose vendors, bees and beekeeping equipment carefully. In order to prevent the exposure of the colony to certain diseases, quarantine procedures should be used.
3. Use a distinctive numerical code/sign to identify each hive.
4. Keep track of each beehive, check colony output and overall wellness.
5. Throughout the year, periodically check on the colonies' state of health. The time of year will determine how frequently hives are inspected. Accessing the beehive should be generally restricted throughout the winter and under inclement weather circumstances.
6. Ensure healthy colony quality in the same apiary, check that the beehive size is adequate to deter overcrowding, and prevent robbing by eradicating very damaged or diseased colonies from the apiary – as they are more likely to be attacked and killed.
7. Use the proper methods to safeguard the well-being of bee colonies, particularly the younger or weaker ones; feed colonies without food reserves or in the event of inclement weather conditions, such as in the autumn, snowy, and extremely chilly or rainy springs; ensure better wintering; or provide sufficient water supplies, notably during hot weather.
8. Take care to avoid using toxic materials in hives, such as hazardous dyes, harsh chemicals for wood, or disinfectants.

9. If the colony's health is unknown, avoid moving honeycombs from one colony to another. The apiary should be cleared of sick colonies, and if required, they should be wiped out.
10. Keep your beekeeping supplies organized and hygienic. Renew the materials as needed. In the event of anomalies, seek professional advice.

5.7 CONCLUSION

Honey bees are susceptible to a wide range of illnesses and insect and mite infestations, which reduce the health and vigour of bee colonies. The diseases can be broadly divided into adult and brood diseases depending on the stage at which they are affected. The most dangerous brood illness that affects honey bees is known as American foulbrood. The bacterium that causes it, *Paenibacillus larvae*, produces heat- and drought-resistant spores that can germinate at any time in a favourable environment. A preliminary diagnosis of the illness is made using the "ropy" test, in which the contents of the infected cell pull out as a rope. The European foulbrood develops during the height of colony development. *Melissococcus plutonius*, a non-spore-producing bacterium, is the responsible organism. The less ropy diseased cell and the loosely adherent scales on the cell walls set it apart from American foulbrood. For the control of the foulbrood illnesses, pathogen exclusion is effective. Chalkbrood and stonebrood are two fungi that can affect honey bee colonies; *Ascosphaera apis* is the fungus that causes the former and *Aspergillus flavus* the latter. The brood becomes mummified as a result of both infections. Both fungi weaken the colony and are best controlled by choosing the right apiary location (away from wetlands and with good exposure), using sound management practices (to minimize water intrusion within the hives), and routinely disinfecting beekeeping tools (for example, sterilizing by torch flames). The viral disease of the brood known as sacbrood results in an accumulation of fluid between the skin of the pre-pupal and larval stages. The outer skin hardens, allowing the entire sac to be expelled without any fluid leaking out.

Nosema apis is a fungus that causes disease, affects adult bees and shortens their lifespan. Supersedure may result from it, and the colony's ability to produce honey is affected. Bee paralysis viruses, septicaemia and amoebiasis are some other illnesses that affect adult bees.

Most of the insects and mites that harm honey bees are pests. The trachea is home to the tracheal mite *Acarapis woodi*, which feeds on bee haemolymph. Through natural auto-grooming, genetically resistant breeds of honey bee mites can be controlled. Only the application of a miticide allows control in cases of severe infestation. The *Varroa* mite is a particularly harmful pest of honey bees. It exists externally on the honey bee's body and consumes haemolymph via intersegmental membranes. If left untreated, a severe *Varroa* mite infestation results in colony collapse, and a robust, healthy colony can only persist for 1 year following the mite infestation. Early diagnosis is crucial and can be managed with the use of miticides. *Tropilaelaps* mite is a significant mite issue in beekeeping. Despite being a pest for quarantine, it has the capacity to harm bee colonies by lowering the worker population. The emerging young bees are malformed because it nurtures itself on the blood of fthe brood.

The bee louse, wax moths, SHBs, wasps and hornets are some of the insects that harm honey bees. By feeding on the honey bees' hive contents, such as regurgitated nectar and wax combs, bee louse and wax moths have an indirect impact. By attacking and killing worker bees while they are in flight, hornets and wasps decrease the population of worker honey bees. Weak colonies' eggs and brood are consumed by the tiny hive beetle. The various management strategies are effective against various insect pests. The best way to control bee louse, wax moths and small hive beetles is with appropriate hive cleaning. Wasp and hornet attacks in the region are decreased by flapping in the apiary.

A management module that addresses the biotic stressors that honey bees encounter continues to be the cornerstone for comprehending apiary dynamics and identifying issues early on.

REFERENCES

Abrol, D. P. and Kakroo, S. K. 1998. Studies on seasonal activity and control of predatory wasps attacking honey bee colonies. *Indian Bee Journal* 60:15–19.

Adlakha, R. L. and Sharma, O. P. 1975. *Apis mellifera* vs. *Apis indica*. *Glean Bee Culture* 103(5):160.

Aronstein, K. A. and Murray, K. D. 2010. Chalkbrood disease in honey bees. Journal of Invertebrate Pathology 103:20–29.

Bailey, L. 1983. *Melissococcus pluton*, the cause of European foulbrood of honeybees (*Apis* ssp.). *Journal of Applied Bacteriology* 55:65–69.

Bailey, L., Gibbs, A. J. and Woods, R. D. 1964. Sacbrood virus of the larval honey bee (*Apis mellifera* Linnaeus). *Virology* 23:425–429.

Batra, L. R., Batra, S. W. T. and Bohart, G. E. 1973. The mycoflora of domesticated and wild bees (Apoidea). *Mycopathologia et Mycologia Applicata* 49(1):13–44.

Borum, A. E. and Ulgen, M. 2008. Chalkbrood (*Ascosphaera apis*) infection and fungal agents of honey bees in north-west Turkey. *Journal of Apicultural Research* 47(2):170–171.

Brodsgaard, C., Ritter, W. and Hansen, H. 1998. Response of in vitro reared honey bee larvae to various doses of Paenibacillus larvae spores. *Apidologie* 29(6):569–578.

Cappa, F., Cini, A., Bortolotti, L., Poidatz, J. and Cervo, R. 2021. Hornets and honey bees: A coevolutionary arms race between ancient adaptations and new invasive threats. *Insects*. 12:1037.

Caron, D. E. 1990. In: Morse, R. A. and Nowogrodzki, R. (eds). *Honeybee Pests, Predators and Disease*. Cornell University Press 2nd edition, Ithaca/New York, pp. 171–172.

Chantawannakul, P., de Guzman, L. I., Li, J. and Williams, G. R. 2016. Parasites, pathogens, and pests of honeybees in Asia. *Apidologie* 47(3):301–324.

Coffey, M. F. 2007. Parasites of the honeybee. The Department of Agriculture, Fisheries and Food, Agriculture and Food Development Authority, Ireland.

Flores, J. M., Ruiz, J. A., Ruz, J. M., Puerta, F., Bustos, M., Padilla, F. and Campano, F. 1996. Effect of temperature and humidity of sealed brood on chalkbrood development under controlled conditions. *Apidologie* 27(4):185–192.

Foley, K., Fazio, G., Jensen, A. B. and Hughes, W. O. 2014. Nutritional limitation and resistance to opportunistic *Aspergillus* parasites in honey bee larvae. Journal of Invertebrate Pathology 111(1):68–73.

Genersch, E. 2008. *Paenibacillus larvae* and American foulbrood–long since known and still surprising. Journal für Verbraucherschutz und Lebensmittelsicherheit 3(4):429–434.

Genersch, E., Forsgren, E., Pentikäinen, J., Ashiralieva, A., Rauch, S., Kilwinski, J. and Fries, I. 2006. Reclassification of *Paenibacillus larvae* subsp. *Pulvifaciens* and *Paenibacillus larvae* subsp. *larvae* as *Paenibacillus larvae* without subspecies differentiation. *International Journal of Systematic and Evolutionary Microbiology* 56(3):501–511.

Gulati, R. and Kaushik, H. D. 2004. Enemies of honeybees and their management–a review. *Agricultural Reviews* 25(3):189–200.

Lee, K. 2012. Purple brood-blog. Bee Informed Partnership, Blogspot. https://beeinformed.org/2012/11/20/purple-brood.

Lopes, L. Q. S., Quatrin, P. M., De-Souza, M. E., De-Almeida, V. R. and Vianna, S. R. C. 2015. Fungal infections in honey bees. *The Journal Fungal Genomics and Biology* 4:102–118.

Meixner, M. D., Francis, R. M., Gajda, A., Kryger, P., Andonov, S. and Uzunov, A. 2014. Occurrence of parasites and pathogens in honey bee colonies used in a European genotype-environment interactions experiment. *Journal of Apicultural Research* 53(2):215–219.

Morse, R. A. and Nowogrodzki, R. (eds). 1990. *Honeybee Pests, Predators and Diseases*, 2nd edn. Comstock Publishing Associates/Cornell University Press, Ithaca.

Murilhas, A. M. 2002. *Varroa destructor* infestation impact on *Apis mellifera carnica* capped worker brood production, bee population and honey storage in a Mediterranean climate. *Apidologie* 33(3):271–281.

Ramachandaran, S. and Mahadevan, V. 1951. The wax moths and their control. *Indian Bee Journal* 13(7/8):85–88.

Rana, B. S., Kumar, A. and Gupta, J. K. 2000. Influence of environmental factors on the population of predatory wasps on honeybees and evaluation of methods for controlling the wasps. *Indian Bee Journal* 62(1–2):47–54.

Ratnieks, F. L. W. 1992. American foulbrood: The spread and control of an important disease of the honeybee. *Bee World* 73(4):177–191.

Rhodes, J. 2005. Non-infectious disorders of honey bees. *Profitable and Sustainable Industries Primefact* :40–45.

Riessberger-Galle, U., Von Der, O. W. and Crailsheim, K. 2001. Adult honeybee's resistance against *Paenibacillus larvae* larvae, the causative agent of the American foulbrood. *Journal of Invertebrate Pathology* 77(4):231–236.

Ritter, W. and Akratanakul, P. 2006. Honey bee diseases and pests: A practical guide agricultural and food engineering technical reports. FAO, Food and Agriculture Organization of the United Nations, Rome.

Sanford, M. T. 1987. Diseases and pests of the honey bee. University of Florida Cooperative Extension Service. *Institute of Food and Agriculture Sciences*, EDIS.

Sarwar, M. 2016. Fungal diseases of honey bees (Hymenoptera: Apidae) that induce considerable losses to colonies and protocol for treatment. *International Journal of Zoology Studies* 1(1):8–13.

Schäfer, M. O., Horenk, J. and Wylezich, C. 2022. Molecular detection of *Malpighamoeba mellificae* in honey bees. *Veterinary Sciences* 9(3):148.

Seitz, N., Traynor, K. S., Steinhauer, N., Rennich, K., Wilson, M. E. and Ellis, J. D. 2015. A national survey of managed honey bee 2014–2015 annual colony losses in the USA. *Journal of Apicultural Research* 54(4):292–304.

Seyedmousavi, S., Guillot, J., Arné, P., De Hoog, G. S., Mouton, J. W., Melchers, W. J. G. and Paul, E. Verweij 2015. Aspergillus and aspergilloses in wild and domestic animals: a global health concern with parallels to human disease. *Medical Mycology* 53:765–797.

Shah, F. A. and Shah, T. A. 1991. *Vespa velutina*, a serious pest of honey bees in Kashmir. *Bee World* 72(4):161–164.

Sharma, P. P. and Deshraj, R. 1985. Diversity of Bee Flora in Kangra Shivaliks and its impact on bee keeping. *Indian Bee Journal* 47(1–4):21–23.

Smith, J. I. B. and Caron, D. M. 1984. Distribution of the bee louse *Braula coeca* Nitzsch in honeybee colonies and its preferences among workers, queens and drones. *Journal of Apicultural Research* 23(3):171–176.

Spiltoir, C. F. and Olive, L. S. 1955. A reclassification of the genus *Pericystis betts*. *Mycologia* 47(2):238–244.

Taber, S. 1999. The small hive beetle, as described by A. E. Lundie in 1940. *The American Bee Journal* 139:450–451.

Viraktamath, S. 1989. Incidence of greater wax moth *Galleria mellonella* (L.) in three species of honey bees. *Indian Bee Journal* 51:139–140.

White, G. F. 1917. Sacbrood. *US Department of Agriculture Bulletin* 431:1–55.

Zaghloul, O. A., Mourad, A. K., El-Kady, M. B., Nemat, F. M. and Morsy, M. E. 2005. Assessment of losses in honey yield due to the chalkbrood disease, with reference to the determination of its economic injury levels in Egypt. Communications in Agricultural and Applied Biological Sciences 70(4):703–714.

6 Honey

Otilia Bobiş, Syeda Tasmia Asma, Ulas Acaroz, Syed Rizwan Ali Shah, Damla Arslan-Acaroz, Sevgi Kolayli and Gulzar Ahmad Nayik

6.1 INTRODUCTION

Honey is a naturally sweet product collected from nature by honey bees (*Apis mellifera*) to meet their food and energy needs. It possesses low water activity and can be stored in hives or human habitations for many years without spoiling.

Written sources show that honey has been an important food source for humans dating back as far as 4000–5000 BC. Written tablets also describe honey being used as an ointment. Ancient cultures and sacred texts also refer to honey as a natural source of food and healing (Samarghandian et al. 2017).

The composition of honey is affected by the characteristic flora of the geographical region where it is collected and the manner of production. Honey exhibits a broad range of biologically active properties and is used in apitherapy due to its antioxidant, antimicrobial, antiviral, anti-inflammatory, antitumoral, and antidepressant characteristics (Otmani et al. 2021). Honey is an essential ingredient of folk medicine, together with other bee products such as bee pollen, royal jelly, propolis, and bee venom, and its use in modern medicine is known as apitherapy (Mijanur Rahman et al. 2014). Apitherapeutic honeys are always untreated raw honeys and are selected from those with high levels of biological activity. Honeys rich in polyphenols, such as forest honeys, dark-colored and fresh honeys, and those not easily crystallized, constitute the class of apitherapeutic honeys.

6.2 TYPES OF HONEY

According to the raw material from where the bees gather nectar and pollen to produce honey, there are two major classes: floral honeys and honeydew honeys. The bees make floral honey from the nectar of flowering plants, enriched with flower pollen and their own substances added in the process of production.

6.2.1 Floral Honey

Floral diversity is the basis of production for a very high number of floral honeys. Whether they are monofloral or multifloral, floral honeys are characterized by specific taste, color, and properties. Unifloral honeys are regarded as more valuable because of their plant-specific flavor and properties (Mădaş et al. 2019; Bobiş et al. 2020, 2021).

A recent review (Mărgăoan et al. 2021) made a clear classification of world unifloral honeys based on plant taxa, chemical composition, mineral content, and phenolic compounds. Useful chemical markers for specific unifloral honeys are presented, mainly from the class of polyphenols, along with the extraction technique, solvent, and method of detection.

Phenolic compounds may be important markers of unifloral honeys, in addition to the chemical composition and palynological analysis. Although it was proposed to use acacetin as a botanical marker of authenticity for acacia (*Robinia pseudoacacia*) honey and quercetin for sunflower honey

DOI: 10.1201/9781003354116-6

(*Helianthus annuus*) (Bobiş et al. 2021), these compounds have been identified in other honey types (Mărgăoan et al. 2021). Resveratrol, an important antioxidant present in different plants and in propolis, was identified in sage honey (*Salvia officinalis* L.) and proposed as a botanical marker (Gašić et al. 2015). Methyl syringate and leptosin are considered botanical markers for Manuka honey (Alvarez-Suarez et al. 2014), and their amount may establish the purity of this type of honey, though they have been identified in other monofloral honeys (Mărgăoan et al. 2021).

6.2.2 Honeydew Honey

Bees produce honeydew honey by collecting sweet substances from the green parts of the plant (extrafloral), either released by the plant in dry periods and with high humidity, or excreted by plant-sucking insects (*Aphides*) that feed on tree sap (Escuredo et al. 2012) (Figure 6.1).

Honeydew honey is easily recognized due to its dark color, high viscosity, and specific taste. Also, the chemical composition is different compared with floral honey, with high pH and electrical conductivity, higher content of di- and trisaccharides (the presence of melezitose), and lower amounts of the monosaccharides fructose and glucose (Seraglio et al. 2019). The presence of melezitose and raffinose, which are not found in floral honey, is an indicator of honeydew honey authenticity (Kaškoniene et al. 2010). Honeydew honey presents higher amounts of bioactive compounds (polyphenols, amino acids) and minerals. Consequently, it exhibits higher antioxidant and antibacterial activity (Bogdanov et al. 2004; Escuredo et al. 2012; Seraglio et al. 2019).

As stated before, honeydew honey presents high amounts of minerals, hence the increased electrical conductivity. It is known that minerals play an important role in the human body, forming structural components of tissues and regulating the metabolic system (Seraglio et al. 2019). As with all components of honey, which are more bioaccessible, minerals from honeydew honey are subject to the same principle. The most abundant mineral in honeydew honey was found to be potassium (K), corresponding to more than 70% of the total quantified. Other minerals reported in honeydew honey are sodium (Na), calcium (Ca), magnesium (Mg), and phosphorus (P), with concentrations exceeding 1% (Seraglio et al. 2019). Microelements that may also be present in honeydew honey are Cu, Fe, Zn, Mn, Al, Cr, Ni, S, and Cd (Escuredo et al. 2012).

Besides sensory and chemical analysis applied to honeydew honey, microscopic analysis is very important in determining the authenticity of this type of honey. In the sediment of honeydew honey, as well as pollen grains, there are different fungal spores from plant pathogens (Escuredo et al.

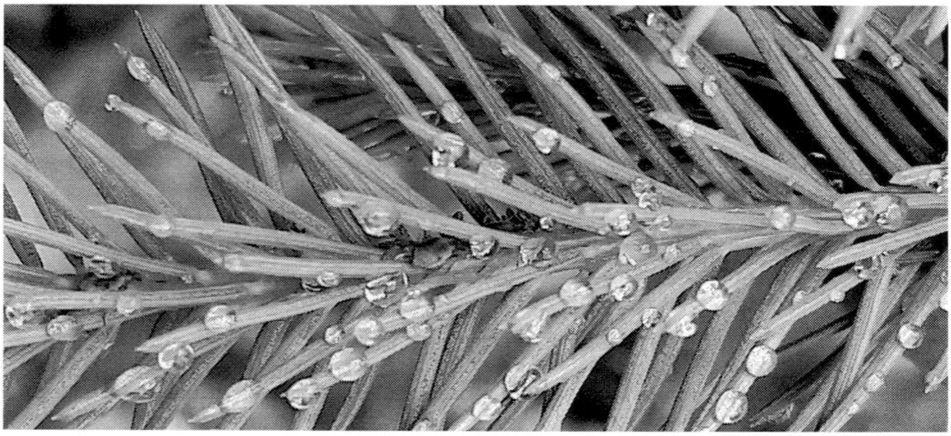

FIGURE 6.1 Honeydew sap and aphids on pine tree. (Original image provided by beekeeper Mihai Grama, Romania)

2012), harvested together with honeydew. The most frequent fungal elements in honeydew honey are *Cladosporium* conidia, *Alternaria*, and *Torula*.

6.3 GENERAL COMPOSITION OF HONEY

Honey comprises about 200 different components, including water, minerals, vitamins, carbohydrates, amino acids, enzymes, pigments, organic acids, and volatile compounds (Alvarez-Suarez et al. 2013; da Silva et al. 2016; Asma et al. 2022). The nutritional and medicinal compounds in honey vary with respect to the nectar and pollens collected by bees during its production stage. Moreover, the variability in the chemical composition depends upon the spatial territories, the altitude at which the bees make colonies, seasonal and ecological aspects, and its supply chain maintenance (Battino et al. 2019).

6.3.1 Carbohydrates

According to Alvarez-Suarez et al. (2013), honey comprises 70–75% simple sugars, of which almost 35–40% is fructose and 30–35% is glucose. Other than simple sugars, honey also contains more complex sugars. Among disaccharides, the highest amounts are represented by maltose (almost 7% in some honeys) and sucrose (up to 5% in authentic honey). Other disaccharides, including maltulose, nigerose, turanose, trehalose, kojibiose, and isomaltose, are found in trace amounts. Among trisaccharides, the most common ones are melezitose (specific to honeydew honey) and maltotriose.

Maltotriose comprises three units of glucose. First, it is converted to maltose, which is hydrolyzed by glucosidase into readily available glucose (Soldatkin et al. 2013). The chemistry of sugars in honey varies with respect to flower nectar, geological lineage, atmospheric influences, and refining and storing constraints (Escuredo et al. 2014). For the identification of monofloral or multifloral honeys, glucose and fructose are valuable biomarkers. Because of the high amounts of simple sugars in honey, the bioavailability of carbohydrates in honey is much higher than in other sweet products. Hence, it serves as an immediate energy booster for its consumers. Impressively, 100 g of honey provides 304 kcal of energy, and one dessert spoon covers 3% of the healthy energy intake daily (Singh et al. 2022).

6.3.2 Water

It is essential to analyze the moisture content in honey to avoid degradation and impairment of the nutritional profile and quality of honey, because fermentation could start. It is crucial to control and check the amount of water in honey before storage, and any variation from the recommended level is not accepted. Moreover, viscosity and glucose crystallization depend upon the amount of water in honey (Bogdanov et al. 2004). The regulatory authority, called the International Honey Commission (IHC), recommends that honey containing 20 g of water per 100 g is acceptable to trade and is suitable for human consumption.

6.3.3 Organic Acids

Honey is slightly acidic because it contains 0.57% of organic acids that are made during the conversion of nectar to honey. Organic acids are important indicators for identifying honey's floral or geological lineage. Besides, organic acids enhance the durability of honey against contamination and have bactericidal actions (da Silva et al. 2016). Among different organic acids, gluconic acid is considered one of the major components compared with others such as α-ketoglutaric acid, acetic acid, aspartic acid, glutamic acid, butyric acid, galacturonic acid, quinic acid, citric acid, glutaric acid, formic acid, pyruvic acid, malonic acid, α-hydroxyglutaric acid, succinic acid, propionic acid, fumaric acid, oxalic acid, malic acid, 2-hydroxybutyric acid, tartaric acid, glyoxylic acid,

2-oxopentanoic acid, isocitric acid, shikimic acid, lactic acid, and methylmalonic acid, which are present in varying concentration ranges (Prica et al. 2014).

6.3.4 Proteins, Enzymes, and Amino Acids

The protein content varies depending on the species of honey bee. Honey produced by *Apis cerana* contains 0.11–3.31% protein, while honey produced by *Apis mellifera* contains 0.21–1.60% protein (Zhang et al. 2019). The concentration of amino acids in honey is approximately 1% w/w. The main amino acid in honey is proline, which constitutes almost 50–80% of the amino acid profile; it is produced mainly by the salivary gland while converting nectar to honey (Truzzi et al. 2014). Proline concentration is the primary marker for identifying the maturation of honey during the harvesting stage, and occasionally, it also indicates the adulteration of honey with external sugars. For instance, in unadulterated honey, the minimum value of proline should be 180 mg/kg of honey (Czipa et al. 2012). Other than the major amino acid proline, honey also contains enzymes that are included in the list of proteins. Invertases, including α-glucosidases, β-glucosidases, and sucrase, are the enzymes present in the highest amounts. Diastases, including α-amylase and β-amylase, are the enzymes that convert polysaccharides into disaccharides and are the primary marker for identifying the quality of the honey. The Honey with remarkable quality contains a higher quantity of diastases. Honey also has bactericidal action because it contains glucose oxidase, which hydrolyzes glucose into gluconic acid and H_2O_2 (Bogdanov et al. 2008).

Honey's composition includes a number of essential and non-essential amino acids, including lysine, alanine, tryptophan, leucine, methionine, tyrosine, arginine, threonine, histidine, glycine, glutamic acid, ornithine, asparagine, glutamine, valine, β-alanine, phenylalanine, α-alanine, phenylalanine, and serine (Asma et al. 2022).

6.3.5 Vitamins and Minerals

The major constituent of vitamins in honey is the vitamin B complex, comprising biotin, riboflavin, thiamine, pantothenic acid, nicotinic acid, folic acid, and pyridoxine (Asma et al. 2022). Vitamin C is also present in honey but is mostly oxidized by enzymatic and chemical reactions (Bogdanov et al. 2008). Minerals in honey vary according to the type of honey; the content is approximately 0.041–0.21%. The mineral content varies with respect to the type of soil and geological background from where the honey bees harvested the nectar. Among all the minerals, potassium is the main element, and it comprises one-third of the total mineral composition.

6.3.6 Phenolic Compounds

Phenolic compounds are present in many different plants with different ratios according to plant type (Acaroz et al. 2019). The bioactive properties of honey are caused mainly by the plant's secondary metabolites, which are delivered to the honey bees via nectar. Phenolic acids and flavonoids are the major phenols in honey. Several new phenolic compounds have been identified in different types of honey; some of these, along with their antioxidant potential, are listed in Table 6.1.

6.3.6.1 Flavonoids

Flavonoids are determined as major functional components of honey. The chemical structure of flavonoids includes benzene and pyran rings. Moreover, based on these rings, they are classified as flavanones (such as naringenin, hesperetin, and pinocembrin), flavonols (such as kaempferol, myricetin, rutin, and quercetin), flavanols (such as gallocatechin and catechin), and flavones (such as apigenin, chrysin, and luteolin). The maximum level of flavonoids in honey is 19.99 mg/kg, and it varies based on the floral lineage of the honey (Alvarez-Suarez et al. 2010). The highest number of flavonoids is reported in Manuka honey (monofloral honey originating from New Zealand),

TABLE 6.1
Some of the Recently Identified Phenolic Compounds Along with Their Antioxidant Activities in Selected Honey Samples

Types of honey	Origin with floral source	Phenolic acid	Flavonoids	Antioxidant capacity	References
Thyme	Greek, monofloral	Caffeic acid Protocatechuic acid *p*-Coumaric acid Vanillic acid *p*-Hydroxybenzoic acid	Quercetin Apigenin Galangin Myricetin Kaempferol Chrysin	ORAC (415–692 μmol TE/kg)	Nousias et al. 2018
Manuka	New Zealand, monofloral	Caffeic acid Gallic acid Syringic acid *p*-Coumaric acid 4-Hydroxybenzoic acid *t*-Cinnamic acid *t*-Ferulic acid	Apigenin Chrysin Galangin Isorhamnetin Kaempferol Leptosin Pinocembrin Quercetin	FRAP (0.14 ± 0.00 mmol TE/100 g) DPPH (0.06 ± 0.01 mmol TE/100 g) TEAC (0.22 ± 0.00 mmol TE/100 g)	Alvarez-Suarez et al. 2016; Afrin et al. 2018
Acacia	Romania, monofloral	*t*-Cinnamic acid *p*-Hydroxybenzoic acid *p*-Coumaric acid Ferulic acid	Kaempferol Naringenin Apigenin Pinocembrin Chrysin Acacetin	FRAP (0.652–0.770 mM Fe^{2+}) DPPH (3.97–4.80 mmol Trolox/100 g)	Bobiş et al. 2021
Rosemary	Spanish, heterofloral	Chlorogenic acid Caffeic acid phenethyl ester Caffeic acid	Quercetin Kaempferol Isorhamnetin	DPPH (EC_{50}), (202 mg/mL)	Combarros-Fuertes et al. 2019

DPPH - 2,2-Diphenyl-1-picrylhydrazyl; EC_{50} - extract concentration (mg/mL) providing 50% of radical scavenging activity; FRAP – Ferric Reducing Antioxidant Power Assay; ORAC – Oxygen Radical Antioxidant Capacity: TE – Trolox Equivalents; TEAC – Trolox Equivalent Antioxidant Capacity.

buckwheat honey, and tualang honey (polyfloral honey originating from Malaysia). However, the lowest amount of flavonoids is found in acacia and gelam honey (Porcza et al. 2016).

6.3.6.2 Phenolic acids

The phenolic acids in honey are hydroxycinnamic and hydroxybenzoic acids (da Silva et al. 2016). All hydroxybenzoic acids differ in methylation and hydroxylation, including salicylic acid, ellagic acid, syringic acid, vanillic acid, syringic acid, and benzoic acid. Hydroxycinnamic acids mainly include ferulic acid, *p*-coumaric acid, caffeic acid, and sinapic acid. Some other phenolic acids, such as chlorogenic acid and *p*-hydroxycinnamic acid, are also present in honey in trace amounts based on the nectar collected by honey bees (da Silva et al. 2016).

6.3.7 Aromatic and Volatile Compounds

The flavor of honey is produced by volatile components and varies according to nectar type and after-harvesting conditions. Monofloral honey has a particular flavor due to the bee's restricted nectar collection and specific volatile components (Castro-Vázquez et al. 2007). A complex study on Romanian acacia honey was performed by Mădaş et al. (2019), and 79 volatile compounds were identified: alcohols, aldehydes, esters, ketones, sulfur compounds, nitrogen compounds, carboxylic and aromatic acids using solid-phase microextraction and gas chromatography/mass spectrometry (GC/MS) determination. The volatile profile, together with the sugar spectrum of the investigated samples, allows the differentiation of the geographical origin of acacia honey samples (Mădaş et al. 2019).

6.4 HONEY PROCESSING

Honey is processed via heating at a particular temperature for purification, filtration, and moisture removal at optimized conditions. It is essential to increase honey's shelf life by delaying crystallization and inhibiting microbial growth. Honey is more prone to the emergence of spontaneous crystallization due to the presence of highly saturated sugars. Crystallization causes an unwanted appearance, resulting in the formation of two different phases that can be visually distinguishable (liquid phase at the top and crystallized phase at the bottom). Commercially, honey is processed by heating (at a specific temperature for purification), filtration, and moisture reduction (to a standard level). The processing is essentially required to extend the shelf life of honey by inhibiting microbial growth and delaying crystallization in honey. Different processing methods have been used for honey processing: conventional processing, spray-drying, high pressure, thermal treatment, microwave processing, and thermosonication.

6.4.1 Conventional Processing

Conventional processing mainly includes six steps for honey processing, from raw honey to bottled honey, after removing suspended particles, microorganisms, and moisture. In conventional methodology, honey is preheated at 40°C, strained and filtered, and then immediately delivered to cold chain management after heating at 60–65°C (for 20–30 minutes) to maintain its natural color with improved taste and nutritional profile (Wakhle 1995). Investigations have reported that heat treatment of honey at 68, 65, and 63°C for 7.5, 25, and 35 minutes, respectively, can significantly destroy the yeast cells in honey (Wakhle 1996). An investigation has documented that hidroxymethylfurfural (HMF) is formed in different types of honey, and its formation varies with respect to temperature (Singh and Bath 1998). The formation of HMF is directly related to the processing temperature and the type of honey; high-temperature treatment for a long time results in increased formation of HMF.

Visquert (2004) studied the impact of thermal treatment on the storage and quality of honey. Honey samples were treated at different temperatures ranging from 35 to 85°C for different periods of time to determine the moisture content, HMF level, electrical conductivity, and acidity of honey samples. The conclusions were that thermal treatment enhanced the level of HMF formation in honey, and on the other hand, moisture content, acidity, and electrical conductivity remained unaffected in thermally treated honey samples. Nowadays, consumer preferences for better-quality honey are increasing, so alternatives to conventional technology are being explored. There have been other techniques compared with conventional heat treatment, and every methodology has pros and cons.

6.4.2 Microwave Processing

In the last few years, heating via microwave has commonly been used for food treatment through heating. It is not only being used at the domestic scale but also being applied at the industrial scale. Microwave processing is advantageous over conventional processing because of its short heating time, space and energy saving, and easy usage (Dimiņš et al. 2019). Microwaves work using the mechanism of non-ionizing radiation by electromagnetic waves with a frequency range from 300

MHz to 300 GHz. The energy conversion to heat in food by these electromagnetic waves is quite effective. A study revealed that the thermal treatment of honey using microwaves, even for a few seconds (10 s), might negatively affect the honey's quality. Escalation in HMF content and reduction in invertase activity cause honey quality to deteriorate (Bartáková et al. 2011).

Honey contains many enzymes, such as amylase, catalase, invertase, and glucose, oxidase, which are a measure of honey quality. Enzymes act as catalysts for various biological reactions and control the pH, composition, and antimicrobial characteristics (Dimiņš et al. 2019). Invertase is a marker for the heat treatment of honey (Sahin et al. 2020). Amylase is more stable to heat as compared with invertase. Hence, the enzymatic action of invertase is lost more quickly as compared with that of amylase. Microwave heating is preferred over conventional heating because of the controlled temperature and time availability. Microwaves produce heat by penetrating the food cells, which is a more effective way of honey processing.

6.4.3 Thermal Treatment

Honey was conventionally processed for storage by using thermal treatment. As in modern processing methods, honey was processed by thermal treatment for color homogenization (for the satisfaction of consumers) (Abu-Jdayil et al. 2002), reducing the moisture content in honey to avert fermentation (Subramanian et al. 2007), facilitating the packaging of honey with low viscosity (Anklam 1998), and destroying the yeast growth by dissolving sugar crystals to impede granulation and consequently increasing the shelf life of honey (Escriche et al. 2009).

Honey undergoes indirect heat treatment (for 25–30 min at 60–65°C, followed by rapid cooling) followed by filtration, straining, and preheating treatment (at 40°C) in the process of thermal treatment. Previously performed investigations revealed that heat treatments at different temperatures, such as 63, 65, and 68°C, for different exposure times (such as 35, 25, and 7.5 minutes, respectively) significantly inhibited the growth of yeast cells (Subramanian et al. 2007). The range for thermal treatment differs little from the mild–high-temperature range. Thermal processing techniques used in industry for the liquefaction and pasteurization of honey mostly use 45–80°C temperatures, respectively (Escriche et al. 2009). Heating can also cause some adverse effects in honey. These adverse effects are directly proportional to the applied temperature and exposure time; the higher the heat and the longer the time, the higher will be the adverse effects. However, the thermal processing method can be used to protect the honey from fermentation by altering its consistency, but the high temperature can adversely affect the honey's quality and biological properties.

6.4.4 Spray-Drying Process

Several foods can be dried into powder form via the spray-drying technique to maintain the quality of food by retaining the activity of particular bioactive components and improving shelf life. The spray-drying technique can be employed to preserve heat-sensitive compounds like lactic acid microorganisms and enzymes (Obón et al. 2020). It provides efficient drying in a short time at a low temperature with a fast evaporation effect. On the other hand, freeze-drying or freezing are preferable on an industrial scale. The drying conditions and the matrix preparation should be properly optimized to avoid any heat-induced damage or enzymatic loss (Shishir and Chen 2017). The heat-sensitive compounds should be maximally retained (Suhag and Nanda 2016).

The processing parameters for honey spray-drying are similar to those for other food materials, with a 150–200°C to 70–106°C temperature range for the inlet to outlet air temperatures, respectively (Samborska 2019). Reducing the outlet temperature and residence interval to a certain level has been observed to improve the probiotic viability and enhance the retention time of enzymatic activity (Silva et al. 2011). An extremely low outlet temperature range can increase the moisture content, damaging the probiotic viability during powder storage. An increase in residence time may disrupt the bioactive compounds. Some other issues that can influence the drying process are related to the configuration of the spray dryer, including chamber design, air flow positioning, type of nozzle, and feed injection (Santivarangkna et al. 2007).

The processing of honey from the liquid state to the powder state is challenging and may induce stickiness in dried honey (Muzaffar et al. 2015), which is characterized by high hygroscopicity, low melting point, and reduced glass transition temperature (*Tg*). To facilitate the spray-drying process of honey, it is essentially required to make some modifications to the drying parameters to increase the *Tg* temperature above the temperature of honey. The addition of a carrier substance (such as Arabic gum, maltodextrin, or starch) can be used to modify the drying process (Jittanit et al. 2010).

6.4.5 High-Pressure Processing

Over the last few years, hydrostatic pressure has been widely applied for the high-quality production of several foods. High-pressure processing (HPP) extends the shelf life of several foods by inactivating microbial growth (Sehrawat et al. 2021). HPP can be considered a practical alternate approach to thermal treatment to enhance high-quality food production (Muntean et al. 2016). At different pressure levels, the properties and level of components are directly related to the duration of treatment. HPP of honey for 20 min at 500 MPa exhibited the highest content of antioxidants and phenols (Chaikham and Prangthip 2015). HPP achieved a considerable increase in phenols and antioxidants. It was further discovered that the antioxidant activity and phenolic compounds can be increased by pressurizing honey samples at 600 MPa (10 min), which exhibited the highest values of DPPH radical scavenging and phenolic content, with an increase of 30% and 47%, respectively, over non-pressurized honey samples (Akhmazillah et al. 2013). The enhancement of these activities or properties in pressurized honey might be because of pollen disintegration inducing an increase in antioxidant compounds.

6.4.6 Thermosonication

Processing of honey using ultrasound integrated with heat is termed thermosonication. This method provides significant processing, negligibly affecting the quality parameters, while ultrasound treatment alone affected honey quality parameters (Aadil et al. 2015; Chong et al. 2017). Ultrasonication is a method with minimum processing and great potential to inhibit microbial growth and promote dehydration in several foods, consequently preserving the food with enhanced sensorial and nutritional properties (Bhargava et al. 2021). The transmission of ultrasound waves in liquid honey may induce cavitation by forming cavities or bubbles, which quickly escalate in size and collapse under the impact of the acoustic field. The collapse of the bubbles results in extreme local hot spots or heating, quickly altering the pressure and temperature in the surrounding medium. It has been documented that when combined with other preservation techniques, e.g., high temperature (30–50°C), ultrasounds can be more significant in improving microbial and enzymatic inactivation. The use of ultrasound waves may not be sufficient for microbial destruction, so thermosonication is a practical alternative processing approach to increase microbial destruction.

Thermosonication decreases the moisture content and water activity by 16.6% and 7.9%, respectively, while conventional heating reduces them by 6.9% and 3.5% (Chong et al. 2017). Several studies have investigated the effect of high temperature, high pressure, and ultrasonic treatments on honey's antioxidant activities and bioactive compounds (Chaikham and Prangthip 2015; Yalcin 2021). The latter group have investigated the antioxidant level, total flavonoids, and phenolic compounds in honey with ultrasonic treatment for 5–20 min (with 20–60% amplitudes at 20 kHz); these were slightly enhanced compared with untreated honey (Chaikham and Prangthip 2015). An increase in treatment duration can significantly enhance the treated honey's antioxidant level and biological properties. Plant cell breakdown is mainly responsible for the escalation in bioactive components in honey samples with ultrasonic treatment. The rupture of cell membranes and organelles releases the enzymes from vacuoles, and these released enzymes can influence the antioxidant properties and phenolic contents in the treated honey samples (Akhmazillah et al. 2013). In a fact-based consideration, no intact cells are present in honey; a possible justification for the observed escalation in antioxidant properties and phenolic compounds in treated honey is pollen disintegration caused by stimulation during ultrasonic treatment (Chaikham and Prangthip 2015).

6.4.7 Membrane Processing

Membrane processing is an alternative method to conventional heat processing. The complete destruction of microbes in honey is challenging using the conventional thermal processing techniques employed by several industries. Thermal processing can also affect or decrease the activity of enzymes. The membrane processing method can provide several advantages, such as decreased viscosity, no murkiness or granulation/sedimentation, and consistent sterile quality can be achieved at the commercial scale (Barhate et al. 2003). The National Honey Board has studied numerous applications of ultra-filtered honey in pharmaceutics, cosmetics, and gel preparations, irrespective of their food and beverage applications (Longmont 1991). The ultra-filtered honey also has the drawback that it may lack some proteins and enzymes and therefore, cannot be recommended for food applications.

6.5 PROPERTIES OF HONEY

More than 300 different types of honey have been recognized (Lay-Flurrie 2008). It is described as a viscous supersaturated sugar solution, which over time can take on a semi-solid or solid state, named the crystallized or granulated state. Generally, the crystallization is dependent on the glucose concentration. The higher the glucose concentration, the faster the crystallization. Chemically, the glucose molecule spontaneously precipitates by losing water, taking the form of a crystal. More crystals are organized as structures, immobilizing the other components of the honey, resulting in a semi-solid or solid state (crystallization). Types of honey differ broadly in physical properties (Table 6.2), taste, and composition. The physicochemical properties of honey are mainly associated with texture, flavor, storage quality, granulation, and nutritional and medicinal properties (Asma et al. 2022). Honey exhibits different colors, with immense variation from colorless to light–dark amber or black (Bodor et al. 2021). The freshly extracted honey is viscous in nature, mostly with an amber or yellow color. Appearance and color vary depending on the environmental and surrounding conditions, such as floral source, pollens, colloidal substances, beehive age, and final storage conditions (Bertoncelj et al. 2007). Some monofloral honeys (where a botanical source of nectar and pollen

TABLE 6.2
Physical Properties of Honey

Characteristics	Value
Specific gravity (250°C, 17% moisture)	1.423
Specific heat (200°C, 17.4% moisture)	2.26 kJ/kg/K
Thermal conductivity	5.36×10^{-5} W/MK
(210°C, 17% moisture)	5.95×10^{-5} W/MK
(710°C, 17% moisture)	
Viscosity	150 P
Freezing point (15% solution)	−1.42 to −1.53°C
Water activity (A_w)	0.5–0.6

From Rashid, S.A. et al., in *Therapeutic Applications of Honey and Its Phytochemicals*, Springer, Singapore, 2020.

is predominant over the others present) have quite a distinctive color, and many times, beekeepers denominate their honey following this color. Some commonly used types of honey include acacia and canola (in liquid state) having very light yellow colors; honeydew, sunflower, eucalyptus, and chestnut honey with pale greenish, bright yellow, light amber, and reddish color tones, respectively (Bobiş et al. 2020, 2021, Bodor et al. 2021). Honey color can even be black-brown for buckwheat (*Fagopyrum aesculentum*) (Nešović et al. 2020) and Japanese knotweed (*Fallopia japonica*) (Bobiş et al. 2019).

The hygroscopicity of honey shows its ability to absorb and retain moisture content . Honey's viscosity differs depending on its composition and the amount of water present in it. Generally, honey contains approximately 18.8% water. The viscosity is directly related to the concentration of high sugars present in honey. The colloidal substances in honey contribute to its viscosity and surface tension, producing foaming in honey (Olaitan et al. 2007). The surface tension varies depending upon the nature and origin of the colloidal substances of honey composition. The pH of the honey is acidic, ranging between 3.2 and 4.5. The acidic pH is due to the organic acids (acetic acid, formic acid, citric acid, and gluconic acid) present in honey, preventing the growth of several bacteria (Olaitan et al. 2007). Honey also has buffering properties due to the presence of mineral salts, phosphates, and carbonates. Upon crystallization, monohydrate crystals are produced by the glucose present in honey. The production of white-colored crystals changes the color of honey to light. Additionally, the crystallization process is quicker, with a high percentage of glucose and lower water content (Al-Habsi et al. 2013).

Standard regulations state that the water content of honey must be below 20% to prevent fermentation (with the exception of tropical and heather honeys). In honey, water content is called moisture content and is considered responsible for fermentation. Actually, the parameter water activity (aw) is responsible for this process, as has been demonstrated for other foods, as most chemical and microbiological reactions are controlled by free water from the product (Prior 1979). In terms of honey, aw values between 0.49 and 0.65 preserve this product against microbial contamination (Cavia et al. 2004). This parameter is dependent on honey type (floral or honeydew), physical state (liquid or crystallized), and botanical origin within the floral category (Cavia et al. 2004; Gleiter et al. 2006).

6.6 HEALTH BENEFITS OF HONEY

6.6.1 Apitherapeutic Components of Honey

Honey is a product containing more than 300 components. Its composition consists of carbohydrates, water, vitamins, minerals, organic acids, polyphenols, alkaloids, anthraquinone glycosides, flavonoids, and volatile compounds, the proportions varying depending on the nectar source (Mijanur Rahman et al. 2014). The dry weight of honey consists of 96% carbohydrates, 85% of which is represented by fructose and glucose. The other sugars are various mono-, di-, and polysaccharides, such as maltose, sucrose, melebiose, melezitose, rhamnose, arabinose, cellobiose, ribose, and trehalose. The remaining 5% of honey consists of various amino acids, peptides and proteins, polyphenols, vitamins, organic acids, volatile compounds, and minerals. The differences between honeys in terms of their authentic structures are due to the secondary metabolites they contain. Aroma and volatile components and polyphenols are the most important natural components that differentiate honey species from one another. The presence, diversity, and quantitative values of these components determine the biological active value of honey, together with its sensory properties (Mijanur Rahman et al. 2014). Bee pollen, the indicator of botanical properties of honey, contains protein-rich macromolecules and consists of vitamins, polyphenols, minerals, and fatty acids. The morphological structures and proportions of bee pollen in honeys are determined based on melissopalynological analyses and their botanical and monofloral or heterofloral properties.

In general, honeys with high polyphenol content exhibit high biological activity. Studies have claimed that honeys with high polyphenol content also exhibit high antioxidant, anti-inflammatory, antimicrobial, and anticarcinogenic activities. Accordingly, honeys with high polyphenol content are included in the dark honey class. Dark-colored honeys are therefore frequently employed in apitherapy applications. However, HMF and some Maillard reaction products increase with caramelization in honey due to heat treatment, and the color of honey thus becomes darker. HMF, diastase activity, and invertase activities are used as indicators to distinguish these honeys. If a honey is to be used in apitherapy, it should comply with the honey codex parameters, have a low HMF value, and exhibit high diastase activity (Sahin et al. 2020).

Depending on the botanical origin of honey, some secondary metabolite agents may also lead to different biological activity loadings. For example, rhododendron honey (*Rhododendron ponticum*), which is particularly grown in the Black Sea region and known as mad honey, or medicinal honey, contains diterpenoids (Grayanotoxins), which lower blood pressure and stop the pulse (Malkoç et al. 2020). Manuka honey (*Leptospermum scoparium*), produced from a plant grown in New Zealand, is another honey with high antimicrobial activity due to its high methylglyoxal content (Kwakman and Zaat 2012).

It is recognized that the darker the honey, the higher the bioactive properties are. Dark-colored honeys include chestnut (*Castania sativa*) (Can et al. 2015), oak (*Quercus* sp.) (Kolayli et al. 2018), heather (*Calluna vulgaris*) (Waś et al. 2011), pine (*Pinus* sp.) (Coskun and Karabulut Dirican, 2019), black cumin (*Nigella sativa*) (Ali and Blunden 2003), Manuka (Mavric et al. 2008), and forest honeys (Belay et al. 2015).

6.6.2 The Wound-Healing Effect of Honey

Honey has been used as a natural wound healer for many hundreds of years. The wound-healing characteristic of honey derives from both its antimicrobial activity and its anti-inflammatory and immunomodulatory properties (Saikaly and Khachemoune 2017). Several *in vitro* and clinical trials have shown that honey is an effective wide-spectrum antibacterial agent (Al-Waili et al. 2011).

Later studies concluded that the antimicrobial activity of honey depends on four basic factors. Due to its viscous structure, honey prevents bacterial growth, since it causes reverse osmosis in bacteria (Osato et al. 1999). Due to the acidic nature of honey, its pH value varies between 3.5 and 5.5, and since this pH prevents the proliferation and growth of bacteria, it also constitutes an important antimicrobial barrier (Mandal and Mandal 2011). Hydrogen peroxide (H_2O_2), which occurs in low amounts in honey due to the glucose oxidase enzyme present in honey, also permits honey to act as a good disinfectant and antimicrobial source. These three factors are common features in almost all honeys and are not particularly affected by the honey flora. The fourth factor, which depends on the authentic structure of honey, consists of the polyphenols and volatile components in its structure. The antimicrobial effect of honey increases in line with the qualitative and quantitative value of the flavonoids it contains (Al-Waili et al. 2011). Studies have shown that polyphenolic agents such as gallic acid, caffeic acid, chrysin, and pinocembrin found in honey are good antimicrobial agents (Yupanqui Mieles et al. 2022). Clinical studies show that the topical use of honey is effective in the treatment of wounds, bedsores, and diabetic wounds due to its antimicrobial activity.

Honey phytochemicals are also reported to provide inhibition against opportunistic pathogens, exhibit anti-quorum sensing activity on *Pseudomonas aeruginosa*, and prevent biofilm formation (Singh et al. 2017; Liaqat et al. 2022).

Irrespective of its origin, honey certainly exhibits antimicrobial and anti-inflammatory effects. Studies of the care of hospital-acquired pressure injury in critically ill children by the same team reported that the use of medicated (Manuka) honey dressings shortened wound healing times (Sankar et al, 2021). Another recent randomized clinical study showed that honey-based ophthalmic drops are capable of accelerating corneal epithelial defect healing in patients with foreign body–induced corneal ulcers (Nejabat et al. 2021). In terms of the anti-inflammatory effect of honey on wound healing, clinical evidence shows that it can be effective in eliminating postoperative complications following tonsillitis surgeries (Hatami et al. 2018).

Due to the antimicrobial, antiseptic, and antiviral properties of honey and its immune system–enhancing effect, it also played an important role in the prevention and treatment of Covid-19. The consumption of honey also increased during the pandemic (Abedi et al. 2021). Despite the large number of studies already performed, further research is still needed to evaluate the role of honey in the treatment of acute and chronic wounds.

6.6.3 The Hepatoprotective Effect of Honey

Several studies have shown that different kinds of honey exhibit hepatoprotective effects in experimental animals exposed to various toxicants, such as CCl_4 (Saral et al. 2016), acetaminophen (Mahesh et al. 2009), and hydrocarbons (Achuba and Nwokogba 2015). These experiments demonstrated that the liver enzymes alanine transaminase, aspartate transaminase, and γ-glutamyl transaminase increased significantly in animals exposed to toxic agents, while significant decreases were observed in groups receiving honey. A study on rabbits reported that honey prevented fatty liver disease to the same extent as the medical drug Atorvastatin in non-alcoholic hepatitis (Alfarisi et al. 2020). The liver is also the most important organ in the body in terms of detoxification, and there is no drug capable of directly treating liver damage. Artichoke and milk thistle extracts, which contain the flavonoids silibinin and silymarin, are used in the treatment of hepatic disorders, and different species of honey have been shown to exhibit hepatoprotective activity (Saral et al. 2016). Although these findings show that honey is an important agent in protecting the liver, further direct randomized clinical studies are needed on this subject.

6.6.4 Antitumoral Effects of Honey

Honey is not generally recommended for cancer patients due to its high sugar content. However, there are no direct studies showing that honey increases the growth of cancer cells. Nonetheless, several mechanisms suggested that the polyphenols in honey may inhibit cell proliferation and induce apoptosis and cell-cycle arrest. No studies have directly examined the effects of honey as an apitherapeutic agent in patients. However, indirect studies of different floral honey ingredients, especially flavonoids, have reported inhibitory effects and apoptotic, tumor necrosis factor–inhibition, antiproliferative, immunomodulatory, anti-mutagenic, anti-inflammatory, and estrogenic effects in both cell culture and experimental animal studies (Waheed et al. 2019). A study investigating renal cell carcinoma reported that honey caused the death of cancer cells by inducing apoptosis (Samarghandian et al. 2011). Numerous studies on different cancer cell lines have shown that honey strengthens the immune system, prevents cell proliferation, induces apoptosis, inactivation of tumor suppressor genes, and deregulation of cell signaling pathways, and changes the progression of the cell cycle.

Research has shown that some polyphenols of honey components, such as chrysin, caffeic acid, caffeic acid phenyl ester, rutin, ellagic acid, quercetin, hesperetin, and apigenin, which vary depending on the honey's botanical characteristics, exhibit antitumoral activity (Waheed et al. 2019). These ingredients of honey cause an antioxidant effect by reducing the formation of oxidative stress agents or by scavenging the formed oxidative stress mediators.

Different studies also show that honey is effective in eliminating the side-effects of chemotherapy and radiotherapy (Charalambous et al. 2018; Gholami et al. 2018). For example, a study conducted using thyme honey reported that this reduced the formation of oral mucositis due to radiation in head and neck cancer patients (Charalambous et al. 2018). Cell culture studies involving pinocembrin, one of the principal components of honey, have shown that it reduced the proliferation of prostate (Chen et al. 2013), breast cancer (Zhu et al. 2021), and colorectal cancer (Jiang et al. 2022) cells, and reduced apoptosis. Ellagic acid, another important component of honey, has been reported to be effective in protecting the testes against the side-effects of bisulfan, a chemotherapy drug, in experimental animals (Rostami et al. 2022).

6.6.5 Honey and Neurodegenerative Disease

Diseases such as Alzheimer's disease, Parkinson's disease, amyotrophic lateral sclerosis, and multiple sclerosis are neurodegenerative conditions involving damage to the central nervous system. Several studies using the natural bee product honey have identified neuroprotective and neurodegenerative activities in terms of apoptosis, necrosis, neuroinflammation, synaptic plasticity, and behavior-modulating neural circuitry (Mijanur Rahman et al. 2014).

The exact etiology of neurodegenerative diseases is still not known, although environmental factors and oxidative stress are thought to play a major role, together with genetic factors. Endogenous antioxidants exhibit a very significant effect in preventing the formation of oxidative stress. Honey's high antioxidant contents are important in preventing neurodegenerative damage. Acetylcholine esterase and monoamine oxidase are two major enzymes that play a role in the central nervous system. Inhibitors of these enzymes are used in the treatment of the diseases listed in the previous paragraph (Hossen et al. 2017). In addition, numerous studies have shown that honey flavonoids play an important role in the inhibition of these enzymes and prevent oxidative activity. A study conducted with chestnut honey reported that this exhibited good inhibition of monoamine oxidase enzyme *in vitro* (Yildiz et al. 2014). Another study reported an increase in pro-inflammatory mediators when lipopolysaccharide-stimulated N13 microglia were used in flavonoid-rich honey extract (Candiracci et al. 2012).

6.6.6 Gastrointestinal Disorders

The gastrointestinal tract, or digestive system, consists of the mouth, esophagus, stomach, duodenum, bile, pancreas, liver, and large intestine. Honey is the most commonly used natural product in the treatment of diseases caused by bacterial infections such as gastrointestinal system diseases, and particularly reflux, ulcer, and gastritis. Most of these studies have focused on *Helicobacter pylori*, which settles in the gastrointestinal tract. *H. pylori* is a slow-growing bacterial strain adapted to microaerophilic environments. The presence of *H. pylori* in the stomach causes numerous diseases, including chronic gastritis, gastric and duodenal peptic ulcer, gastric adenocarcinoma, and gastric mucosa-associated lymphoid tissue lymphoma (Yazbek et al. 2015). Although this Gram-negative bacterium cannot survive in an acidic environment, it is capable of surviving in the extracellular environment thanks to the urease enzyme it secretes. This converts urea into carbon dioxide and ammonia, and together with the resulting ammonia, the bacteria attach themselves to the stomach wall by providing an alkaline or basic environment (Mohtashami et al. 2015; Yazbek et al. 2015). Numerous *in vitro* and *in vivo* studies have shown that different types of honey, such as Turkish honey (Kolayli et al. 2020), Manuka honey (Matongo and Nwodo 2014), Korean acacia honey (Kim et al. 2017), South African honeys (Manyi-Loh et al. 2010), Greek oregano honey (Voidarou et al. 2021), and *Nigella sativa* honey (Mohtashami et al. 2015) exhibited powerful inhibition of *H. pylori* strains. *In vitro* studies have shown that honey causes non-peroxidic inhibition against *H. pylori* and results in the inhibition of its urease enzyme (Kolayli et al. 2017).

REFERENCES

Aadil, R.M.; Zeng, X.A.; Zhang, Z.H. et al. 2015. Thermosonication: A potential technique that influences the quality of grapefruit juice. *International Journal of Food Science and Technology* 50(5): 1275–1282. https://doi.org/10.1111/ijfs.12766.

Abedi, F.; Ghasemi, S.; Farkhondeh, T.; Azimi-Nezhad, M.; Shakibaei, M. 2021. Possible potential effects of honey and its main components against Covid-19 infection. *Dose-Response* 19(1). https://doi.org/10.1177/1559325820982423.

Abu-Jdayil, B.; Ghzawi, A.A.-M.; Al-Malah, K.I.; Zaitoun, S. 2002. Heat effect on rheology of light- and dark-colored honey. *Journal of Food Engineering* 51(1): 33–38. https://doi.org/10.1016/S0260-8774(01)00034-6.

Acaroz, U.; Arslan-Acaroz, D.; Ince, S. 2019. *A Wide Perspective on Nutrients in Beverages*, 1–39. Elsevier. https://doi.org/10.1016/B978-0-12-816842-4.00001-0.

Achuba, F.I.; Nwokogba, C.C. 2015. Effects of honey supplementation on hydrocarbon-induced kidney and liver damage in Wistar albino rats. *Biokemistri* 27(1): 50–55.

Afrin, S.; Giampieri, F.; Gasparrini, M. et al. 2018. The inhibitory effect of Manuka honey on human colon cancer HCT-116 and LoVo cell growth. Part 1: the suppression of cell proliferation, promotion of apoptosis and arrest of the cell cycle. *Food and Function* 9(4): 2145–2157. https://doi.org/10.1039/c8fo00164b.

Akhmazillah, M.; Farid, M.; Silva, F. 2013. High pressure processing (HPP) of honey for the improvement of nutritional value. *Innovative Food Science and Emerging Technologies* 20: 59–63. https://doi.org/10.1016/j.ifset.2013.06.012.

Alfarisi, H.A.H.; Bin Ibrahim, M.; Mohamed, Z.B.H.; Azahari, N.; Hamdan, A.H.B.; Mohamad, C.A.C. 2020. Hepatoprotective effects of a novel trihoney against nonalcoholic fatty liver disease: A comparative study with atorvastatin. *The Scientific World Journal*: 4503253. https://doi.org/10.1155/2020/4503253.

Al-Habsi, N.A.; Davis, F.J.; Niranjan, K. 2013. Development of novel methods to determine crystalline glucose content of honey based on DSC, HPLC, and viscosity measurements, and their use to examine the setting propensity of honey. *Journal of Food Science* 78(6): E845–E852. https://doi.org/10.1111/1750-3841.12103.

Ali, B.H.; Blunden, G. 2003. Pharmacological and toxicological properties of *Nigella sativa*. *Phytotherapy Research* 17(4): 299–305. https://doi.org/10.1002/ptr.1309.

Alvarez-Suarez, J.M.; Gasparrini, M.; Forbes-Hernandez, T.Y.; Mazzoni, L.; Giampieri, F. 2014. The composition and biological activity of honey: A focus on Manuka honey. *Foods* 3(3): 420–432. https://doi.org/10.3390/foods3030420.

Alvarez-Suarez, J.M.; Giampieri, F.; Battino, M. 2013. Honey as a source of dietary antioxidants: Structures, bioavailability and evidence of protective effects against human chronic diseases. *Current Medicinal Chemistry* 20(5): 621–638. https://doi.org/10.2174/092986713804999358.

Alvarez-Suarez, J.M.; Giampieri, F.; Cordero, M. et al. 2016. Activation of AMPK/Nrf2 signalling by Manuka honey protects human dermal fibroblasts against oxidative damage by improving antioxidant response and mitochondrial function promoting wound healing. *Journal of Functional Foods* 25: 38–49. https://doi.org/10.1016/j.jff.2016.05.008.

Alvarez-Suarez, J.M.; Tulipani, S.; Romandini, S.; Bertoli, E.; Battino, M. 2010. Contribution of honey in nutrition and human health: A review. *Mediterranean Journal of Nutrition and Metabolism* 3(1): 15–23. https://doi.org/10.1007/s12349-009-0051-6.

Al-Waili, N.; Salom, K.; Al-Ghamdi, A.A. 2011. Honey for wound healing, ulcers, and burns; data supporting its use in clinical practice. *The Scientific World Journal* 11: 766–787. https://doi.org/10.1100/tsw.2011.78.

Anklam, E. 1998. A review of the analytical methods to determine the geographical and botanical origin of honey. *Food Chemistry* 63(4): 549–562. https://doi.org/10.1016/S0308-8146(98)00057-0.

Asma, S.T.; Bobiş, O.; Bonta, V. et al. 2022. General nutritional profile of bee products and their potential antiviral properties against mammalian viruses. *Nutrients* 14(17): 3579. https://doi.org/10.3390/nu14173579.

Barhate, R.; Subramanian, R.; Nandini, K.; Hebbar, H.U. 2003. Processing of honey using polymeric microfiltration and ultrafiltration membranes. *Journal of Food Engineering* 60(1): 49–54. https://doi.org/10.1016/S0260-8774(03)00017-7.

Bartáková, K.; Dračková, M.; Borkovcová, I.; Vorlová, L. 2011. Impact of microwave heating on hydroxymethylfurfural content in Czech honeys. *Czech Journal of Food Sciences* 29(4): 328–336. https://doi.org/10.17221/110/2009-cjfs.

Battino, M.; Forbes-Hernández, T.Y.; Gasparrini, M. et al. 2019. Relevance of functional foods in the Mediterranean diet: The role of olive oil, berries and honey in the prevention of cancer and cardiovascular diseases. *Critical Reviews in Food Science and Nutrition* 59(6): 893–920. https://doi.org/10.1080/10408398.2018.1526165.

Belay, A.; Solomon, W.K.; Bultossa, G.; Adgaba, N.; Melaku, S. 2015. Botanical origin, color, granulation, and sensory properties of the Harenna forest honey, Bale, Ethiopia. *Food Chemistry* 167: 213–219. https://doi.org/10.1016/j.foodchem.2014.06.080.

Bertoncelj, J.; Doberšek, U.; Jamnik, M.; Golob, T. 2007. Evaluation of the phenolic content, antioxidant activity and colour of Slovenian honey. *Food Chemistry* 105(2): 822–828. https://doi.org/10.1016/j.foodchem.2007.01.060.

Bhargava, N.; Mor, R.S.; Kumar, K.; Sharanagat, V.S. 2021. Advances in application of ultrasound in food processing: A review. *Ultrasonics Sonochemistry* 70: 105293. https://doi.org/10.1016/j.ultsonch.2020.105293.

Bobiş, O.; Bonta, V.; Cornea-Cipcigan, M.; Nayik, G.A.; Dezmirean, D.S. 2021. Bioactive molecules for discriminating Robinia and Helianthus honey: High performance liquid chromatography – Electron Spray ionization- mass spectrometry polyphenolic profile and physicochemical determinations. *Molecules* 26(4): 4433. https://doi.org/10.3390/molecules26154433.

Bobiş, O.; Dezmirean, D.S.; Bonta, V. et al. 2019. Japanese knotweed (*Fallopia japonica*): Landscape invasive plant versus high quality honey source. Scientific Papers. Series D. *Animal Science* 62(1): 231–235.

Bobiş, O.; Moise, A.R.; Ballestros, I. et al. 2020. Eucalyptus honey: Quality parameters, chemical composition and health–promoting properties. *Food Chemistry* 325: 126870. https://doi.org/10.1016/j.foodchem.2020.126870.

Bodor, Z.; Benedek, C.; Urbin, A.; Szabo, D.; Sipos, L. 2021. Colour of honey: Can we trust the Pfund scale? – An alternative graphical tool covering the whole visible spectra. *LWT- Food Sciednce and Technology* 149: 111859. https://doi.org/10.1016/j.lwt.2021.111859.

Bogdanov, S.; Jurendic, T.; Sieber, R.; Gallmann, P. 2008. Honey for nutrition and health: A review. *Journal of the American College of Nutrition* 27(6): 677–689. https://doi.org/10.1080/07315724.2008.10719745.

Bogdanov, S.; Ruoff, K.; Oddo, L.P. 2004. Physico-chemical methods for the characterisation of unifloral honeys: A review. *Apidologie* 35(Suppl. 1): S4–S17. https://doi.org/10.1051/apido:2004047.

Can, Z.; Yildiz, O.; Sahin, H.; Turumtay, E.A.; Silici, S.; Kolayli, S. 2015. An investigation on Turkish honeys: Their physic-chemical properties, antioxidant capacities and phenolic profile. *Food Chemistry* 180: 133–141. https://doi.org/10.1016/j.foodchem.2015.02.024.

Candiracci, M.; Piatti, E.; Dominguez-Barragán, M. et al. 2012. Anti-inflammatory activity of a honey flavonoid extract on lipopolysaccharide-activated N13 microglial cells. *Journal of Agricultural and Food Chemistry* 60(50): 12304–12311. https://doi.org/10.1021/jf302468h.

Castro-Vázquez, L.; Díaz-Maroto, M.; Pérez-Coello, M. 2007. Aroma composition and new chemical markers of Spanish citrus honeys. *Food Chemistry* 103(2): 601–606. https://doi.org/10.1016/j.foodchem.2006.08.031.

Cavia, M.M.; Fernández-Muiño, M.A.; Huidobro, J.F.; Sancho, M.T. 2004. Correlation between moisture and water activity of honeys harvested in different years. *Journal of Food Science* 69(5): 368–370.

Chaikham, P.; Prangthip, P. 2015. Alteration of antioxidative properties of longan flower-honey after high pressure, ultra-sonic and thermal processing. *Food Bioscience* 10: 1–7. https://doi.org/10.1016/j.fbio.2015.01.002.

Charalambous, M.; Raftopoulos, V.; Paikousis, L. et al. 2018. The effect of the use of thyme honey in minimizing radiation-induced oral mucositis in head and neck cancer patients: A randomized controlled trial. *European Journal of Oncology Nursing* 34: 89–97. https://doi.org/10.1016/j.ejon.2018.04.003.

Chen, Z.; Rasul, A.; Zhao, C. et al. 2013. Antiproliferative and apoptotic effects of pinocembrin in human prostate cancer cells. *Bangladesh Journal of Pharmacology* 8(3): 255–262. https://doi.org/10.3329/bjp.v8i3.14795.

Chong, K.; Chin, N.; Yusof, Y.A. 2017. Thermosonication and optimization of stingless bee honey processing. *Food Science and Technology International* 23(7): 608–622. https://doi.org/10.1177/1082013217713331.

Combarros-Fuertes, P.; Estevinho, L.M.; Dias, L.G. et al. 2019. Bioactive components and antioxidant and antibacterial activities of different varieties of honey: A screening prior to clinical application. *Journal of Agricultural and Food Chemistry* 67: 688–698. https://doi.org/10.1021/acs.jafc.8b05436.

Coskun, F.; Karabulut Dirican, L. 2019. Effects of pine honey on the physicochemical, microbiological and sensory properties of probiotic yoghurt. *Food Science and Technology* 39: 616–628. https://doi.org/10.1590/fst.24818.

Czipa, N.; Borbély, M.; Győri, Z. 2012. Proline content of different honey types. *Acta Alimentaria* 41(1): 26–32. https://doi.org/10.1556/AAlim.2011.0002.

da Silva, P.M.; Gauche, C.; Gonzaga, L.V.; Costa, A.C.O.; Fett, R. 2016. Honey: Chemical composition, stability and authenticity. *Food Chemistry* 196: 309–323. https://doi.org/10.1016/j.foodchem.2015.09.051.

Dimiņš, F.; Miķelsone, V.; Augšpole, I.; Niklāvs, A. 2019. Microwave facilities for thermal treatment of honey. *Key Engineering Materials*. https://doi.org/10.4028/www.scientific.net/KEM.800.103.

Escriche, I.; Visquert, M.; Juan-Borrás, M.; Fito, P. 2009. Influence of simulated industrial thermal treatments on the volatile fractions of different varieties of honey. *Food Chemistry* 112(2): 329–338. https://doi.org/10.1016/j.foodchem.2008.05.068.

Escuredo, O.; Dobre, I.; Fernández-González, M.; Seijo, M.C. 2014. Contribution of botanical origin and sugar composition of honeys on the crystallization phenomenon. *Food Chemistry* 149: 84–90. https://doi.org/10.1016/j.foodchem.2013.10.097.

Escuredo, O.; Fernández-González, M.; Seijo, M.C. 2012. Differentioation of blossom honey and honeydew honey from Northwest Spain. *Agriculture* 2(1): 25–37. https://doi.org/10.3390/agriculture2010025.

Gašić, U.M.; Natić, M.M.; Mišić, D.M. et al. 2015. Chemical markers for the authentication of unifloral *Salvia officinalis* L. honey. *Journal of Food Composition and Analysis* 44(12): 128–138. https://doi.org/10.1016/j.jfca.201508.008.

Gholami, M.; Abbaszadeh, A.; Baharvans, P.; Hasanvand, A.; Hasanvand, A.; Gharravi, A.M. 2018. Protective effects of Persian honey, *Apis mellifera Meda Skorikov* on side effects of chemotherapy and ischemia/reperfusion induced testicular injury. *Journal of Complementary and Integrative Medicine* 15(4): 20160035. https://doi.org/10.1515/jcim.2016-0035.

Gleiter, R.A.; Horn, H.; Isengard, H.D. 2006. Influence of type and state of crystallization on the water activity of honey. *Food Chemistry* 96(3): 441–445. https://doi.org/10.1016/j.foodchem.2005.03.051.

Hatami, M.; Mirjalili, M.; Ayatollahi, V.; Vaziribozorg, S.; Zand, V. 2018. Comparing the efficacy of peritonsillar injection of tramadol with honey in controlling post-tonsillectomy pain in adults. *Journal of Craniofacial Surgery* 29(4): e384–e387. https://doi.org/10.1097/SCS.0000000000004393.

Hossen, M.S.; Ali, M.Y.; Jahurul, M.; Abdel-Daim, M.M.; Gan, S.H.; Khalil, M.I. 2017. Beneficial roles of honey polyphenols against some human degenerative diseases: A review. *Pharmacological Reports* 69(6): 1194–1205. https://doi.org/10.1016/j.pharep.2017.07.002.

Jiang, L.; Yang, Y.; Feng, H.; Zhou, Q.; Liu, Y. 2022. Pinocembrin inhibits the proliferation, migration, invasiveness, and epithelial-mesenchymal transition of colorectal cancer cells by regulating LACTB. *Cancer Biotherapy and Radiopharmaceuticals* 37(7): 527–536. https://doi.org/10.1089/cbr.2020.4052.

Jittanit, W.; Niti-Att, S.; Techanuntachaikul, O. 2010. Study of spray drying of pineapple juice using maltodextrin as an adjunct. *Chiang Mai Journal of Science* 37(3): 498–506.

Kaškoniené, V.; Venskutonis, P.R.; Čeksterytė, V. 2010. Carbohydrate composition and electrical conductivity of different honeys from Lithuania. *LWT – Food Science and Technology* 43(5): 801–807. https://doi.org/10.1016/j.lwt.2010.01.007.

Kim, S.G.; Hong, I.P.; Woo, S.O.; Jang, H.R.; Pak, S.C.; Han, S.M. 2017. Isolation of abscisic acid from Korean acacia honey with anti-*Helicobacter pylori* activity. *Pharmacognosy Magazine* 13(Suppl. 2): S170–S173. https://doi.org/10.4103/0973-1296.210166.

Kolayli, S.; Baltas, N.; Sahin, H.; Karaoglu, S. 2017. Evaluation of anti-*Helicobacter pylori* activity and urease inhibition by some Turkish authentic honeys. *Journal of Food Science and Engineering* 7(2): 67–73. https://doi.org/10.17265/2159-5828/2017.02.001.

Kolayli, S.; Can, Z.; Cakir, H.E.; Okan, O.T.; Yildiz, O. 2018. An investigation on Trakya region Oak (Quercus spp.) honeys of Turkey: Their physico-chemical, antioxidant and phenolic compounds properties. *Turkish Journal of Biochemistry*, 43(4): 362–374. https://doi.org/10.1515/tjb-2017-0174.

Kolayli, S.; Palabiyik, I.; Atik, D.S.; Keskin, M.; Bozdeveci, A.; Karaoglu, S.A. 2020. Comparison of antibacterial and antifungal effects of different varieties of honey and propolis samples. *Acta Alimentaria* 49(4): 515–523. https://doi.org/10.1556/066.2020.49.4.18.

Kwakman, P.H.S.; Zaat, S.A.J. 2012. Antibacterial components of honey. *IUBMB Life* 64(1): 48–55. https://doi.org/10.1002/iub.578.

Lay-Flurrie, K. 2008. Honey in wound care: Effects, clinical application and patient benefit. *British Journal of Nursing* 17(Suppl.5): mS30-S36. https://doi.org/10.12968/bjon.2008.17.sup5.29649.

Liaqat, I.; Gulab, B.; Hanif, U. et al. 2022. Honey potential as antibiofilm, antiquorum sensing and dispersal agent against multispecies bacterial biofilm. *Journal of Oleo Science* 71(3): 425–434. https://doi.org/10.5650/jos.ess21199.

Longmont, N. 1991. *Definition of Honey and Honey Products; Ultrafiltered Honey*. National Honey Board.

Mădaş, N.M.; Mărghitaş, L.A.; Dezmirean, D.S. et al. 2019. Volatile profile and Physico-chemical analysis of Acacia honey for geographical origin and nutritional value determination. *Foods* 8(10): 445. https://doi.org/10.3390/foods8100445.

Mahesh, A.; Shaheetha, J.; Thangadurai, D.; Rao, D.M. 2009. Protective effect of Indian honey on acetaminophen induced oxidative stress and liver toxicity in rat. *Biologia* 64(6): 1225–1231. https://doi.org/10.2478/s11756-009-0205-5.

Malkoç, M.; Yaman, S.Ö.; Imamoğlu, Y. et al. 2020. Anti-inflammatory, antioxidant and wound-healing effects of mad honey in streptozotocin-induced diabetic rats. *Journal of Apicultural Research* 59(4): 426–436. https://doi.org/10.1080/00218839.2019.1689036.

Mandal, M.D.; Mandal, S. 2011. Honey: Its medicinal property and antibactgerial activity. *Asian Pacific Journal of Tropical Biomedicine* 1(2): 154–160. https://doi.org/10.1016/S2221-1691(11)60016-6.

Manyi-Loh, C.E.; Clarke, A.M.; Munzhelele, T.; Green, E.; Mkwetshana, N.F.; Ndip, R.N. 2010. Selected South African honeys and their extracts possess *in vitro* antiHelicobacter pylori activity. *Archives of Medical Research* 41(5): 324–331. https://doi.org/10.1016/j.arcmed.2010.08.002.

Mărgăoan, R.; Topal, E.; Balkanska, R.. et al. 2021. Monofloral honeys as a potential source of natural antioxidants, minerals and medicine. *Antioxidants* 10(7): 1023. https://doi.org/10.3390/antiox10071023.

Matongo, F.; Nwodo, U.U. 2014. *In vitro* assessment of *Helicobacter pylori* ureases inhibition by honey fractions. *Archives of Medical Research* 45(7): 540–546. https://doi.org/10.1016/j.arcmed.2014.09.001.

Mavric, E.; Wittmann, S.; Barth, G.; Henle, T. 2008. Identification and quantification of methylglyoxal as the dominant antibacterial constituent of Manuka (*Leptospermum scoparium*) honeys from New Zealand. *Molecular Nutrition and Food Research* 52(4): 483–489. https://doi.org/10.1002/mnfr.200700282.

Mieles, Y.J.; Vyas, C.; Aslan, E.; Humphreys, G.; Diver, C.; Bartolo, P. 2022. Honey: An advanced antimicrobial and wound healing biomaterial for tissue engineering applications. *Pharmaceutics* 14(8): 1663. https://doi.org/10.3390/pharmaceutics14081663.

Mohtashami, R.; Huseini, H.F.; Heydari, M. et al. 2015. Efficacy and safety of honey based formulation of *Nigella sativa* seed oil in functional dyspepsia: A double blind randomized controlled clinical trial. *Journal of Ethnopharmacology* 175: 147–152. https://doi.org/10.1016/j.jep.2015.09.022.

Muntean, M.-V.; Marian, O.; Barbieru, V. et al. 2016. High pressure processing in food industry–characteristics and applications. *Agriculture and Agricultural Science Procedia* 10: 377–383. https://doi.org/10.1016/j.aaspro.2016.09.077.

Muzaffar, K.; Nayik, G.A.; Kumar, P. 2015. Stickiness problem associated with spray drying of sugar and acid rich foods: A mini review. *Journal of Nutrition and Food Sciences* S12: 1. https://doi.org/10.4172/2155-9600.s12-003.

Nejabat, M.; Soltanzadeh, K.; Yasemi, M.; Daneshamouz, S.; Akbarizadeh, A.R.; Heydari, M. 2021. Efficacy of honey-based ophthalmic formulation in patients with corneal ulcer: A randomized clinical trial. *Current Drug Discovery Technologies* 18(3): 457–462. https://doi.org/10.2174/1570163817666200214113055.

Nešović, M.; Gašić , U.; Tosti, T. et al. 2020. Polyphenol profile of buckwheat honey, nectar and pollen. *Royal Society Open Science* 7(12): 201576. https://doi.org/10.1098/rsos.201576.

Nousias, P.; Karabagias, I.; Riganakos, K. 2018. Deep inside polyphenols of Hellenic thyme honey. *Austin Journal of Nutrition and Food Sciences* 6(1): 1098.

Obón, J.; Luna-Abad, J.; Bermejo, B.; Fernández-López, J. 2020. Thermographic studies of cocurrent and mixed flow spray drying of heat sensitive bioactive compounds. *Journal of Food Engineering* 268: 109745. https://doi.org/10.1016/j.jfoodeng.2019.109745.

Olaitan, P.B.; Adeleke, O.E.; Iyabo, O. 2007. Honey: A reservoir for microorganisms and an inhibitory agent for microbes. *African Health Sciences* 7(3). https://doi.org/10.5555/afhs.2007.7.3.159.

Osato, M.S.; Siddharta, G.R.; Graham, D.Y. 1999. Osmotic effect of honey on growth and viability of *Helicobacter pylori*. *Digestive Diseases and Sciences* 44(3): 462–464.

Otmani, A.; Amessis-Ouchemoukh, N.; Birinci, C. et al. 2021. Phenolic compounds and antioxidant and antibacterial activities of Algerian honeys. *Food Bioscience* 42: 101070. https://doi.org/10.1016/j.fbio.2021.101070.

Porcza, L.M.; Simms, C.; Chopra, M. 2016. Honey and cancer: Current status and future directions. *Diseases* 4(4): 30. https://doi.org/10.3390/diseases4040030.

Prateesha; Singh, B.R.; Shoeb, M. et al. 2017. Scaffold of selenium nanovectors and honey phytochemicals for inhibition of *Pseudomonas aeruginosa* quorum sensing and biofilm formation. *Frontiers in Cellular and Infection Microbiology* 7: 93. https://doi.org/10.3389/fcimb.2017.00093.

Prica, N.; Baloš, M.Ž.; Jakšić, S. et al. 2014. Moisture and acidity as indicators of the quality of honey originating from Vojvodina region. *Archives of Veterinary Medicine* 7(2): 99–109. https://doi.org/10.46784/e-avm.v7i2.135.

Prior, B.A. 1979. Measurement of water activity in foods: A review. *Journal of Food Protection* 42(8): 668–674. https://doi.org/10.4315/0362-028X-42.8.668.

Rahman, M.M.; Gan, S.H.; Khalil, I. 2014. Neurological effects of honey: Current and future prospects. *Evidence-Based Complementary and Alternative Medicine* ID 958721. https://doi.org/10.1155/2014/958721.

Rashid, S.A.; Rashid, S.M.; Amin, I. et al. 2020. Role of honey for enhancing performance in endurance sports. In *Therapeutic Applications of Honey and Its Phytochemicals*. ed. M.U. Rehman; S. Majid, Springer Singapore. https://doi.org/10.1007/978-981-15-6799-5-20.

Rostami, A.; Vakili, S.; Koohpeyma, F. et al. 2022. Ellagic acid effects on testis, sex hormones, oxidative stress, and apoptosis in the relative sterility rat model following busulfan administration. *BMC Complementary Medicine and Therapies* 22(1): 1–14. https://doi.org/10.1186/s12906-022-03650-w.

Sahin, H.; Kolayli, S.; Beykaya, M. 2020. Investigation of variations of invertase and glucose oxidase degrees against heating and timing options in raw honeys. *Journal of Chemistry*. https://doi.org/10.1155/2020/5398062.

Saikaly, S.K.; Khachemoune, A. 2017. Honey and wound healing: An update. *American Journal of Clinical Dermatology* 18(2): 237–251. https://doi.org/10.1007/s40257-016-0247-8.

Samarghandian, S.; Afshari, J.T.; Davoodi, S. 2011. Honey induces apoptosis in renal cell carcinoma. *Pharmacognosy Magazine* 7(25): 46–52. https://doi.org/10.4103/0973-1296.75901.

Samarghandian, S.; Farkhondeh, T.; Samini, F. 2017. Honey and health: A review of recent clinical research. *Pharmacognosy Research* 9(2): 121–127. https://doi.org/10.4103/0974-8490.204647.

Samborska, K. 2019. Powdered honey–drying methods and parameters, types of carriers and drying aids, physicochemical properties and storage stability. *Trends in Food Science and Technology* 88: 133–142. https://doi.org/10.1016/j.tifs.2019.03.019.

Sankar, J.; Lalitha, A.V.; Rameshkumar, R.; Mahadevran, S.; Kabra, S.K.; Lodha, R. 2021. Use of Honey versus Standard care for hospital-acquired pressure injury in critically ill children: A multicenter randomized controlled trial. *Pediatric Critical Care Medicine* 22(6): e349–e362. https://doi.org/10.1097/PCC.0000000000002611.

Santivarangkna, C.; Kulozik, U.; Foerst, P. 2007. Alternative drying processes for the industrial preservation of lactic acid starter cultures. *Biotechnology Progress* 23(2): 302–315. https://doi.org/10.1021/bp060268f.

Saral, O.; Yildiz, O.; Aliyazicioglu, R. et al. 2016. Apitherapy products enhance the recoveryof CCl_4 – Induced hepatic damages in rats. *Turkish Journal of Medical Sciences* 46: 194–202. https://doi.org/10.3906/sag-1411-35.

Sehrawat, R.; Kaur, B.P.; Nema, P.K.; Tewari, S.; Kumar, L. 2021. Microbial inactivation by high pressure processing: Principle, mechanism and factors responsible. *Food Science and Biotechnology* 30(1): 19–35. https://doi.org/10.1007/s10068-020-00831-6.

Seraglio, S.K.; Silva, B.; Bergamo, G. et al. 2019. An overview of physicochemical characteristics and health-promoting properties of honeydew honey. *Food Research International* 119: 44–66. https://doi.org/10.1016/j.foodres.2019.01.028.

Shishir, M.R.I.; Chen, W. 2017. Trends of spray drying: A critical review on drying of fruit and vegetable juices. *Trends in Food Science and Technology* 65: 49–67. https://doi.org/10.1016/j.tifs.2017.05.006.

Silva, J.; Freixo, R.; Gibbs, P.; Teixeira, P. 2011. Spray-drying for the production of dried cultures. *International Journal of Dairy Technology* 64(3): 321–335. https://doi.org/10.1111/j.1471-0307.2011.00677.x.

Singh, N.; Bath, P.K. 1998. Relationship between heating and hydroxymethylfurfural formation in different honey types. *Journal of Food Science and Technology (Mysore)* 35(2): 154–156.

Singh, I.; Singh, S.; Hans, V.; Singh, J. 2022. Development of inclined plate honey moisture reduction system. *Journal of Food Science and Technology* 59(2): 615–624. https://doi.org/10.1007/s13197-021-05049-w.

Soldatkin, O.; Peshkova, V.; Saiapina, O. et al. 2013. Development of conductometric biosensor array for simultaneous determination of maltose, lactose, sucrose and glucose. *Talanta* 115: 200–207. https://doi.org/10.1016/j.talanta.2013.04.065.

Subramanian, R.; Umesh Hebbar, H.; Rastogi, N. 2007. Processing of honey: A review. *International Journal of Food Properties* 10(1): 127–143. https://doi.org/10.1080/10942910600981708.

Suhag, Y.; Nanda, V. 2016. Optimization for spray drying process parameters of nutritionally rich honey powder using response surface methodology. *Cogent Food and Agriculture* 2(1): 1176631. https://doi.org/10.1080/23311932.2016.1176631.

Truzzi, C.; Annibaldi, A.; Illuminati, S.; Finale, C.; Scarponi, G. 2014. Determination of proline in honey: Comparison between official methods, optimization and validation of the analytical methodology. *Food Chemistry* 150: 477–481. https://doi.org/10.1016/j.foodchem.2013.11.003.

Visquert, M. 2004. Changes in the quality parameters of honey caused by thermal processes. *Food, Equipments and Tecnology* 23(188): 87–92.

Voidarou, C.; Rozos, G.; Alexopoulos, A. et al. 2021. *In vitro* screening potential antibacterial properties of the Greek oregano honey against clinical isolates of *Helicobacter pylori*. *Foods* 10(7): 1568. https://doi.org/10.3390/foods10071568.

Waheed, M.; Bilal Hussain, M.; Javed, A. et al. 2019. Honey and cancer: A mechanistic review. *Clinical Nutrition* 38: 2499–2503. https://doi.org/10.1016/j.clnu.2018.12.01.9.

Wakhle, D. 1995. Design for honey processing unit Part I. *Indian Bee Journal* 57: 144–146.

Wakhle, D. 1996. Design for honey processing Unit Part II. *Indian Bee Journal* 58: 5–9.

Waś, E.; Rybak-Chmielewska, H.; Szczesna, T.; Kachaniuk, K.; Teper, D. 2011. Characteristics of Polish unifloral honeys. III. Heather honey (*Calluna vulgaris* L.). *Journal of Apicultural Science* 55(1): 129–136.

Yalçin, G. 2021. Effects of thermal treatment, ultrasonication, and sunlight exposure on antioxidant properties of honey. *Turkish Journal of Pharmaceutical Sciences* 18(6): 776. https://doi.org/10.4274/tjps.galenos.2021.53810.

Yazbek, P.B.; Trinidade, A.B.; Chin, C.M.; dos Santos, J.L. 2015. Challenges to the treatment and new perspectives for the eradication of *Helicobacter pylori. Digestive Diseases and Sciences* 60(10): 2901–2912. https://doi.org/10.1007/s10620-015-3712-y.

Yildiz, O.; Karahalil, F.; Can, Z.; Sahin, H.; Kolayli, S. 2014. Total monoamine oxidase (MAO) inhibition by chestnut honey, pollen and propolis. *Journal of Enzyme Inhibition and Medicinal Chemistry* 29(5): 690–694. https://doi.org/10.3109/14756366.2013.843171.

Zhang, Y.-Z.; Chen, Y.-F.; Wu, Y.-Q. et al. 2019. Discrimination of the entomological origin of honey according to the secretions of the bee (*Apis cerana* or *Apis mellifera*). *Food Research International* 116: 362–369. https://doi.org/10.1016/j.foodres.2018.08.049.

Zhu, X; Li, R; Wang, C. et al. 2021. Pinocembrin inhibits the proliferation and metastasis of breast cancer via suppression of the PI3K/AKT signaling pathway. *Frontiers in Oncology* 11: 661184. https://doi.org/10.3389/fonc.2021.661184.

7 Royal Jelly

Md. Shahidullah Kayshar, Md. Fahad Jubayer, Md. Anisur Rahman Mazumder, Afzal Rahman, Dayanand Peter and Thottiam Vasudevan Ranganathan

7.1 INTRODUCTION

Royal jelly (RJ) is an intricately structured bee product. Young nurse bees (aged 5 to 14 days) create royal jelly in the hypopharyngeal and mandibular salivary glands. It is a white-yellow, gelatinous, viscous, sour-tasting glandular fluid with a faintly recognizable phenol scent (Fujita et al. 2013). This substance is produced in the stomachs of worker bees as a result of the worker bees' inability to fully digest honeydew (Moselhy et al. 2013). RJ is the only sustenance for bee larvae from the time they hatch until the third day of their lives. After 3 days, worker bee larvae consume worker jelly, which is constituted largely of honey and pollen. RJ is fed to larvae that are selected to become queens up until the fifth day of their life. RJ is the only nourishment the queen bee consumes for the remainder of her life once the cell is operculated. Overall, the composition of RJ is very steady at the macro level, despite the fact that the precise composition of RJ may vary depending on factors including bee nutrition, bee species, climate, season, harvesting technique, and the age of the bee larvae. Proteins, fatty acids, carbohydrates, hormones, vitamins, and minerals are all abundant in royal jelly and are essential for the growth of larvae. The lifespan of a bee is also significantly impacted by RJ. A worker bee lives for about 45 days, whereas a queen bee can live for up to 5 years and can lay many eggs every day (approximately 2000–3000 eggs per day). RJ must be stored in specific ways for human work in order to keep its properties. Whether RJ's properties are maintained depends on how it is preserved. RJ is oxidized when it comes into contact with air directly, and it is light- and heat-sensitive (Scarselli et al. 2005; Liu et al. 2008; Sabatini et al. 2009; Kheyri et al. 2012; Buttstedt et al. 2013; Zhang et al. 2019). Additionally, RJ is a type of widely used functional food (Ramadan and Al-Ghamdi, 2012) and dietary supplement that has numerous biological effects, including anti-tumor, anti-allergy, anti-inflammatory, and immunological modulation (Pasupuleti et al. 2017).

7.2 HISTORICAL VIEW

RJ intake by people is first recorded in antiquity, in ancient Greece. RJ was thought by the ancient Greeks to have contributed to the production of "ambrosia," the nectar that granted immortality to the Gods of Olympus. This belief was recorded for the first time in recorded history. According to historical accounts, before the benefits of honey, larvae, propolis, pollen, and RJ were fully understood, they were often eaten straight from the honeycombs (Mraz 1995). The value of RJ to bee society was initially recognized by Aristotle. He came to the conclusion that RJ use boosted physical strength and, more crucially, enhanced mental capacity by seeing its effects on queen bees. The only items served for breakfast at his school were honey and RJ (Molan 1999; Domerego 2012). Ancient Egyptians used RJ as a cosmetic, and Cleopatra is famous for using it as one of her own beauty secrets. Additionally, because the Pharaohs frequently ate RJ during that time, it started to be considered a representation of their power and majesty (Fratini et al. 2016). Since the beginning of Chinese civilization, RJ has been used in traditional Chinese medicine. This by-product of beekeeping, which was produced in certain royal gardens, was thought to be connected to the sexual potency and longevity of the early Chinese emperors, even in old age (Cherbuliez et al. 2003). The Dutch naturalist, microscopist, and entomologist Jan Swammerdam (1637–1680) was the first

 DOI: 10.1201/9781003354116-7

to define the nutritional makeup of the royal cell. Swammerdam also revealed that the "beehive chief" is actually a queen, rather than a king, as was previously thought to be the case (Viel and Doré 2003). The name "Royal Jelly" was created by the French scientist René Antoine de Réaumur (1683–1757) to describe the food that queen bees eat and to explain the remarkable development of the queen (Molan 1999; Cherbuliez et al. 2003). The first chemical analysis of RJ was performed in 1852 by Reverend Langstroth, known as the "Father of American Beekeeping," albeit his methods were unable to ensure results that were applicable to science (Domerego 2012).

In the 1950s, Langstroth actively promoted the use of RJ as a commercial product, particularly in regions where the production of honey was unprofitable (Viel and Doré 2003). Since the invention of "Apitherapy" in the early 1960s, research has been conducted on the use of RJ as a functional chemical and health booster. RJ's characteristics were later recognized, and it has frequently been used to treat both humans and bees (Molan 1999).

The majority of RJ's biological components are RJ peptides. These compounds have a wide range of biological properties, including actions that are antibacterial, immunomodulatory, anti-aging, antioxidant, neuroprotective, anti-inflammatory, and antihypertensive, as well as actions that improve immune system performance (Moradi et al. 2013; Fratini et al. 2016; Ibrahim and Kosba, 2018; Maghsoudlou et al. 2019; Zhang et al. 2019). Along with its antioxidant, antineoplastic, antibiotic, and antidiarrheal properties, regular consumption of RJ is generally thought to benefit a number of medical conditions, including Alzheimer's, depression, infertility, digestive problems, aging, anemia, stress-related illnesses, and weakened immune function (Elnagar 2010; Zhang et al. 2019). RJ has additionally been recommended as a dietary supplement to assist with various reproductive problems (Elnagar 2010; Mahdivand et al. 2019).

7.3 COMPOSITION

Comparing royal jelly from different colonies, bee breeds, and temperature ranges reveals that its chemical composition is remarkably stable. However, there may be a difference between honey and royal jelly in terms of composition. Honey is a natural food made mostly of sugars and water, with traces of additional nutrients such vitamins, minerals, amino acids, and organic acids as well as aromatic compounds, flavonoids, and other phenolic compounds. This composition is prone to major variations based on storage time and conditions, especially the minor constituents, which are connected to the botanical and geographic origin (Santos-Buelga and González-Paramás 2017).

Royal jelly also contains 3–6% fats, vitamins, minerals, and free amino acids in addition to 60–70% water, 12–15% crude protein, and 10–16% total protein in royal jelly sugar (Table 7.1). In contrast to honey, the pH lies between 3.6 and 4.2 (Guo et al. 2021).

The water activity (a_w) of royal jelly exceeds 0.92. Although the ash and pH values alter with the seasons, the protein content rarely changes. The moisture and carbohydrate percentages are highest during the monsoon season, while the lipid content is highest during the hot season (Wongchai and Ratanavalachai 2002).

However, currently, there are no universally accepted specifications for royal jelly production. On the other hand, some countries have set national standards, and these include countries like Brazil, Bulgaria, Japan, and Switzerland. The International Honey Commission has formed a working group to develop a global benchmark (Moselhy et al. 2013). According to these commissions, the composition of royal jelly is presented in Table 7.2. Most of the carbohydrates in royal jelly are simple sugars like glucose and sucrose, but some complex sugars like maltose, trehalose, melibiose, ribose, and erlose in very small amounts can also be found.

7.3.1 Proteins, Peptides, and Amino Acids

Major royal jelly proteins (MRJPs), which are largely comprised of MRJP1–9 and make up more than 80% of the proteins in royal jelly, share 60–70% of their amino acid sequences. MRJP1–3

TABLE 7.1
Key Active Components of Royal Jelly (RJ)

	Active elements
Proteins	MRJPs (MRJP1–9); Jelleines (JelleineI–IV); aspirin, royalactina, apolipophorin-III-like protein, and royalsin
Free amino acids	Lysine, cystine, aspartic acid, and proline
Lipids/fats	Sebacic acid (SA), 3,10-dihydroxydecanedioic acid (3,10-HDecDA), 9-hydroxy-2-decenoic acid (9-HDA), 8-hydroxyoctanoic acid (8-HOC), 2-decene-1,10-dioic acid (2-DecDA), and 3-hydroxydecanoic acid (3-HHDA)
Sugars	Sucrose, glucose, and fructose
Flavonoids	Naringenin, chrysin, acacetin, formononetin, isorhamnetin glycosides, etc.
Phenols	Pinobanksin, dodecanoic acid, octanoic acid, 1,2-benzenedicarboxylic acid and their esters
Vitamins	Vitamins E, niacin, vitamin B6, pantothenic acid, thiamine, riboflavin, folate, inositol, and biotin
Minerals	Fe, Na, Zn, Ca, Mg, Cu, Cr, Cd, K, P
Others	Prolactin, estradiol, testosterone, N1 oxide of adenosine monophosphate, and flavonoids

(Guo et al. 2021)

TABLE 7.2
Royal Jelly (RJ) Composition

Components (%)	Fresh	Lipolyzed
Water/moisture	60–70	<5
Lipid	3–8	8–19
Protein	9–18	27–41
Glucose	4–8	–
Fructose	3–13	–
Sucrose	0.5–2	–
Ash	0.8–3	2–5
Acidity in milliliters 0.1N NaOH/g	3–6	–

(Bogdanov 2011)

and 5 make up 31%, 16%, 26%, and 9%, respectively, of the royal jelly protein, which accounts for 82% of the total royal jelly protein (Guo et al. 2021). These complex proteins of the MRJP family are abundant in the amino acids required to support both the queen bee and the growing larvae (Ramanathan et al. 2018). MRJP1 is extensively studied for its role as a "queen maker," even though the other MRJPs also have a big impact on how larvae develop. The endurance of the organism is affected and improved by all MRJPs, not just MRJP1, according to recent research on *Drosophila melanogaster* by Xin et al. (2016). *D. melanogaster*'s lifespan, reproduction, and feeding rates were all improved by royal jelly. The MRJPs' antioxidant capabilities were principally responsible for the increased life expectancy. It was revealed that the levels of the enzyme superoxide dismutase increased when *D. melanogaster* was given royal jelly (Xin et al. 2016). Royal jelly also has trace amounts of royalsin, jelleines, aspirin, and royalactina in addition to MRJPs. Common antimicrobial peptides include royalsin and jelleine. Apolipophorin-III-like protein, another lipid-binding protein found in royal jelly, acted as an antibacterial agent by forming a protein–lipid complex and dispersing lipids into the surrounding water (Fratini et al. 2016; Guo et al. 2021). In royal jelly, lysine is the most abundant free amino acid, followed by proline, cystine, and aspartic acid. There are also amino acids with concentrations lower than 5 mg/100 g, such as valine, glutamic acid, serine, glycine, cysteine, threonine, and alanine (Guo et al. 2021). Nine distinct genes encode the nine members of the royal jelly protein family, MRJP1, MRJP2, MRJP3, MRJP4, MRJP5, MRJP6, MRJP7, MRJP8, and MRJP9. Figure 7.1 provides a clear illustration of the roles and classification of MRJPs.

7.3.2 Lipids

Lipids, which make up about 3% to 19% of the dry weight of royal jelly, are the second most significant active component after proteins. Fatty acids make up 80–85% of the lipids, followed by waxes, steroids (3–4%), phenolic lipids (4–10%), and phospholipids, which make up 0.4–0.8%. Most of the organic acids are present in nature in their free state, and they have an unusual structure made up of mono- and dihydroxy acids as well as dicarboxylic acids with 8 and 10 carbon atoms (Maghsoudlou et al. 2019). Fresh royal jelly contains 0.77 to 0.93 g of 10-hydroxy-2-decenoic acid (10-HDA), which is the primary fatty acid. The quality of royal jelly can be determined using 10-HDA. Royal jelly also contains traces of sterols, as well as sebacic acid (SA), 8-hydroxyoctanoic acid (8-HOC), 3,10-dihydroxydecanedioic acid (3,10-HDecDA), 9-hydroxy-2-decenoic acid (9-HDA), 3-hydroxydecanoic acid (3-HHDA), and 2-decene-1,10-dioic acid (2-DecDA) (Ahmad et al. 2020; Guo et al. 2021).

7.3.3 Carbohydrates and Sugars

About 30% of the dry weight of royal jelly is made up of carbohydrates. While there is little variation in the components' overall quality, there is a wide range in their actual numerical values. Like honey, the predominant sugars here are the monosaccharides fructose and glucose. Although there is a wide range of acceptable levels, the presence of a small amount of sugar serves as a great indicator of adulteration because the quality of the primary carbohydrates is so consistent (Ramadan and Al-Ghamdi, 2012). There are also some oligosaccharides that can be detected, including trehalose, maltose, gentiobiose, isomaltose, raffinose, erlose, and melezitose. These oligosaccharides, despite their extremely low concentrations, are helpful for identifying a pattern similar to that of honey and in some cases, are a marker of a product's authenticity (Sabatini et al. 2009). Ninety percent of the sugar content in royal jelly comes from glucose and fructose, while only 0.8% to 3.6% comes from sucrose. Glucose outweighs fructose in terms of total content, and sucrose concentration tends to fluctuate widely. Sugars like maltose, trehalose, and others can be found in royal jelly in trace amounts. Whether or not lactose is present is currently unknown and requires additional investigation (Guo et al. 2021).

7.3.4 Vitamins and Minerals

The vitamin concentrations in royal jelly span a broad spectrum, with riboflavin, thiamine, niacin, and folic acid exhibiting relatively uniform values. Vitamin C is present in trace amounts, in contrast to the absence of the fat-soluble vitamins A, D, E, and K (Bogdanov 2011). The vitamin and mineral content of royal jelly can be found in Table 7.3. Since worker bees gather their vitamin C and other nutrients from the pollen of flowers, royal jelly's vitamin content fluctuates with the seasons. Additionally, royal jelly contains a number of minor chemical constituents, such as biopterine, neopterine, and heterocyclic compounds. In addition to phosphates, ATP, ADP, AMP, acetylcholine, and gluconic, benzoic, malic, citric, and lactic acids, royal jelly also includes free nucleotides (adenosine, guanosine, cytidine, and iridine). Although their purposes are as yet unknown, it is thought that these components are derived from nurse bees (Fratini et al. 2016).

7.3.5 Phenols and Flavonoids

Royal jelly contains 23.3 mg/g of phenolics and 1.28 mg/g of flavonoids. Phenolic compounds include pinobanksin, organic acids (such as octanoic acid, dodecanoic acid, and 1,2-benzenedicarboxylic

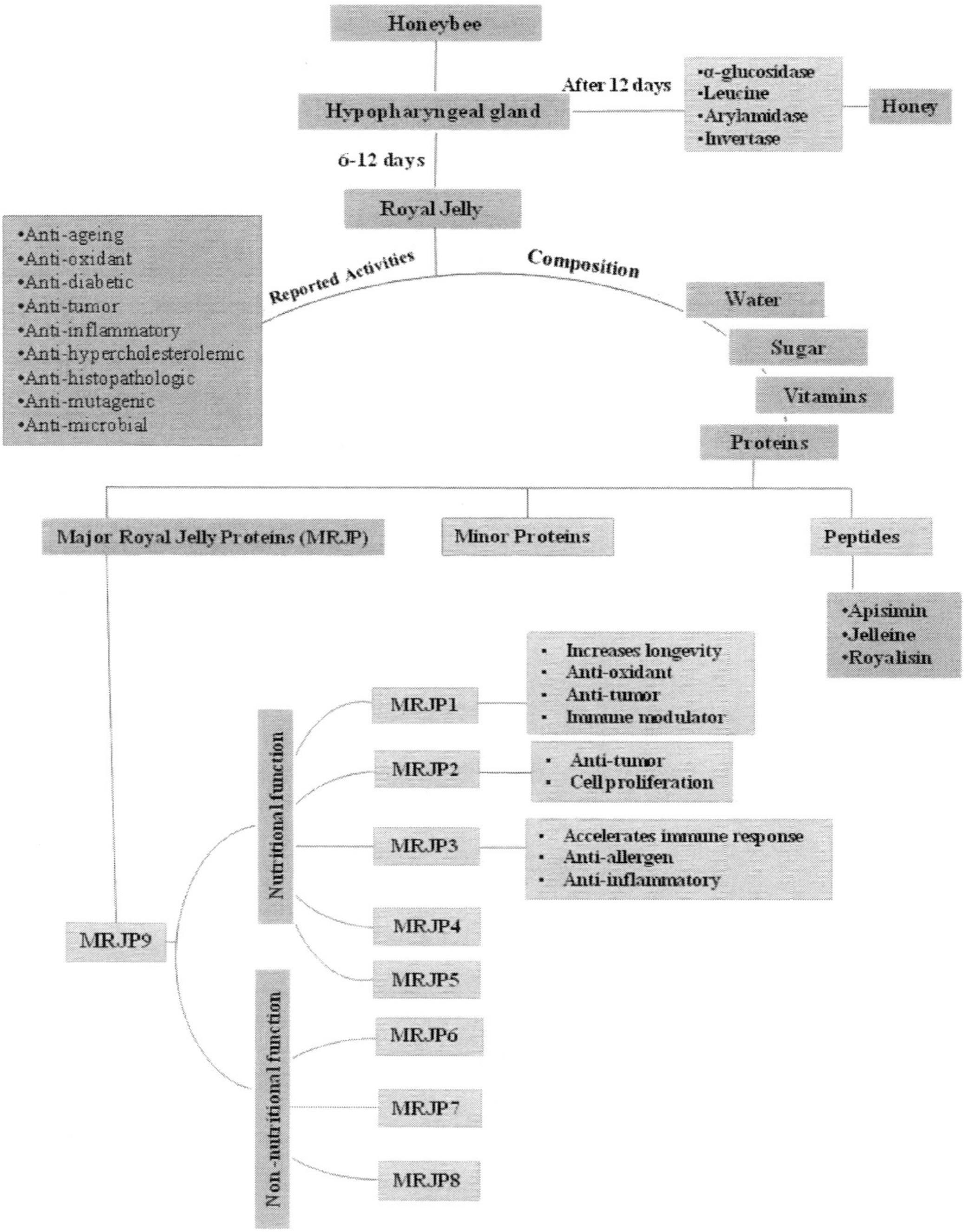

FIGURE 7.1 Royal jelly proteins and their functions. (From Ramanathan, A.N.K.G. et al., *J. Funct. Foods*, 44, 255–264, 2018.)

TABLE 7.3
Vitamin and Mineral Content of Royal Jelly

Vitamin or mineral	Amount (mg/100g)
Potassium (K)	200.0–1000.0
Magnesium (Mg)	20.0–100.0
Calcium (Ca)	25.0–85.0
Iron (Fe)	1.0–11.0
Zinc (Zn)	0.7–8.0
Copper (Cu)	0.33–1.6
B1 (Thiamin)	0.1–1.7
B2 (Riboflavin)	0.5–2.5
B3 (Niacin)	4.5–19.0
B5 (Pantothenic acid)	3.6–23.0
B6 (Pyridoxin)	0.2–5.5
Biotin	0.15–0.55
Folic acid	0.01–0.06

(Bogdanov 2011; Maghsoudlou et al. 2019)

acid), and their esters. Hesperetin, naringenin, isosakuranetin, chrysin, acacetin, luteolin, and its glycoside are examples of flavanones. Flavonols, which include kaempferol and isorhamnetin glycosides, are examples of isoflavonoids (coumestrol, genistein, and formononetin) (López-Gutiérrez et al. 2014; Nabas et al. 2014; Kunugi and Ali 2019).

7.3.6 Other Ingredients

Adenosine monophosphate and adenosine monophosphate N1 oxide are also present in royal jelly. This is the end result of adenosine oxidation at the adenine base moiety's N1 position. There is no other place in nature where it can be discovered besides royal jelly. The N1 oxide of adenosine monophosphate performs an essential physiological function. It can activate the mitogen-activated protein kinase (MAPK)/extracellular signal-regulated kinases pathway via the A2A adenosine receptor to promote neurite outgrowth and start pheochromocytoma PC12 cell differentiation (Hattori et al. 2007; Hattori et al. 2010). Pure royal jelly contains between 27.0 and 50.0 mg/g of adenosine, whereas royal jelly used as a dietary supplement contains between 2.0 and 173.0 mg/g. Acetylcholine (ACh), hormones (testosterone, flavonoids, prolactin, and estradiol), and other bioactive substances are also present in royal jelly (Kim and Lee 2011).

7.4 PROPERTIES OF ROYAL JELLY

7.4.1 Rheological Properties

Due to its bioactive properties, royal jelly has become an increasingly important nutritional product in recent years, attracting more and more consumers. The degradation of royal jelly's bioactive properties can be accelerated by processing temperature and storage conditions. Royal jelly has a creamy, milk-like consistency and is rich in proteins and carbohydrates. As a result, the rheological categorization of royal jelly can provide useful information for a variety of purposes, including process equipment design and energy utilization, as well as product structure and freshness and quality for consumers (Saricaoglu et al. 2019). Royal jelly is much more viscous than regular honey. It is highly adaptive for honeybees to have two different, unique head glands controlling the release of royal jelly. Hypopharyngeal and mandibular gland secretions are responsible for lowering pH to the degree where oligo MRJP1/apisimin complexes

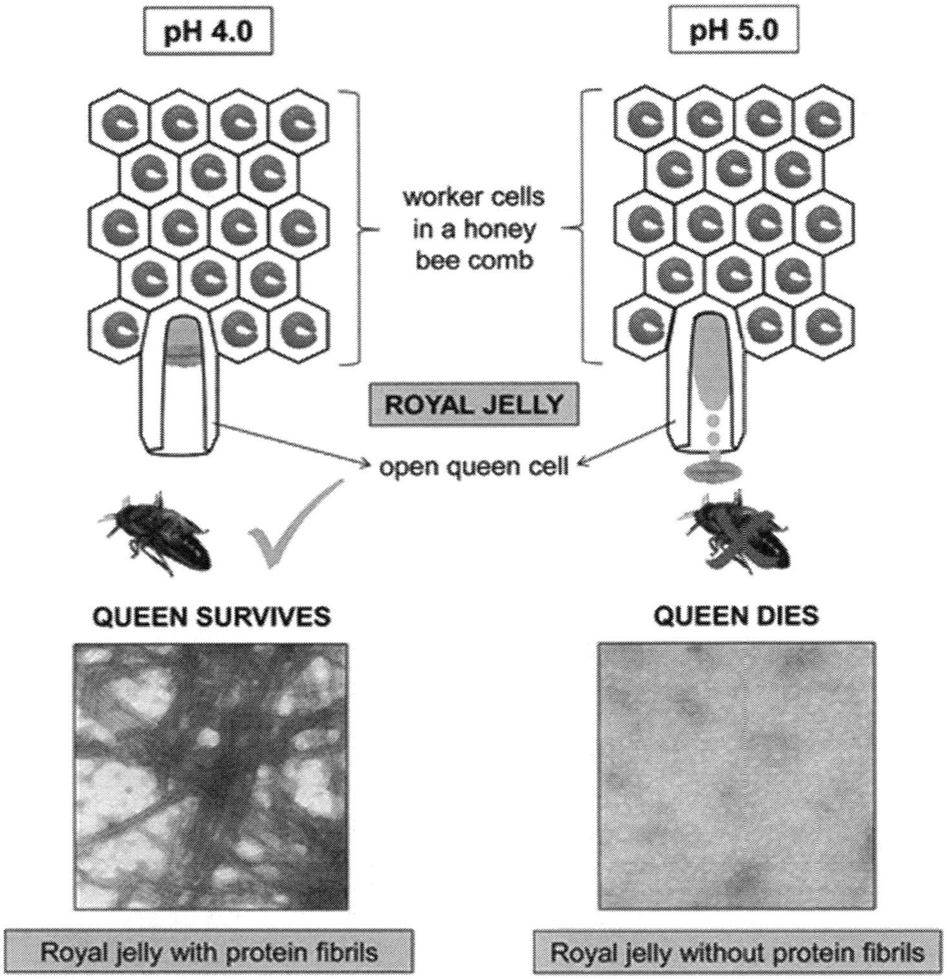

FIGURE 7.2 Viscosity of royal jelly is determined by a two-protein fibril network. (From Buttstedt, A. et al., *Curr. Biol.* 28, 1095–1100, 2018.)

form fibril networks to generate the necessary RJ viscosity. Nursing bees blend the secretions of two glands to control RJ's viscosity when filling a queen cell (Buttstedt et al. 2018). Queen honeybee larvae are grown in special, vertically oriented, downward-opening cells. Nurse bees leave protein-rich RJ on the cell ceiling as sustenance for the larvae. Two of these proteins are crucial for the growth of queens. They regulate the viscosity of the royal jelly to prevent the larva from escaping (Buttstedt et al. 2018) (Figure 7.2). Fresh royal jelly has the consistency of a thick jelly. This substance is partially soluble in water. It starts off looking somewhat pale and then turns more yellow as it ages. It has a foul smell, but it tastes delicious. Sensory qualities are crucial evaluation indicators. A deeper color and a rotten flavor characterize royal jelly that has been incorrectly preserved for a long time. As a result, keeping it frozen preserves its quality. The viscosity increases with time and water content, and it thickens up when held at room temperature (Kausar and More, 2019).

7.4.2 Authenticity

Royal jelly is seen as a pricey product that could be adulterated in an effort to reduce production costs as well as an alluring functional food. Other factors can also have an impact on the ultimate

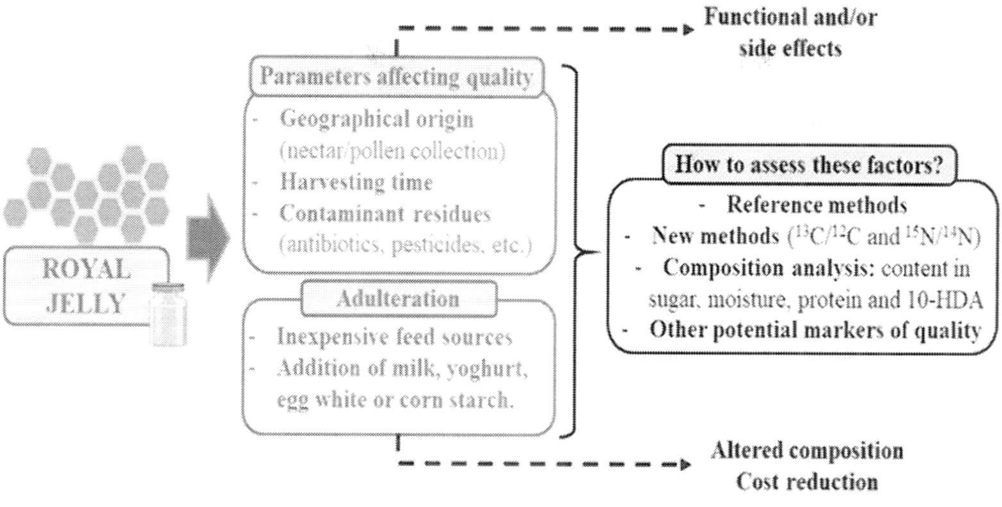

FIGURE 7.3 Overview of the essential factors influencing the quality, adulteration techniques, and indicators to evaluate these factors of royal jelly. (From Collazo, N. et al., *Nutrients*, 13, 543, 2021.)

quality of royal jelly. Using reference methods to examine the composition and assess these elements is necessary to ensure the product's quality and its intended physicochemical and organoleptic features (Collazo et al. 2021). An overview of royal jelly authenticity determination is given in Figure 7.3. Royal jelly's quality and authenticity are often assessed using sugar, moisture, protein, and 10-hydroxy-2-decenoic acid (10-HDA). Specific amino acids, amines, carbohydrates, and vitamins have been identified as possible biomarkers through additional research. The most crucial marker of real royal jelly is 10-hydroxy-2-decenoic acid (Sabatini et al. 2009; Virgiliou et al. 2020).

On the other hand, because bees' diets are affected by their surroundings, the final composition of royal jelly will depend on its place of origin. Recent research has revealed that environmental factors (such as proximity to the ocean or the interior) can alter royal jelly's chemical composition. Additionally, the distinctive sugar content appeared to be linked to free radical scavenging activity. Pollen studies can also be used to infer a plant's likely origin. As with other foods, the qualities of royal jelly can change depending on when it was harvested (Ramadan and Al-Ghamdi, 2012; Mokaya et al. 2020).

7.4.3 Health-Promoting and Biological Properties of Royal Jelly

Royal jelly is one of the most popular functional foods. The dietetic and cosmeceutical benefits of this product were its main marketing points. But, it has been revealed that royal jelly has numerous pharmacological qualities that can either prevent or treat a wide range of illnesses (Collazo et al. 2021). Today, scientists are looking at the different ways that royal jelly can improve health. According to a meta-analysis, royal jelly treatments can improve blood levels of high-density lipoprotein (HDL) and lower total cholesterol. The expression of cholesterol 7-hydroxylase (CYP7A1) may also increase as a result of RJ, which may then increase the activity of hepatic receptors implicated in the production of extremely low-density lipoproteins (LDL) (Kamakura et al. 2006). Additionally, in the denuded aortic ring, RJ might inhibit the extracellular Ca^{2+} influx caused by elevated K^+ and the intracellular Ca^{2+} release caused by norepinephrine (NE). RJ has the potential to enhance the generation of nitric oxide (NO) in isolated aortic rings through the activation of cyclic guanosine monophosphate (cGMP). NO production is associated with RJ's antihypertensive effects, and muscarinic receptor agonists cause vasodilation via the NO/cGMP pathway and calcium channels (Pan et al. 2019). Mice fed a high-fat diet supplemented with royal jelly had greater thermogenic capacity in their brown adipose tissue than those fed a normal diet, and this effect was shown even in the absence of changes in food or energy intake (Yoneshiro et al. 2018). In contrast, earlier studies found conflicting results regarding royal jelly's impact on lipid profiles (Collazo et al. 2021). Ascorbic acid, short peptides, flavonoids, α-tocopherol, and enzymes including glucose oxidase, catalase, and peroxidase are just a few of the components of honey that may have antioxidant properties. For this reason, royal jelly could replace the use of synthetic sulfites in the preservation of fruits and vegetables and the production of juices and preserves. In addition to their aesthetic appeal, the high antioxidant content of royal jelly and propolis makes them promising candidates for usage as functional foods (Viuda-Martos et al. 2008). Royal jelly has been shown to inhibit the growth of Lewis lung carcinoma and colorectal adenocarcinoma cells (Filipič et al. 2015).

Finding new antibiotics to combat the current widespread antibiotic resistance is crucial in preventing the spread of various infectious diseases. Antimicrobial peptides (AMPs) have emerged as a potential new class of antibiotics in this investigation. Royal jelly substances such as jelleines, royalisin, MRJPs, royalactin, and apisimin are examples of AMPs. Moreover, among the AMPs, there are additional royal jelly compounds with antibacterial activity, such as 10-HDA, which inhibits the growth of certain bacteria (Šedivá et al. 2018). Research revealed that royal jelly and its primary constituents have potent antibacterial activity against Gram-positive bacteria and to a lesser extent, Gram-negative bacteria. The potential of royal jelly as an anti-biofilm agent was also investigated. The possibility of consumer illness and contamination makes biofilm formation a significant issue in the food and beverage industry. In the year 2018, a study conducted in Europe investigated the efficacy of royal jelly in combating *Listeria monocytogenes*, a prevalent bacterium typically found in food biofilms. (Authority, 2021).

Along with a healthy diet, functional foods can help one feel better and possibly help to avoid certain diseases. If royal jelly were added to foods, it might make up for or even improve the nutritional value of the diet. It's a terrific cure that helps add more vital nutrients to the body. Research is currently being done on novel functional foods focused on royal jelly, including analyses of their characteristics and pharmacological functions in the human body (Núnez-González et al. 2011). Inoculating skimmed milk with varying concentrations of royal jelly, for instance, has been proposed as a method for the management of human disorders such as hyperglycemia (type 2 diabetes), hypertension, and various cancers (including breast and skin cancers) (Alu'datt et al. 2015). In addition, royal jelly has numerous cosmeceutical applications. Royal jelly has estrogen-like actions; a royal jelly cream improved postmenopausal women's quality of life more effectively than Premarin cream and lubricant. In another approach, 10-HDA may be effective for addressing skin barrier disorders. It was determined that its synthetic counterpart, Hydroxydecine®, is excellent at restoring the skin barrier, reducing irritation, and moisturizing dry skin (Collazo et al. 2021). Lesions on desquamated skin took less time to heal after receiving royal jelly. This means that it can reduce inflammation and speed up the healing process after a wound. Royal jelly may be helpful in the prevention and treatment of periodontal disorders, according to recent studies. These are chronic inflammatory disorders that are brought on by bacterial infection and lead to the loss of bone that holds teeth in place (Melliou and Chinou 2014).

RJ increases the activity of DAF-16, the *Caenorhabditis elegans* homolog of the human Forkhead Box O transcription factor (FOXO), by inhibiting insulin/insulin-like growth factor-1 signaling (IIS). In addition to improving connections with its associated transcriptional coregulator proteins, host cell factor (HCF-1), sirtain homolog (SIR-2.1), and FTT-2, activation of DAF-16/FOXO increases its translocation from the cytoplasm to the nucleus (a 14-3-3 protein). The formation of a protein complex within the cells, the upregulation of several genes that promote longevity, and the regulation of the dietary restriction signaling (which lowers food intake) and the mechanistic target of rapamycin (mTOR) pathway's downstream processes all contribute to the extension of the lifespan of the control unc-24/+ mutants. DNA repair, autophagy, antioxidant properties, anti-inflammatory action, stress tolerance, and cell proliferation are all bolstered by the interplay of many genes engaged in IIS, mTOR, and dietary restriction

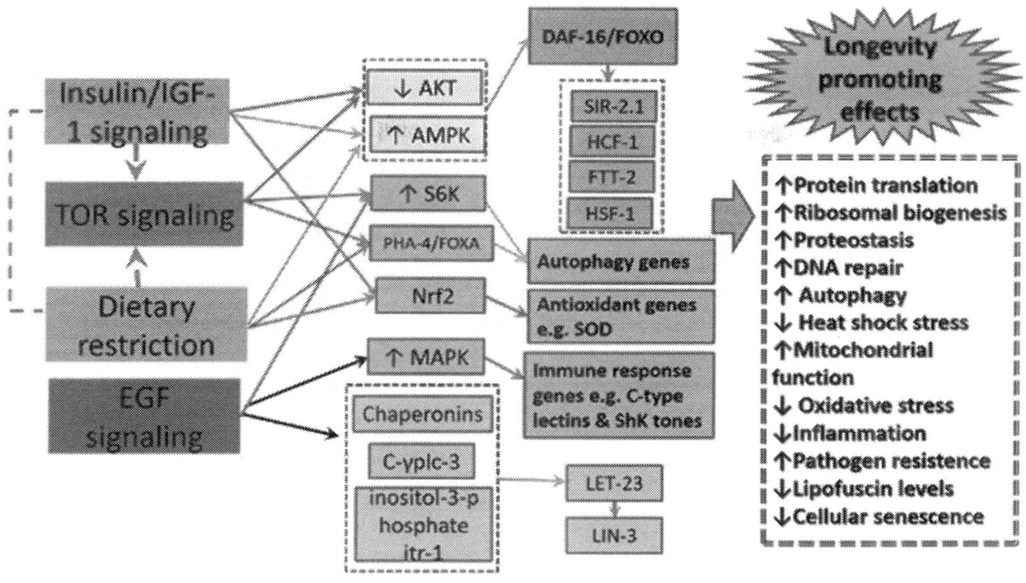

FIGURE 7.4 Possible ways that royal jelly (RJ) and its components increase longevity. (From Kunugi, H. and Mohammed Ali, A., *Int. J. Mol. Sci.*, 20, 4662, 2019.)

signaling. RJ, on the other hand, predominantly stimulates epidermal growth factor receptor (EGFR) signaling *via* activating the receptor (LET-23).

Elongation factors and chaperonins are upregulated by EGFR through the downstream phospholipase C-plc-3 and inositol-3-phosphate receptor itr-1, which boosts protein translation and proteasome activity. Rebuilding cellular components, increasing cellular detoxification, ribosomal function, and muscle maintenance, and lowering lipofuscin levels are all part of this system. A detailed scenario can be found in Figure 7.4.

7.5 HEALTH BENEFITS

Today, more people are turning to natural therapies to treat their health problems. It has been demonstrated that these products reduce the risk of chronic illnesses and cut healthcare costs (Jenkhetkan et al. 2017). An increasing number of people have been turning to natural bee products for treatment, in both alternative and mainstream medicine, in recent years. RJ, a bee by-product, is used for human health maintenance as well (Pasupuleti et al. 2017). Royal jelly has a number of benefits that are comparable to those of functional foods, which have grown in popularity in recent years. Numerous studies have been done on the pharmacological qualities and health-promoting effects of RJ in both animal models and people. RJ, a dietary modification of honey and bee bread, is widely sold in Asia as a health food and cosmeceutical, in China and Japan in particular (Fratini et al. 2016; Clarke and McDonald 2017). RJ is also used to research additional drug applications and conventional ingestion as "remedies" for people and animals (Pasupuleti et al. 2017). Recent studies have examined the genesis and function of RJ, as well as its MRJPs, for larval development, antibacterial qualities, therapeutic value, proteins and peptides, prospective usefulness for the treatment of cancer, and healthy aging and lifespan (Bogdanov 2011; Buttstedt et al. 2014; Fratini et al. 2016; Pasupuleti et al. 2017; Miyata and Sakai, 2018; Ramanathan et al. 2018; Kunugi and Ali 2019). The beneficial effects of RJ on human health are the primary focus of this section. The bioactive ingredients and pharmaceutical application of RJ are also discussed. These biological activities include hepato-renal protection, neuroprotection, estrogenic action, and fertility, as well as those

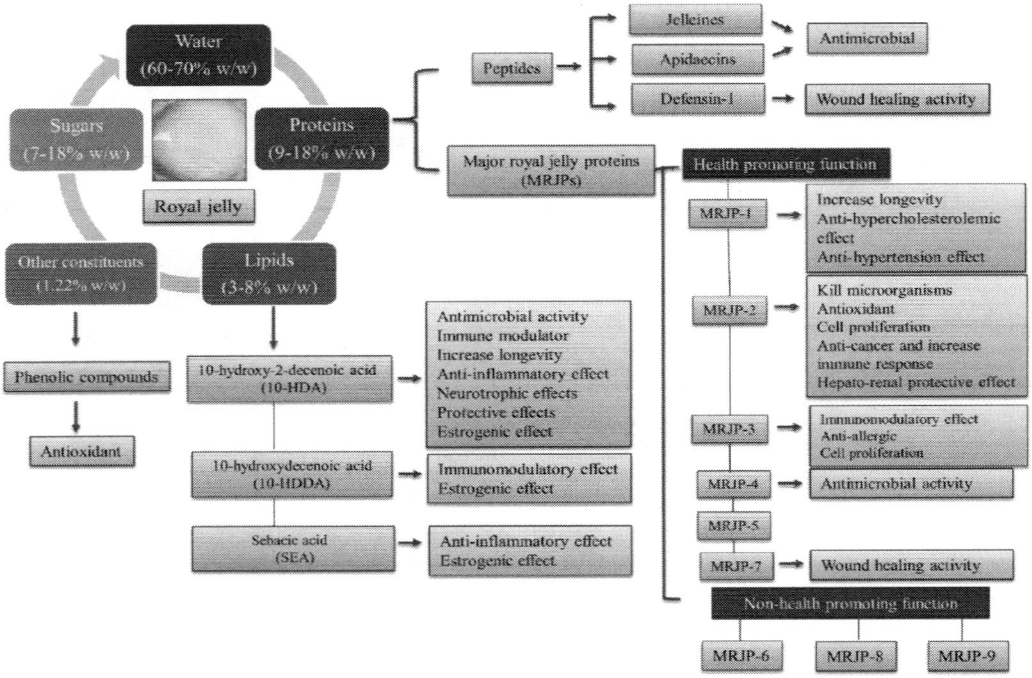

FIGURE 7.5 Schematic illustration of the primary biological components of RJ and their biological functions. (Ahmad et al. 2020).

that are immunomodulatory, wound healing, antimicrobial, antioxidant, anti-inflammatory, anti-aging, anti-cancer, anti-diabetic, anti-hyperlipidemic, anti-hypertension, and so on.

7.5.1 Bioactive Agents

According to Alu'datt et al. (2018), RJ is a rich source of nutrients and bioactive compounds, which may be essential for biological processes and therapeutic applications. It has been demonstrated that proteins, peptides, lipids, phenolics, and flavonoids are the main bioactive components responsible for the variety of pharmacological properties of RJ (Figure 7.5). The natural variation of bioactive compounds is determined by the diversity of flora species found in diverse environments (Alu'datt et al. 2018).

Laboratory animals (rabbits, mice, rats, and hamsters), microorganisms (bacteria, fungi, viruses, and nematodes), farm animals (ewes and buffalos), and clinical trials have all been used to study the biological functions of RJ and its use (Table 7.4). This research lays the foundation for future developments of RJ's pharmaceutical effects (human disease treatment). Several biological activities have been connected to the active ingredient concentration of RJ.

7.5.2 Health Beneficiary Functions of Royal Jelly

RJ is a functional food that affects several biological processes, including the immune system, aging, memory, digestion, blood sugar, obesity, and antimicrobial activities, and has advantages against cancer. RJ has positive impacts on both human and bee health. It is a natural antibiotic that effectively aids in the development of brood cell larvae and retains ovulatory features for the whole of the bee's lifespan. Additionally, RJ contains antioxidants that may lower the chance of developing cancer, hypertension, diabetes, and cardiovascular illnesses (Aslan and Aksoy 2015; Malekinejad et al.

TABLE 7.4
Biological Effects of Royal Jelly's Bioactive Ingredients

Bioactive compound	Biological effects
RJ (Royal Jelly)	a) Influence on diabetes b) Genoprotective impact c) Enhances the triglyceride, lipoprotein, and cholesterol levels in serum d) Reduces glucose levels and boosts insulin levels e) Elevation of the BCL2/BAX (BCL2 associated X protein) ratio is necessary for cell endurance f) Improvement in hTERT/BAX with age g) For the antioxidative response, NRF2/BAX are elevated
RJ and Purified RJ Peptides	a) Neuroprotective b) Boost estrogen levels, cholinergic response, and antioxidant capabilities c) Reduce the production of β-secretase to lower cholesterol and prevent the synthesis of amyloid-beta d) Improve autonomic nervous system and blood–brain barrier
MRJPs (Major Royal Jelly Proteins)	a) Boost memory b) Lengthen life expectancy by boosting antioxidant and EGF (Epidermal Growth Factor) production c) Affecting the metabolism of cysteine, taurine, and energy will help older rats with their spatial memory
MRJP1 (Major Royal Jelly Protein-1) oligomer	a) Defend against intestinal inflammation b) Higher level of HDL-c, reduced levels of LDL-c and total cholesterol c) Encourage the growth of IEC-6 cells d) Effect on cholesterol levels
MRJP2 (Major Royal Jelly Protein-2) and isomer X1	a) Anti-cancer impact b) Lower levels of TNF-a, MDA, glomerular diameter, adhesion molecule-1 expression, leukocyte infiltration, blood urea, and MDA (tumor necrosis factor) c) Encourage caspase-dependent apoptosis and prevent HepG2 cells from expressing Bcl-2 and p53 d) Increased levels of SOD (superoxide dismutase) and GPx (glutathione peroxidase)
MRJP3 (Major Royal Jelly Protein-3)	a) Effect on immune regulation b) Reduces antigenicity and prevents the generation of IL-4 (interleukin-4), IL-2, and IFN-γ (interferon-gamma)
10-HDA (10-hydroxydecanoic acid)	a) Prolonging life b) Increases *Caenorhabditis elegans* lifetime c) Anti-diabetic effect d) Anti-cancer impact e) Defends against intestinal inflammation f) Impact on neurons g) Enhances insulin resistance and hyperglycemia by promoting PGC-1α expression h) Possesses anti-inflammatory properties in WiDr cells from human colorectal cancer i) Prevented pigmentation of mouse skin by decreased melanin synthesis, TRP1, TRP2, and MITF levels in B16F1 melanoma cells j) Reduces glutathione levels while increasing lipid peroxidation with MDA (malondialdehyde) k) Inhibited oligodendrocyte, astrocyte, and neuronal differentiation l) Improves collagen formation by upregulating TGF-β1 expression to shield it from photo-aging
3,10-DDA (3,10-dihydroxy-decanoic acid)	a) Immunoregulatory effect b) IL-12 and IL-18 production should be increased, while IFN-U production should be stimulated

(Continued)

TABLE 7.4 (CONTINUED)
Biological Effects of Royal Jelly's Bioactive Ingredients

Bioactive compound	Biological effects
HPO-DAEE (4-Hydroperoxy-2-decenoic acid ethyl ester)	a) Anticancer activity b) The encouragement of apoptosis in human lung cancer A549 cells via the ROS (Reactive Oxygen Species)-ERK (Extracellular Regulated protein Kinases)-p38 pathway and to a lesser degree, the CHOP pathway
Royalisin	a) Antibacterial b) Prolonging life c) Damages cell walls and cell membranes to inhibit Gram-positive bacteria d) Increases *Caenorhabditis elegans* lifespan, encourages epidermal growth factor signaling
Jelleines I–III	a) Yeasts and Gram-positive and Gram-negative bacteria are suppressed b) Antibacterial
ACh	a) Defends against intestinal swelling b) M receptor (muscarinic receptor) should be used as a mediator in the stimulation of intestinal smooth muscle
Apolipophorin-III-like protein	a) Effect on immune regulation b) Improves immunological response upon phosphorylation
AMP-N1 oxide as well as its analogs	a) Anti-cancer impact b) Activate the integrin signaling, adenosine A2A receptor, and MAPK (mitogen-activated protein kinase) signaling pathways to promote the generation of axons in rat pheochromocytoma PC12 cells

(Ahmad et al. 2020)

2016; Al-Kushi et al. 2018). Furthermore, RJ has an impact on the morphological traits, development, learning, size, and form variations in a variety of species, including honey bees, mice, and people (Shi et al. 2018).

7.5.2.1 Antimicrobial Activity

RJ has potent antibacterial activities against a variety of diseases (Han et al. 2014; Coutinho et al. 2018; Kim et al. 2019). RJ may also treat periodontal pathogens such as *Porphyromonas gingivalis*, *Aggregatibacter actinomycetemcomitans*, *Prevotella intermedia*, and *Fusobacterium nucleatum* (Coutinho et al. 2018). Gram-negative *Escherichia coli* is resistant to the antibacterial effects of MRJPs (2–5 and 7). By including positively charged lysine, arginine, and histidine residues, antibacterial peptides are able to interact with the anionic phospholipids of the cell membrane, causing the membrane to collapse and killing the bacteria. The three intramolecular disulfide connections between cysteine residues provide royalisin with its potent antibacterial activity against a wide range of Gram-positive and Gram-negative microorganisms. Natural jelleines can also decrease microorganisms that are both Gram-positive and Gram-negative. MRJPs 2 and 4 serve as antimicrobial agents with a broad spectrum of activity against bacteria (Gram-positive and Gram-negative), fungi, and yeasts. Recombinant MRJP-2 and MRJP-4 can kill microorganisms by attaching to the cell wall of fungi, yeast, and bacteria and changing the structure of the cell wall (Kim et al. 2019; Park et al. 2019). *Aspergillus fumigatus*, *Syncephalastrum racemosum*, and *A. niger* have all shown resistance to RJ's antifungal effects. Additionally, it has been suggested that royalisin shows an antifungal response to necrotrophic fungi such as *Botrytis cinerea* (Bíliková et al. 2001). The natural jelleine-Il protein has an inhibitory effect on *Candida albicans* (Fontana et al. 2004; Han et al. 2014). RJ also performs well against *C. albicans* and as a backup agent. Fatty acids like 3,10-HDA, 11S, 10-HDA, and 10-acetooxy-2-DEA may significantly suppress the growth of yeasts such as *Candida tropicalis*, *C. albicans*, and *C. glabrata* (Melliou and Chinou, 2005).

7.5.2.2 Antioxidant Activity

Various chronic and degenerative disorders might potentially be prevented and treated using the antioxidant activity of RJ. Researchers have recently looked at the antioxidant capabilities of RJ and its enzymatic hydrolysates. We may infer that RJ's antioxidant activity is influenced by phenolic chemicals, proteins, and peptides. RJ affects the expression of proteins and scavenges reactive oxygen species inside the cell. RJ regulates oxidative stress and apoptosis in the liver and kidneys and possesses potent antioxidant and free radical scavenging properties. Additionally, it minimized the negative effects of chemical agents and the oxidative stress brought on by fumonisin (Nabas et al. 2014; Guo et al. 2009). Twenty-nine antioxidant peptides extracted from RJ hydrolysate were tested for their ability to scavenge hydroxyl radicals and hydrogen peroxide, with 12 short peptides containing 2–4 residues exhibiting the greatest activity. Additionally, three dipeptides in RJ exhibit considerable scavenging activity as a result of their phenolic hydroxyl group's donation of a hydrogen atom (Guo et al. 2009).

7.5.2.3 Prolonging Lifespan/Healthy Longevity

From the beginning of recorded history, people have believed that RJ may prevent DNA damage and reduce oxidative stress, hence extending the human lifespan (Pasupuleti et al. 2017). RJ is linked to an increase in longevity in various different species, including queen honey bees (Fratini et al. 2016). Mice, honey bees, crickets, silkworms, nematodes, and other creatures can all live longer thanks to the protein and lipid components of RJ. These can also prevent the senescence of human tissues in cell cultures by upregulating epidermal growth factor and downregulating insulin-like growth factors (Kunugi and Ali. 2019). RJ increases oxidative phosphorylation and respiration, enhancing physiological functions (Han et al. 2011). It affects the cellular activities of estrogen and gonadotropin. RJ lowers conjunctivitis, corneal burn, blepharitis, and excessive ocular blood pressure in the elderly. Additionally, by moisturizing the skin and preventing the production of melanin, it reduces the appearance of wrinkles and spots (Tatsuhiko et al. 2011).

7.5.2.4 Wound-Healing Activity

Healing wounds is a big health issue, and RJ seems to play a significant role. RJ is suitable for wound healing because of its fatty acids, proteins, and antimicrobial characteristics (Majtan et al. 2010). When used on hamsters with severe oral mucositis caused by the chemotherapy drug 5-fluorouracil, dose-dependent RJ speeds up the healing process. Additionally, RJ at a dosage of 5 g/mL alters the levels of various lipids and raises the number of sphingolipids, which facilitates the migration of fibroblasts in humans (Kim et al. 2019). A helpful technique to assist persons with diabetic foot ulcers is to use an RJ dressing in addition to other conventional treatments. Additionally, this procedure results in vasodilation around the lesion, which may enlarge blood vessels to increase blood flow and prevent the entry of additional bacteria into the wound (Siavash et al. 2011). The defensin-1 peptide in RJ aids in skin regeneration and cutaneous wound healing by increasing matrix metalloproteinase-9 synthesis and keratinocyte migration (Bucekova et al. 2017). RJ increases granulation tissue development and wound healing capacity while speeding up the healing of desquamated skin lesions and minimizing exudation (Galaly et al. 2014).

7.5.3 Immunomodulatory Function

Immunomodulatory action is one of the many pharmacological effects of RJ that have been extensively researched. Allergies, cancer, and inflammation are all caused by an immune modulatory response, which increases the production of antibodies or decreases the activity of white blood cells (Labro 2012). After 3 months of treatment, the effect of RJ therapy in systemic lupus erythematosus (SLE) indicated a noteworthy improvement. This is according to the findings of the first human study on childhood systemic SLE (Nwobi et al. 2008). RJ combats respiratory disorders

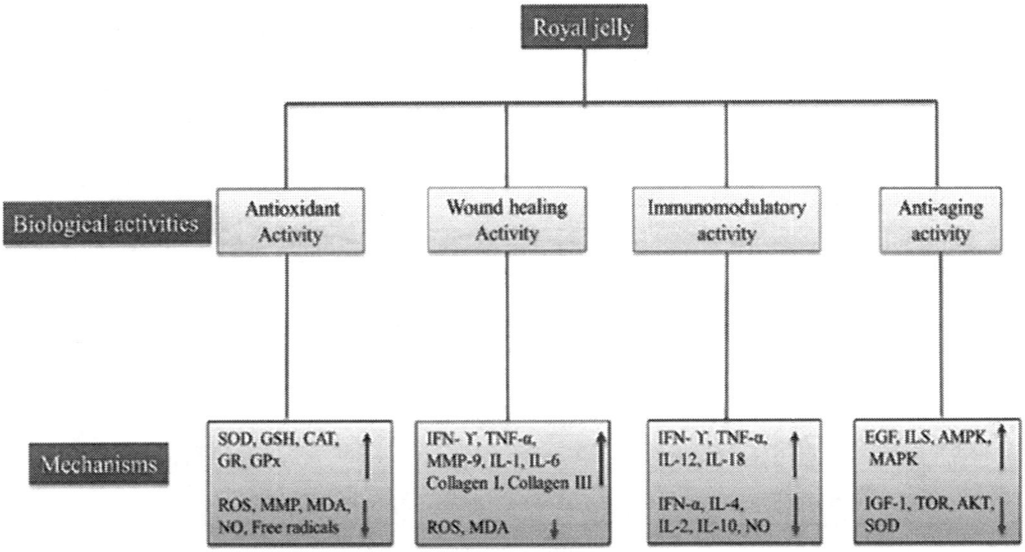

FIGURE 7.6 The biological activities of royal jelly and its mechanism. (From Ahmad, S. et al., *Int. J. Mol. Sci.*, 21, 382, 2020.)

and asthma. It inhibits type I allergy responses by restoring macrophage activity and Th1/Th21 cytokine responses. The immune system is supported in its fight against infections, cancer, allergies, and inflammation by unsaturated fatty acids, amino and gamma globulin, enzymes, proteins, and vitamins E and A. RJ contains most of these chemicals; hence, it may strengthen the immune system. Metastasis development is inhibited by the action of RJ on interferon-alpha suppression of human colon cancer cell growth (Mannoor 2008). Figure 7.6 depicts RJ's biological activity and any underlying potential pathways (Ahmad et al. 2020).

7.5.3.1 Anti-Cancer Effect

Recent pharmacological studies have demonstrated that RJ has potential activity against cancer, diabetes, cardiovascular disease, and Alzheimer's disease (Premratanachai and Chanchao 2014; Al-Kushi et al. 2018). By inhibiting or boosting the proliferation of cancer cells, RJ is beneficial, since it affects the immune system. The therapeutic effects of RJ in the body of a cancer patient are long-lasting. It is possible that RJ's anti-cancer qualities restrict tumor development and/or dissemination in the liver and/or lung by inhibiting tumor-induced angiogenesis and/or by promoting immune function (Kimura 2008). Crude RJ can protect human breast cancer cells from growth triggered by bisphenol A. RJ therapy administered to individuals with benign prostatic hyperplasia for 3 months has superior results in lowering prostatic-specific antigen and enhancing quality of life (Pajovic et al. 2016). Figure 7.6 depicts the biological actions of RJ and their putative underlying mechanisms.

7.5.3.2 Anti-Diabetic Effect

Many medications exist to treat diabetes and its complications, but novel strategies are required to give patients the greatest therapeutic benefit with the fewest adverse effects. After taking RJ, blood glucose levels dropped noticeably in healthy trial participants (Münstedt et al. 2009). RJ supplementation significantly decreased serum levels of fasting blood glucose (FBG) and glycosylated hemoglobin by increasing insulin concentration, which may help with the management of diabetes (Pourmoradian et al. 2014). Another study found that type 2 diabetes patients receiving RJ therapy have higher overall antioxidant capacities and lower insulin resistance homeostasis model evaluations. In RJ, the oxidation of glucose for the creation of energy and maintenance of the optimal blood sugar level is facilitated by the presence of insulin-like peptides, chromium, sulfur, and vitamin B3 (Ashry and Elkady 2014). RJ also reduces alloxan-induced diabetes, which damages the structure of insulin. It functions in the liver as a potent free radical scavenger, preventing or reducing the negative effects of medications (Li et al. 2011). RJ is anti-diabetic and protects diabetic individuals from insulin resistance and hypercholesterolemia. RJ also significantly affects blood glucose levels, apolipoprotein A1 (ApoA-1) levels, and (ApoB/ApoA-1) ratios, which may lower the risk of cardiovascular disease in those with type 2 diabetes (Khoshpey et al. 2016).

7.5.3.3 Effect on Cardiovascular Health

RJ reduces the lipid, cholesterol, and total serum protein levels in rats and rabbits. Without changing blood sugar or total cholesterol levels, RJ therapy for rats lowers plasma triglyceride and insulin levels and has a propensity to lower systolic blood pressure (Zamami et al. 2008). However, oral treatment for RJ lowers systolic and diastolic blood pressure while raising NO levels. RJ causes vasorelaxation in isolated aortic rings by inhibiting L-NAME (a nitric oxide synthase inhibitor), indomethacin (a cyclooxygenase inhibitor and an antagonist of the atropine-M3 receptor), and methylene blue (a guanylate cyclase inhibitor) (Pan et al. 2019). RJ proteins may reduce systolic blood pressure by inhibiting the angiotensin 1-converting enzyme (ACE) activity that is produced by the hydrolysis of gastrointestinal enzymes (Matsui et al. 2002). Vascular smooth muscle cells (VSMCs) show diminished cell migration, proliferation, and contraction as well as blood pressure with MRJP-1 protein (Fan et al. 2016).

7.5.3.4 Effects Against Inflammation

A large number of biological and chemical factors, such as cytokines, pro-inflammatory enzymes, low–molecular weight compounds (eicosanoids), and the enzymatic breakdown of tissues, contribute to the inflammatory process (Libby et al. 2002). In autoimmune diseases, including rheumatoid arthritis and inflammatory bowel disease, RJ may be crucial for improving quality of life (Kohno et al. 2004). In addition, RJ has an anti-inflammatory action and encourages cell regeneration in rats with colitis brought on by acetic acid. Moreover, 10-HDA blocks the creation of lipopolysaccharide (LPS)-induced production of interleukin (IL)-6 and IkB-z, both of which are linked to autoimmune and inflammatory diseases (Sugiyama et al. 2012).

7.5.3.5 Neurotrophic and Pharmaceutical Action

RJ utilizes a neurotrophic factor produced by glial cells to enhance the formation of brain cells from neural stem cells as well as to stimulate and activate the central nervous system. A bioactive compound in RJ, namely, 10-HDA, stimulates neuronal generation (Pyrzanowska et al. 2012). RJ performs neurotrophic and neuroprotective functions in the adult rat hippocampus. RJ promotes the development of neuronal, oligodendrocyte, and astrocyte cells (Ito et al. 2012). While RJ is present in both the peripheral and central nervous systems, it plays a role as a neuromodulator only in the motor branch of the somatic nervous system (Mannoor et al. 2008). RJ was found to improve spatial memory, block stress-related immune depression, and reduce neuronal mortality and neurogenesis in Alzheimer's and Parkinson's diseases (Pasupuleti et al. 2017). Additionally, RJ significantly reduces the body weight, blood lipid level, amyloid-beta, acetyl-cholinesterase, and malondialdehyde (MDA) level while increasing choline acetyltransferase and superoxide dismutase (SOD) levels in brain tissue to significantly improve the behavioral impairments and brain image structure of cholesterol-fed rabbits (Zhang et al. 2019). Pharmaceutical studies demonstrate that the bioactive substances in RJ, including proteins, peptides, lipids, phenolics, and flavonoid chemicals, are responsible for different actions. In contemporary pharmaceutical research, RJ has shown promise against cancer, diabetes, cardiovascular disease, and Alzheimer's disease. As is shown in Figure 7.7, RJ and its constituents have pharmacological actions that promote health by influencing a variety of biological processes.

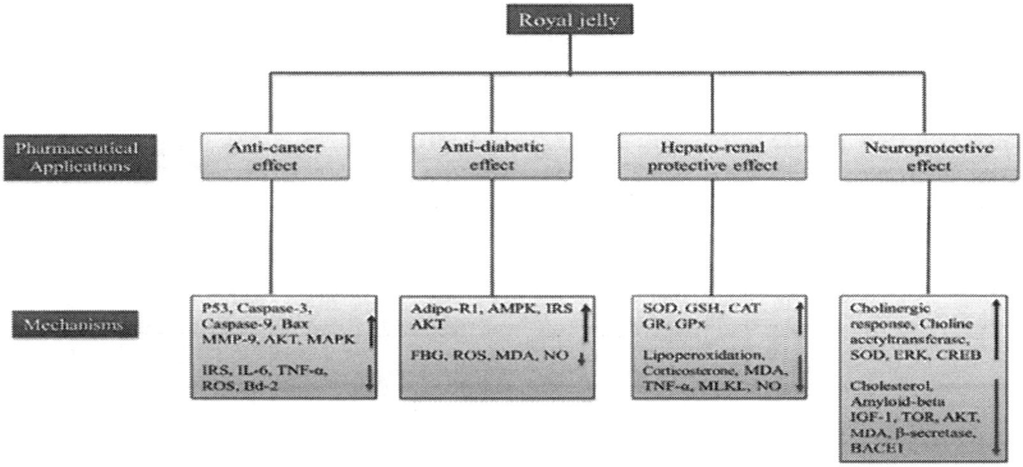

FIGURE 7.7 The mechanisms and pharmacological effects of royal jelly. (From Ahmad, S. et al., *Int. J. Mol. Sci.*, 21, 382, 2020.)

7.5.3.6 Effects on Fertility and the Reproductive System

Infertility and sexual dysfunction are common clinical issues with few effective treatments. According to the evidence outlined in this section, RJ increases both male and female fecundity. Because of its interaction with estrogen receptors, RJ affects gene expression and stimulates cell growth in both in vivo and in vitro settings (Mishima et al. 2005). Combining exogenous progesterone with intramuscular RJ injections and oral intake enhances the estrous impact and pregnancy rate in Awassi ewes (Husein and Kridli 2002). Additionally, it effectively treats sexual and urinary problems, and it enhances postmenstrual women's quality of life (Seyyedi et al. 2016). RJ treatment may improve the post-thaw quality and in vitro fertilization potential of Nili-Ravi buffalo bull sperm. Oocyte and cumulus cell RJ therapy has been shown to increase glycolysis, the pentose phosphate pathway, and antioxidant enzyme activity, which may lead to improved oocyte maturation, fertilization, and blastocyst formation (Eshtiyaghi et al. 2016). Infertile men who receive RJ therapy have significantly higher levels of testosterone, luteinizing hormones, and active sperm motility. In male hamsters, long-term RJ usage increases testosterone levels and spermatogenesis by correcting age-related testicular function loss (Kohguchi et al. 2007). Infertility resulting from asthenozoospermia may be successfully treated with honey and RJ (Abdelhafiz and Muhamad, 2008).

In addition, RJ might reduce summer infertility and increase the physiological health of male rabbits subjected to heat stress (El-Hanoun et al. 2014). Under heat-stress situations, co-administration may be an efficient way to boost sperm features, early embryo development, and sperm lipid peroxidation (Mahdivand et al. 2019).

7.6 DOSAGES OF ROYAL JELLY FOR HUMAN USE

Bee products may cause an allergic reaction in certain people. For instance, an allergic reaction may occur if RJ is used in cases of asthma and anaphylaxis (Thien et al. 1996). The recommendation of royal jelly to patients is highly dependent on their unique demands, their health problems, and their age, as well as the results obtained in the laboratory by researchers. Occasionally, very low dosages may be quite beneficial, and other times, massive quantities may be required. Studies indicate that a dosage of 100 mg/kg is the most effective treatment for severe illnesses, and that greater doses are seldom required.

7.6.1 Infants

Supplementation of royal jelly has been reported to improve the condition of infants with nutritional deficiencies and preterm infants. This includes an improvement in their growth and development, immune system, appetite and strengthening of the nervous systems. A dosage of 1 g per day for 5 days has been reported to improve weight gain and blood glucose levels of premature babies. A range of doses from 50 mg to 1 g per day was mentioned for treatment of preterm infants. A gradual increase in the dosage over a period improved the conditions of infants with no side effects reported. Preterm infants were given 0.25 g of uncooked royal jelly every day with no observed side effects. (Mahmoud et al. 1997; Magdalena, 2010; Strant 2017a; Strant 2017b).

7.6.2 Children

For impaired neurological system, weakened immune system (fetal distress, labor problems), lethargy, loss of appetite, anorexia, anemia, etc., royal jelly should be consumed in doses of 0.5 to 1 g daily for children 1 to 5 years old and 0.5 to 1 g daily for children 5 to 12 years old. Royal jelly is given in smaller doses when coupled with other bee products. The suggested dose is 2.5 g of royal jelly per day for children under 5, and 5 g per day for children between 5 and 12, with acute illnesses and colds (Strant 2017a; Strant 2017b).

7.6.3 Adults

Indications include immune system function, insomnia, skin issues, anemia, diminished libido, hormonal imbalance, wounds, premenstrual syndrome, menopause, and osteoporosis. Diabetes, depression, and Hashimoto's disease may be treated with 1 to 2 g per day, and 3–5 g per day can be used to treat arthritis. In our practice, various amounts of royal jelly are used in situations of recent depression. In a few rare cases, higher dosages, such as 10 g/day, are offered for a more rapid, strong response over a shorter time frame (10 days a month for 3 months). In the following situations, 1 g of royal jelly each day, together with other bee products and plants, is ideal: immunity, convalescence, preparation for surgery, autoimmune diseases, cancer, hormonal imbalances, infertility, ovarian cyst, uterine fibroma, thyroid problems, etc. Dosage should be as follows: i) when a cold first starts, up to 10 g/day for 1 to 3 days; ii) for other acute infections, up to 10 g/day for 1 to 3 days; and iii) for post-operative recovery, up to 5 g/day for up to 3 to 5 days. For persons with demanding job situations, 10 grams of royal jelly every day. For the first 1–3 days of a cold, 10 g of cough medicine till it stops, and 10 g for other acute illnesses; to hasten the healing process following surgery, 5–10 g for 3–5 days. The adverse effects of chemotherapy are paresthesia, scorching or uncomfortable fingers, instability when walking, and a feeling of weakness in the legs. Three grams of royal jelly per day may be used to treat infants aged 6 to 8 weeks (Strant et al. 2016; Strant 2017a; Strant 2017b).

7.7 ROYAL JELLY CONTRAINDICATIONS AND ADVERSE REACTIONS

In the acute stages of bronchial asthma in crisis, hypersensitivity to any of its ingredients, Addison's disease, and cancer, royal jelly is contraindicated. Overdosing may result in digestive system problems such as stomach aches, vomiting, diarrhea, and insomnia (especially in women). These side effects will go away if the dosage of royal jelly is decreased or stopped. Major overdoses may result in toxicity. Multiple variables may reduce the efficacy of RJ. The active components in royal jelly cannot be utilized effectively if the body is too weak or unable to digest and/or absorb, or if a person has major structural or genetic problems. More caution should be exercised and a number of factors should be taken into account prior to use in these circumstances (Strant et al. 2019).

7.8 PRODUCTION AND PROCESSING

During the 1940s, Japan was the primary innovator of royal jelly production and processing methods. Typically, bees generate small amounts of royal jelly because they feed a small number of queen larvae. Beekeepers take advantage of the bees' propensity to develop a new queen when the old one dies in order to increase the output of royal jelly. Each colony can produce up to 500 g of royal jelly in a single season in specialized beekeeping units. Continuous and discontinuous production are the two available techniques (Hu et al. 2019). Method, planning, and timing are essential for the successful commercial manufacturing of royal jelly. Attendance every single day is mandatory. The time and effort put into making royal jelly is far higher than that of regular bee products because of the level of knowledge and patience necessary. When honey is in short supply, bee colonies can be put to good use making royal jelly. Royal jelly spoils easily in the elements, so avoid exposing it to too much of anything. Many physical and chemical changes have been documented after long-term preservation; however, there is little data on whether or not these changes affect the biological efficacy for people (Bogdanov 2016).

The method they employed to generate royal jelly was to remove the queen from the colony in order to collect royal jelly directly from the new queen cells, which the worker honey bees constructed as soon as they learned the queen was missing. The production process for royal jelly is given here (Bogdanov 2016):

A) Place larvae that are now two days old into the cups of the queen cell frame, which are fastened to the bar. Typically, there are 30–90 cells in a queen cell frame. The larvae shouldn't have their position changed or be hurt during the grafting process.
B) To ensure that the bees quickly attend the grafted frame, it should be placed in the center of strong queen-right colonies close to the open brood.
C) Improve cell acceptance by feeding sugar syrup (1:1 sugar to water).
D) Take out the grafted-cell frames after 72 hours (3 days), they are taken out to be harvested. The cells are slit open at their narrowest point to make collecting easier and faster.
E) Carefully remove the larvae from the jelly using a pair of delicate forceps so as not to kill the insects and ruin the jelly. Use a small spatula or a sucking device to empty the royal jelly from each cell. The cells are immediately prepared to undergo a new round of proliferation.
F) Royal jelly can be filtered through a fine nylon net to eliminate wax fragments and larvae (nylon stockings work great for this). Filters made of metal should be avoided.
G) To prevent the jelly from drying out, store it in airtight dark glass vials or food-safe plastic containers. It requires immediate refrigeration.
H) Royal jelly requires thorough cleaning and disinfection of all surfaces, including hands, using either high heat or rubbing alcohol.

7.9 CONCLUSION

The market for functional foods, of which royal jelly (RJ) is a part, continues to grow. This has led to large-scale imports in nations whose indigenous supply is insufficient to fulfill demand. RJ's exceptional biological qualities give it broad commercial appeal, and it is now employed in a wide range of industries, from pharmaceuticals and food to cosmeceuticals and industry. RJ or its particular components have been shown to have anti-lipidemic, antioxidant, antiproliferative, antimicrobial, neuroprotective, anti-inflammatory, immunomodulatory, anti-aging, and estrogenic actions. In terms of biological properties, research has concentrated on anti-lipidemic, antioxidant, antimicrobial, anti-inflammatory actions and other benefits such as anti-aging or estrogenic functionality. But, further study is needed to determine how exactly it works. Thus, research capacities must be strengthened to enable both a reliable qualitative and quantitative assessment of the different components, and analytical tests must be implemented on commercially available items to detect possible adulteration. RJ contains various valuable medicinal benefits that have been utilized from ancient times until now. Scientists have just begun to investigate the myriad health advantages of this amazing superfood, and while some of its biological and therapeutic activities have been confirmed, there are only a small number of reliable studies to back up those claims. RJ's chemistry and bioactive components, on the other hand, are not well understood. Further testing (in vitro, animal studies, and clinical trials) and validation would be required to demonstrate any measurable benefit and mechanism of action of natural RJ and isolated compounds. Because of the rapid expansion and increase in demand for RJ, it is evident that the RJ market, which is now controlled by China, has to be regulated. These procedures should aim not only to establish controls on the validity and origin of the RJ, but also to preserve the environment and bees. Future study should concentrate on developing a complete understanding of the pathways of RJ compounds in order to generate new uses and products for the nutraceutical and cosmetic industries, as well as the pharmaceutical business. Knowledge about the composition of newly manufactured RJ is required to create a standard composition, assess the quality of commercial items, and detect the existence of RJ in other products containing it. Furthermore, developing a liquid chromatography–mass spectrometry approach, among other techniques, is critical for studying the complete composition of RJ. RJ has captured scientific attention since the first scientific research on it was published in 1852, and it is still a "hot" natural product. Every year, research groups from all over the world publish a plethora of highly fascinating scientific discoveries in international literature, suggesting that the field will continue to be explored for a long time to come.

ABBREVIATIONS

AKT: protein kinase B
AMPK: 50 AMP-activated protein kinase
BACE1: β-site amyloid precursor protein cleaving enzymes
BAX: BCL2 associated X protein
BCL2: B-cell lymphoma 2
CAT: catalase
CREB: cAMP Response Element-Binding Protein
EGF: epidermal growth factor
ERK: extracellular signal-regulated kinases
GPx: glutathione peroxidase
GR: glutathione reductase
GSH: glutathione
IFN-α: interferon-α
IFN-γ: interferon-gamma
IgE: immunoglobulin E
IGF-1: insulin-like growth factor-1
IL-4: interleukin-4
IRS: insulin receptor substrate 1
MAKL: mixed linkage kinase domain-like
MAPK: mitogen-activated protein kinase
MDA: malondialdehyde
MITF: microphthalmia associated transcription factor
MMP-9: matrix metallopeptidases-9
MRJP-4: major royal jelly protein 4
NO: nitric oxide
NRF2: nuclear factor erythroid 2 related factor 2
RJ: royal jelly
RJPs: purified royal jelly peptides
RJPH: royal jelly protein hydrolysate
ROS: reactive oxygen species

SOD: superoxide dismutase
TNF: tumor necrosis factor
TNF-α: tumor necrosis factor-alpha
TOR: target of rapamycin
10-HDA: 10-hydroxydecanoic acid
10HΔ2DA: 10-hydroxy-delta-2-decenoic acid
3,10-DDA: 3,10-dihydroxy-decanoic acid
2-AF: 2-aminofluorene

REFERENCES

Abdelhafiz, A.T. and Muhamad, J.A. 2008. Midcycle pericoital intravaginal bee honey and royal jelly for male factor infertility. *International Journal of Gynecology and Obstetrics* 101(2): 146–149.

Ahmad, S., Campos, M.G., Fratini, F., Altaye, S.Z. and Li, J. 2020. New insights into the biological and pharmaceutical properties of royal jelly. *International Journal of Molecular Sciences* 21(2): 382.

Al-Kushi, A.G., Header, E.A., ElSawy, N.A., Moustafa, R.A. and Alfky, N.A.A. 2018. Antioxidant effect of royal jelly on immune status of hyperglycemic rats. *Pharmacognosy Magazine* 14(58): 528.

Alu'datt, M.H., Rababah, T., Sakandar, H.A., Imran, M., Mustafa, N., Alhamad, M.N., Mhaidat, N., Kubow, S., Tranchant, C., Al-Tawaha, A.R. and Ayadi, W. 2018. Fermented food-derived bioactive compounds with anticarcinogenic properties: Fermented royal jelly as a novel source for compounds with health benefits. In *Anticancer Plants: Properties and Application*, pp. 141–165. Springer.

Alu'datt, M.H., Rababah, T., Obaidat, M.M., Ereifej, K., Alhamad, M.N., Mhaidat, N., Andrade, J.E., Johargy, A. and Ayadi, W. 2015. Probiotics in milk as functional food: Characterization and nutraceutical properties of extracted phenolics and peptides from fermented skimmed milk inoculated with royal jelly. *Journal of Food Safety* 35(4): 509–522.

Ashry, K. and Elkady, A.A. 2014. Royal jelly modulates hepatotoxic effect of rats treated with cisplatin. *Arab Journal of Nuclear Sciences and Applications* 47(1): 172–180.

Aslan, Z. and Aksoy, L. 2015. Anti-inflammatory effects of royal jelly on ethylene glycol induced renal inflammation in rats. *International Brazilian Journal of Urology* 41(5): 1008–1013.

Bíliková, K., Wu, G. and Šimúth, J. 2001. Isolation of a peptide fraction from honeybee royal jelly as a potential antifoul brood factor. *Apidologie* 32(3): 275–283.

Bogdanov, S. 2011. Royal jelly, bee brood: Composition, health, medicine: A review. *Lipids* 3(8): 8–19.

Bogdanov, S. 2016. Royal jelly and bee brood: Harvest, composition, quality. In *The Royal Jelly Book; Bee Product Science: Muethlethurnen,* Switzerland.

Bucekova, M., Sojka, M., Valachova, I., Martinotti, S., Ranzato, E., Szep, Z., Majtan, V., Klaudiny, J. and Majtan, J. 2017. Bee-derived antibacterial peptide, defensin-1, promotes wound reepithelialisation in vitro and in vivo. *Wound Healing Southern Africa* 10(2): 25–35.

Buttstedt, A., Mureşan, C.I., Lilie, H., Hause, G., Ihling, C.H., Schulze, S.H., Pietzsch, M. and Moritz, R.F. 2018. How honeybees defy gravity with royal jelly to raise queens. *Current Biology* 28(7): 1095–1100.

Buttstedt, A., Moritz, R.F. and Erler, S. 2014. Origin and function of the major royal jelly proteins of the honeybee (*Apis mellifera*) as members of the yellow gene family. *Biological Reviews* 89(2): 255–269.

Buttstedt, A., Moritz, R.F.A. and Erler, S. 2013. More than royal food-Major royal jelly protein genes in sexuals and workers of the honeybee *Apis mellifera. Frontiers in Zoology* 10(1): 1–10.

Cherbuliez, T. and Domerego, R. 2003. L'apithérapie, Médecine des Abeilles. Editions Amyris. *Collect Douce Altern* 255.

Clarke, M. and McDonald, P. 2017. Australian royal jelly-market opportunity assessment based on production that uses new labour saving technology. *Rural Industries Research and Development Corporation* 4: 1–10.

Collazo, N., Carpena, M., Nuñez-Estevez, B., Otero, P., Simal-Gandara, J. and Prieto, M.A. 2021. Health promoting properties of bee royal jelly: Food of the queens. *Nutrients* 13(2): 543.

Coutinho, D., Karibasappa, S.N. and Mehta, D.S. 2018. Royal jelly antimicrobial activity against periodontopathic bacteria. *Journal of Interdisciplinary Dentistry* 8(1): 1–18.

Domerego, R. 2001. Ces abeilles qui nous guérissent. Lattès.

El-Hanoun, A.M., Elkomy, A.E., Fares, W.A. and Shahien, E.H. 2014. Impact of royal jelly to improve reproductive performance of male rabbits under hot summer conditions. *World Rabbit Science* 22(3): 241–248.

Elnagar, S.A. 2010. Royal jelly counteracts bucks' "summer infertility". *Animal Reproduction Science* 121(1–2): 174–180.

Eshtiyaghi, M., Deldar, H., Pirsaraei, Z.A. and Shohreh, B. 2016. Royal jelly may improve the metabolism of glucose and redox state of ovine oocytes matured in vitro and embryonic development following in vitro fertilization. *Theriogenology* 86(9): 2210–2221.

Fan, P., Han, B., Feng, M., Fang, Y., Zhang, L., Hu, H., Hao, Y., Qi, Y., Zhang, X. and Li, J. 2016. Functional and proteomic investigations reveal major royal jelly protein 1 associated with anti-hypertension activity in mouse vascular smooth muscle cells. *Scientific Reports* 6(1): 1–13.

Filipič, B., Gradišnik, L., Rihar, K., Šooš, E., Pereyra, A. and Potokar, J. 2015. The influence of royal jelly and human interferon-alpha (HuIFN-αN3) on proliferation, glutathione level and lipid peroxidation in human colorectal adenocarcinoma cells in vitro. *Arhiv za higijenu rada i toksikologiju* 66(4): 269–274.

Fontana, R., Mendes, M.A., De Souza, B.M., Konno, K., César, L.M.M., Malaspina, O. and Palma, M.S. 2004. Jelleines: A family of antimicrobial peptides from the royal jelly of honeybees (Apis mellifera). *Peptides* 25(6): 919–928.

Fratini, F., Cilia, G., Mancini, S. and Felicioli, A. 2016. Royal jelly: An ancient remedy with remarkable antibacterial properties. *Microbiological Research* 192: 130–141.

Fujita, T., Kozuka-Hata, H., Ao-Kondo, H., Kunieda, T., Oyama, M. and Kubo, T. 2013. Proteomic analysis of the royal jelly and characterization of the functions of its derivation glands in the honeybee. *Journal of Proteome Research* 12(1): 404–411.

Galaly, S.R., Abdella, E.M. and, Mohammed, H.M. 2014. Effects of royal jelly on genotoxicity and nephrotoxicity induced by valproic acid in albino mice. *Beni-Suef University Journal of Basic and Applied Sciences* 3(1): 1–15.

Guo, J., Wang, Z., Chen, Y., Cao, J., Tian, W., Ma, B. and Dong, Y. 2021. Active components and biological functions of royal jelly. *Journal of Functional Foods* 82: 104514.

Guo, H., Kouzuma, Y. and Yonekura, M. 2009. Structures and properties of antioxidative peptides derived from royal jelly protein. *Food Chemistry* 113(1): 238–245.

Han, B., Fang, Y., Feng, M., Lu, X., Huo, X., Meng, L., Wu, B. and Li, J. 2014. In-depth phosphoproteomic analysis of royal jelly derived from western and eastern honeybee species. *Journal of Proteome Research* 13(12): 5928–5943.

Han, S.M., Yeo, J.H., Cho, Y.H. and Pak, S.C. 2011. Royal jelly reduces melanin synthesis through down-regulation of tyrosinase expression. *The American Journal of Chinese Medicine* 39(06): 1253–1260.

Hattori, N., Nomoto, H., Fukumitsu, H., Mishima, S. and Furukawa, S. 2007. Royal jelly-induced neurite outgrowth from rat pheochromocytoma PC12 cells requires integrin signal independent of activation of extracellular signal-regulated kinases. *Biomedical Research* 28(3): 139–146.

Hattori, N., Nomoto, H., Fukumitsu, H., Mishima, S. and Furukawa, S. 2010. AMP N1-oxide, a unique compound of royal jelly, induces neurite outgrowth from pc12 vells via signaling by protein kinase A independent of that by mitogen-activated protein kinase. *Evidence-Based Complementary and Alternative Medicine* 7(1): 63–68.

Hu, F.L., Bíliková, K., Casabianca, H., Daniele, G., Salmen Espindola, F., Feng, M., Guan, C., Han, B., Krištof Kraková, T., Li, J.K. and Li, L. 2019. Standard methods for Apis mellifera royal jelly research. *Journal of Apicultural Research* 58(2): 1–68.

Husein, M.Q. and Kridli, R.T. 2002. Reproductive responses following royal jelly treatment administered orally or intramuscularly into progesterone-treated Awassi ewes. *Animal Reproduction Science* 74(1–2): 45–53.

Ibrahim, S.E.L.M. and Kosba, A.A. 2018. Royal jelly supplementation reduces skeletal muscle lipotoxicity and insulin resistance in aged obese rats. *Pathophysiology* 25(4): 307–315.

Ito, S., Nitta, Y., Fukumitsu, H., Soumiya, H., Ikeno, K., Nakamura, T. and Furukawa, S. 2012. Antidepressant-like activity of 10-hydroxy-trans-2-Decenoic acid, a unique unsaturated fatty acid of royal jelly, in stress-inducible depression-like mouse model. *Evidence-Based Complementary and Alternative Medicine* 2012: 139140.

Jenkhetkan, W., Thitiorul, S., Jansom, C. and Ratanavalachai, T. 2017. Molecular and cytogenetic effects of Thai royal jelly: Modulation through c-MYC, h-TERT, NRF2, HO-1, BCL2, BAX and cyclins in human lymphocytes in vitro. *Mutagenesis* 32(5): 525–531.

Kamakura, M., Moriyama, T. and Sakaki, T. 2006. Changes in hepatic gene expression associated with the hypocholesterolaemic activity of royal jelly. *Journal of Pharmacy and Pharmacology* 58(12): 1683–1689.

Kausar, S.H. and More, V.R. 2019. Royal jelly: Organoleptic characteristics and physicochemical properties. *Lipids* 2: 11–57.

Kheyri, H., Cribb, B.W., Reinhard, J., Claudianos, C. and Merritt, D.J. 2012. Novel actin rings within the secretory cells of honeybee royal jelly glands. *Cytoskeleton* 69(12): 1032–1039.

Khoshpey, B., Djazayeri, S., Amiri, F., Malek, M., Hosseini, A.F., Hosseini, S., Shidfar, S. and Shidfar, F. 2016. Effect of royal jelly intake on serum glucose, apolipoprotein AI (ApoA-I), apolipoprotein B (ApoB) and ApoB/ApoA-I ratios in patients with type 2 diabetes: A randomized, double-blind clinical trial study. *Canadian Journal of Diabetes* 40(4): 324–328.

Kim, J. and Lee, J. 2011. Observation and quantification of self-associated adenosine extracted from royal jelly products purchased in USA by HPLC. *Food Chemistry* 126(1): 347–352.

Kim, B.Y., Lee, K.S., Jung, B., Choi, Y.S., Kim, H.K., Yoon, H.J., Gui, Z.Z., Lee, J. and Jin, B.R. 2019. Honeybee (Apis cerana) major royal jelly protein 4 exhibits antimicrobial activity. *Journal of Asia-Pacific Entomology* 22(1): 175–182.

Kimura, Y. 2008. Antitumor and antimetastatic actions of various natural products. *Studies in Natural Products Chemistry* 34: 35–76.

Kohguchi, M., Inoue, S.I., Ushio, S., Iwaki, K., Ikeda, M. and Kurimoto, M. 2007. Effect of royal jelly diet on the testicular function of hamsters. *Food Science and Technology Research* 10(4): 420–423.

Kohno, K., Okamoto, I., Sano, O., Arai, N., Iwaki, K., Ikeda, M. and Kurimoto, M. 2004. Royal jelly inhibits the production of proinflammatory cytokines by activated macrophages. *Bioscience, Biotechnology, and Biochemistry* 68(1): 138–145.

Kunugi, H. and Mohammed Ali, A. 2019. Royal jelly and its components promote healthy aging and longevity: from animal models to humans. *International Journal of Molecular Sciences* 20(19): 4662.

Labro, M.T. 2012. Immunomodulatory effects of antimicrobial agents. Part I: antibacterial and antiviral agents. *Expert Review of Anti-Infective Therapy* 10(3): 319–340.

Li, C., Mannoor, M.K., Toma, N., Taniguchi, T., Inafuku, M., Yamaguchi, K., Sato, Y. and Watanabe, H. 2011. The efficacy of royal jelly in the restoration of alcoholic liver injury in mouse model. *Biomedical Research* 22(1): 1–8.

Libby, P., Ridker, P.M. and Maseri, A. 2002. Inflammation and atherosclerosis. *Circulation* 105(9): 1135–1143.

Liu, Je-R., Yang, Y.C., Shi, Li-S. and Peng, C.C. 2008. Antioxidant properties of royal jelly associated with larval age and time of harvest. *Journal of Agricultural and Food Chemistry* 56(23): 11447–11452.

López-Gutiérrez, N., del Mar Aguilera-Luiz, M., Romero-González, R., Vidal, J.L.M. and Frenich, A.G. 2014. Fast analysis of polyphenols in royal jelly products using automated TurboFlow™-liquid chromatography–Orbitrap high resolution mass spectrometry. *Journal of Chromatography. B* 973: 17–28.

Magdalena, M. 2010. Effect of royal jelly on breast infant with distrophy and maldevelopment. In *20th Apimondia International Beekeeping Congress in Bukarest*, Romania, pp. 583–585.

Maghsoudlou, A., Mahoonak, A.S., Mohebodini, H. and Toldra, F. 2019. Royal jelly: Chemistry, storage and bioactivities. *Journal of Apicultural Science* 63(1): 17–40.

Mahdivand, N., Najafi, G., Nejati, V., Shalizar-Jalali, A. and Rahmani, F. 2019. Royal jelly protects male rats from heat stress-induced reproductive failure. *Andrologia* 51(3): e13213.

Mahmoud, S.A., El-Banby, A.M., El-Shakankiry, H.M., Abdel-Hamid, K.M. and Hassabel-Naby, M.A. 1997. Effect of diet supplementation with honey or royal jelly on preterms. *Wirkung einer zusätzlichen Ernährung mit Honig oder Weiselfuttersaft bei Frühgeborenen Der XXXV. Internationale Bienenzüchterkongress der Apimondia Antwerpen, Apimondia-Verlag*, p. 432.

Majtan, J., Kumar, P., Majtan, T., Walls, A.F. and Klaudiny, J. 2010. Effect of honey and its major royal jelly protein 1 on cytokine and MMP-9 mRNA transcripts in human keratinocytes. *Experimental Dermatology* 19(8): e73–e79.

Malekinejad, H., Ahsan, S., Delkhosh-Kasmaie, F., Cheraghi, H., Rezaei-Golmisheh, A. and Janbaz-Acyabar, H. 2016. Cardioprotective effect of royal jelly on paclitaxel-induced cardio-toxicity in rats. *Iranian Journal of Basic Medical Sciences* 19(2): 221.

Mannoor, M.K. 2008. The efficacy of royal jelly in the restoration of stress-induced disturbance of lymphocytes and granulocytes. *Biomedical Research* 19: 69–77.

Matsui, T., Yukiyoshi, A., Doi, S., Sugimoto, H., Yamada, H. and Matsumoto, K. 2002. Gastrointestinal enzyme production of bioactive peptides from royal jelly protein and their antihypertensive ability in SHR. *The Journal of Nutritional Biochemistry* 13(2): 80–86.

Melliou, E. and Chinou, I. 2014. Chemistry and bioactivities of royal jelly. *Studies in Natural Products Chemistry* 43: 261–290.

Melliou, E. and Chinou, I. 2005. Chemistry and bioactivity of royal jelly from Greece. *Journal of Agricultural and Food Chemistry* 53(23): 8987–8992.

Mishima, S., Suzuki, K.M., Isohama, Y., Kuratsu, N., Araki, Y., Inoue, M. and Miyata, T. 2005. Royal jelly has estrogenic effects in vitro and in vivo. *Journal of Ethnopharmacology* 101(1–3): 215–220.

Miyata, Y. and Sakai, H. 2018. Anti-cancer and protective effects of royal jelly for therapy-induced toxicities in malignancies. *International Journal of Molecular Sciences* 19(10): 3270.

Mokaya, H.O., Njeru, L.K. and Lattorff, H.M.G. 2020. African honeybee royal jelly: Phytochemical contents, free radical scavenging activity, and physicochemical properties. *Food Bioscience* 37: 100733.

Molan, P.C. 1999. Why honey is effective as a medicine. 1. Its use in modern medicine. *Bee World* 80(2): 80–92.

Moradi, A.R., Malekinejad, H., Farrokhi-Ardabili, F. and Bernousi, I. 2013. Royal jelly improves the sperm parameters of ram semen during liquid storage and serves as an antioxidant source. *Small Ruminant Research* 113(2–3): 346–352.

Moselhy, W.A., Fawzy, A.M. and Kamel, A.A. 2013. An evaluation of the potent antimicrobial effects and unsaponifiable matter analysis of the royal jelly. *Life Science Journal* 2(10): 290–296.

Mraz, C. 1995. *Health and the Honeybee*. Lynchburg: Queen City Publications.

Münstedt, K., Bargello, M. and Hauenschild, A. 2009. Royal jelly reduces the serum glucose levels in healthy subjects. *Journal of Medicinal Food* 12(5): 1170–1172.

Nabas, Z., Haddadin, M.S., Haddadin, J. and Nazer, I.K. 2014. Chemical composition of royal jelly and effects of synbiotic with two different locally isolated probiotic strains on antioxidant activities. *Polish Journal of Food and Nutrition Sciences* 64(3): 171–180.

Núnez-González, E., Serra-Majem, L., Fika-Hernándo, M., Fernández-Vallhonrat, B., Bravo-Martínez, J., Martín-Ferrer, J.M., Chas-Barbeito, C. and Bautista-Castano, I. 2011. Determinants of specific food consumption in the Canary Islands (Spain). *Food and Function* 2(10): 627–632.

Nwobi, O., Abitbol, C.L., Chandar, J., Seeherunvong, W. and Zilleruelo, G. 2008. Rituximab therapy for juvenile-onset systemic lupus erythematosus. *Pediatric Nephrology* 23(3): 413–419.

Pajovic, B., Radojevic, N., Dimitrovski, A., Tomovic, S. and Vukovic, M. 2016. The therapeutic potential of royal jelly in benign prostatic hyperplasia. Comparison with contemporary literature. *The Aging Male* 19(3): 192–196.

Pan, Y., Xu, J., Jin, P., Yang, Q., Zhu, K., You, M., Hu, F. and Chen, M. 2019. Royal jelly ameliorates behavioral deficits, cholinergic system deficiency, and autonomic nervous dysfunction in ovariectomized cholesterol-fed rabbits. *Molecules* 24(6): 1149.

Park, H.G., Kim, B.Y., Park, M.J., Deng, Y., Choi, Y.S., Lee, K.S. and Jin, B.R. 2019. Antibacterial activity of major royal jelly proteins of the honeybee (Apis mellifera) royal jelly. *Journal of Asia-Pacific Entomology* 22(3): 737–741.

Pasupuleti, V.R., Sammugam, L., Ramesh, N. and Gan, S.H. 2017. Honey, propolis, and royal jelly: A comprehensive review of their biological actions and health benefits. *Oxidative Medicine and Cellular Longevity* 2017: 1259510.

Pourmoradian, S., Mahdavi, R., Mobasseri, M., Faramarzi, E. and Mobasseri, M. 2014. Effects of royal jelly supplementation on glycemic control and oxidative stress factors in type 2 diabetic female: A randomized clinical trial. *Chinese Journal of Integrative Medicine* 20(5): 347–352.

Premratanachai, P. and Chanchao, C. 2014. Review of the anticancer activities of bee products. *Asian Pacific Journal of Tropical Biomedicine* 4(5): 337–344.

Pyrzanowska, J., Piechal, A., Blecharz-Klin, K., Graikou, K., Widy-Tyszkiewicz, E. and Chinou, I. 2012. Chemical analysis of Greek royal jelly–Its influence of the long-term administration on spatial memory in aged rats. *Planta Medica* 78(11): PI467.

Ramadan, M.F. and Al-Ghamdi, A. 2012. Bioactive compounds and health-promoting properties of royal jelly: A review. *Journal of Functional Foods* 4(1): 39–52.

Ramanathan, A.N.K.G., Nair, A.J. and Sugunan, V.S. 2018. A review on royal jelly proteins and peptides. *Journal of Functional Foods* 44: 255–264.

Sabatini, A.G., Marcazzan, G.L., Caboni, M.F., Bogdanov, S. and de Almeida-Muradian, L.B. 2009. Quality and standardization of royal jelly. *Journal of Apiproduct and Apimedical Science* 1(1): 1–6.

Santos-Buelga, C. and González-Paramás, A.M. 2017. Chemical composition of honey. In *Bee Products-Chemical and Biological Properties*, pp. 43–82. Springer, Cham. Edited by: José M. Alvarez-Suarez.

Saricaoglu, F.T., Cinar, A., Demircan, H. and Oral, R.A. 2019. Rheological and microstructural characterization of royal jelly at different temperatures. *Journal of Food Process Engineering* 42(8): e13285.

Scarselli, R., Donadio, E., Giuffrida, M.G., Fortunato, D., Conti, A., Balestreri, E., Felicioli, R., Pinzauti, M., Sabatini, A.G. and Felicioli, A. 2005. Towards royal jelly proteome. *Proteomics* 5(3): 769–776.

Šedivá, M., Laho, M., Kohútová, L., Mojžišová, A., Majtán, J. and Klaudiny, J. 2018. 10-HDA, a major fatty acid of royal jelly, exhibits pH dependent growth-inhibitory activity against different strains of Paenibacillus larvae. *Molecules* 23(12): 3236.

Seyyedi, F., Rafiean-Kopaei, M. and Miraj, S. 2016. Comparison of the effects of vaginal royal jelly and vaginal estrogen on quality of life, sexual and urinary function in postmenopausal women. *Journal of Clinical and Diagnostic Research: JCDR* 10(5): QC01.

Shi, J.L., Liao, C.H., Wang, Z.L. and Wu, X.B. 2018. Effect of royal jelly on longevity and memory-related traits of Apis mellifera workers. *Journal of Asia-Pacific Entomology* 21(4): 1430–1433.

Siavash, M., Shokri, S., Haghighi, S., Mohammadi, M., Shahtalebi, M.A. and Farajzadehgan, Z. 2011. The efficacy of topical royal jelly on diabetic foot ulcers healing: A case series. *Journal of Research in Medical Sciences: The Official Journal of Isfahan University of Medical Sciences* 16(7): 904.

Strant, M. 2017, October. A treasure in Apiterapy "royal jelly"-myths and realities. II. In *Marmaris Apitherapy and Apicultural Products Symposium*, pp. 14–15.

Strant, M. 2017. Personal experience and practices for royal jelly. *Cluj Napoca.*

Strant, M., Varadi, A. and Aoşan, C. 2016. The APILARNIL and Queen larvae–studies, utilization, doses, clinical cases. *Cluj Napoca.*

Strant, M., Yücel, B., Topal, E., Puscasu, A.M., Margaoan, R. and Varadi, A. 2019. Use of royal jelly as functional food on human and animal health. *Hayvansal Üretim* 60(2): 131–144.

Sugiyama, T., Takahashi, K., Tokoro, S., Gotou, T., Neri, P. and Mori, H. 2012. Inhibitory effect of 10-hydroxy-trans-2-decenoic acid on LPS-induced IL-6 production via reducing IκB-ζ expression. *Innate Immunity* 18(3): 429–437.

Tatsuhiko, T., Naoko, K. and Yuko, H. 2011. Application of the material of honeybee origin. Application of the consmetic material of the honeybee origin. *Fragrance Journal* 30: 17–24.

Thien, F.C.K., Leung, R., Baldo, B.A., Weinbr, J.A., Plomley, R. and Czarny, D. 1996. Asthma and anaphylaxis induced by royal jelly. *Clinical and Experimental Allergy* 26(2): 216–222.

Viel, C. and Doré, J.C. 2003. Histoire et emplois du miel, de l'hydromel et des produits de la ruche. *Revue d'histoire de la pharmacie* 91(337): 7–20.

Virgiliou, C., Kanelis, D., Pina, A., Gika, H., Tananaki, C., Zotou, A. and Theodoridis, G. 2020. A targeted approach for studying the effect of sugar bee feeding on the metabolic profile of royal jelly. *Journal of Chromatography. A* 1616: 460783.

Viuda-Martos, M., Ruiz-Navajas, Y., Fernández-López, J. and Pérez- Álvarez, J.A. 2008. Functional properties of honey, propolis, and royal jelly. *Journal of Food Science* 73(9): R117–R124.

Wongchai, V. and Ratanavalachai, T. 2002. Seasonal variation of chemical composition of royal jelly produced in Thailand. *Science and Technology Asia* 7(2): 1–8.

Xin, X.X., Chen, Y., Chen, D., Xiao, F., Parnell, L.D., Zhao, J., Liu, L., Ordovas, J.M., Lai, C.Q. and Shen, L.R. 2016. Supplementation with major royal-jelly proteins increases lifespan, feeding, and fecundity in Drosophila. *Journal of Agricultural and Food Chemistry* 64(29): 5803–5812.

Yoneshiro, T., Kaede, R., Nagaya, K., Aoyama, J., Saito, M., Okamatsu-Ogura, Y., Kimura, K. and Terao, A. 2018. Royal jelly ameliorates diet-induced obesity and glucose intolerance by promoting brown adipose tissue thermogenesis in mice. *Obesity Research and Clinical Practice* 12(1): 127–137.

Zamami, Y., Takatori, S., Goda, M., Koyama, T., Iwatani, Y., Jin, X., Takai-Doi, S. and Kawasaki, H. 2008. Royal jelly ameliorates insulin resistance in fructose-drinking rats. *Biological and Pharmaceutical Bulletin* 31(11): 2103–2107.

Zhang, X., Yu, Y., Sun, P., Fan, Z., Zhang, W. and Feng, C. 2019. Royal jelly peptides: Potential inhibitors of β-secretase in N2a/APP695swe cells. *Scientific Reports* 9(1): 1–11.

8 Bee Pollen

Meryem Bakour, Hassan Laaroussi, Asmae El Ghouizi, Driss Ousaaid, Soumaya Touzani and Badiaa Lyoussi

8.1 INTRODUCTION

The flower–bee relationship is a mutualistic relationship in which the flowers reproduce sexually, and the bees enjoy a meal of pollen and nectar (Dibos 2010; Melin 2011). The pollen grains that cover the body of the bee during foraging are collected by rubbing movements and then transferred to the hind legs, moistened, and compacted using regurgitation of the crop until they become stickier and are transformed into a ball called bee pollen (Nicolson et al. 2018; Brodschneider and Crailsheim 2010). The need for bee pollen in a beehive is estimated at 25 kg annually (Lapert 2010). Bee pollen is gaining more and more importance thanks to its rich and varied composition of molecules of biological interest. The composition of bee pollen is discussed in the following subsections.

8.1.1 Free Amino Acids

Amino acids are molecules that have two functional groups, a carboxylic acid (-COOH) and an amine (NH_2); when amino acids combine, they form proteins (Azad 2018). Amino acids actively participate in food taste, including sweetness, umami, bitterness, and sourness. In 1908, it was discovered for the first time that monosodium glutamate, a salt form of glutamic acid, was the key taste of some traditional Japanese seasonings (Ikeda 1908).

From a human diet point of view, amino acids are divided into two categories. The first one includes those that are not synthesized in the organism and must therefore be provided by food. These amino acids are histidine, leucine, valine, methionine, lysine, phenylalanine, threonine, isoleucine, tryptophan, and arginine. The second category includes cysteine, glutamic acid, asparagine, hydroxylysine, alanine, tyrosine, glycine, serine, proline, and aspartic acid. These are synthesized in the organism, and it is not necessary to provide them through food (Akram et al. 2011). Bee pollen samples from Saudi Arabia, South Korea, Spain, Italy, Colombia, Turkey, Brazil, China, Taiwan, and Romania revealed considerable amounts of proline, aspartic acid, arginine, glutamic acid, serine, valine, lysine, alanine, and leucine (Taha, Al-Kahtani, and Taha 2019; Gardana et al. 2018; Themelis et al. 2019; Ghosh and Jung 2017; Hsu et al. 2021; Yang et al. 2013; da Silva et al. 2014). The ratio of proline/total amino acids is an indicator of bee pollen freshness. To categorize a sample as fresh, the proline index [(proline/total free amino acids) × 100] must be less than 80% in fresh bee pollen (Ecem Bayram 2021).

It was shown that bee pollen samples from China and Taiwan have high contents of amino acids; for instance, proline content in samples from China ranged from 7.4 to 59.5 mg/g, and in samples from Taiwan, it ranged from 5.5 to 33.5 mg/g. The aspartic acid content in Chinese samples varied between 11.3 and 25.8 mg/g, and in Taiwanese samples between 11.3 and 25.8 mg/g. In addition, glutamic acid was represented by a content of 11.1 to 28.7 mg/g in Chinese samples and 14.6 to 33.3 mg/g in Taiwanese samples. Cystine represents the lowest concentration of amino acids, ranging from 1.5 to 3 mg/g and 0.3 to 6.9 mg/g in Chinese and Taiwanese samples, respectively. It bears mentioning that the analysed bee pollen samples shared the botanical origins of *Brassica napus*, *Nelumbo nucifera*, and *Zea mays* (Hsu et al. 2021; Yang et al. 2013).

DOI: 10.1201/9781003354116-8

From the standpoint of honey bees' nutrition, it was revealed that the minimum requirement for the following amino acids is: alanine (40 milligrams per gram of protein), isoleucine (40 milligrams per gram of protein), leucine (45 milligrams per gram of protein), lysine (30 milligrams per gram of protein), threonine (30 milligrams per gram of protein), phenylalanine (15 milligrams per gram of protein), arginine (30 milligrams per gram of protein), and histidine (15 milligrams per gram of protein) (Groot 1953).

It was revealed that the free amino acids composition of bee pollen changed according to botanical origins and storage conditions (De-Melo and Almeida-Muradian 2017). In addition, it was found that bee pollen samples collected during two different seasons (spring and winter) were richer in the amino acids required for human and honey bee diets (Al-Kahtani et al. 2020).

8.1.2 Phenolic Compounds

Phenolic compounds, or polyphenols, are secondary metabolites synthesized by plants as a response to biotic and abiotic aggression, such as plant injury, insect attack, and environmental change (Bruni and Sacchetti 2009). A phenolic ring with at least one hydroxyl group is the basic structural unit of polyphenols, and depending on the strength and the number of the phenolic ring in the molecule, we can classify the phenolic compounds into phenolic acids, flavonoids, lignans, lignins, coumarins, and condensed and hydrolysable tannins (Abbas et al. 2017). Interestingly, these compounds present an essential source of active pharmaceuticals thanks to their broad therapeutic spectrum, such as antioxidant, antitumor, antiatherosclerotic, anti-inflammatory, cardioprotective, anti-thrombogenic, and antimicrobial effects (Tungmunnithum et al. 2018; Agrawal 2011; Rasouli, Farzaei, and Khodarahmi 2017).

The absorption of bee pollen flavonoids in the intestine depends on several physicochemical criteria, such as lipophilicity, the size of the molecule, the pK_a, and the solubility. After absorption of flavonoids in the small intestine, their hydroxyl groups are subjected to conjugation with glucuronic acid or sulfate due to specific metabolizing enzymes in the intestine, such as phlorizin hydrolase and UDP-glucuronosyl transferase. These metabolites reach tissues and cells, resulting in multiple health benefits (Thakur and Nanda 2020a; Hollman 2004).

Several reports have shown the richness of bee pollen in phenolic compounds. For instance, the analysis of bee pollen collected from Turkey, Romania, Morocco, and Brazil indicated that many phenolic compounds were found in bee pollen samples, like cinnamic acid, gallic acid, vanillic acid, protocatechuic acid, ferulic acid, caffeic acid, chlorogenic acid, *p*-coumaric acid, syringic acid, rosmarinic acid, and ellagic acid. In addition, more than ten flavonoid compounds were identified, such as myricetin, isorhamnetin, rutin, luteolin, quercetin, kaempferol, hesperidin, and catechin. Recent evidence has shown that flavonoid compounds can be considered fingerprints and chemotaxonomic indicators of botanical origin. For instance, it was revealed that hesperetin and naringin detection could be used as markers of citrus species foraging (Ferreres et al. 1993), 8-methoxy-kaempferol detection is a marker of rosemary pollen, and luteolin is a marker of lavender pollen (Castillo, Benavente, and del Río 1992; Cavazza et al. 2013; Anklam 1998). Evidence in a study published by M. Campos et al. (1997) showed that bee pollen collected from *Ulex europaeus* and *Eucalyptus globulus* shared the same flavonoid profile as the foraged plant. In addition, flavonoid compounds such as rutin were used as a marker of bee quality and the correct conditioning processes (Canale et al. 2016).

The study by Castagna et al. (2020) showed a relationship between the drying processes and storage time of bee pollen and polyphenols, flavonoids, and rutin content. For instance, it was revealed that the flavonoid content decreased significantly in willow bee pollen processed during three different technical drying methods. In addition, Campos et al. (2003) indicated that the antioxidant activity of bee pollen decreased significantly during ageing. For instance, it was found that the free radical scavenging ability changed from 40 ± 0.3 µg/mL in *Eucalyptus globulus* fresh bee pollen to 120 ± 0.3 µg/mL after 3 years of storage, and from 82 ± 0.5 µg/mL in *Metrosideros umbellata* fresh bee pollen to 152 ± 0.9 µg/mL in the same sample after 3 years of storage. In contrast to Campos et al. (2003) and Castagna et al. (2020), Zuluaga-Domínguez, Serrato-Bermudez, and Quicazán (2018) argue that bioactive compounds such as flavonoids and phenolics increased significantly by up to 55% due to the bee pollen drying temperature (at 40, 50, and 60°C). The authors explain this rise by the release of these compounds due to the changes in the outer part of pollen grains contained in bee pollen.

8.1.3 Minerals

Bee pollen is a rich source of minerals. The major mineral elements in bee pollen are potassium, phosphorus, calcium, and magnesium. The concentrations of these minerals vary according to the country of origin, pedoclimatic conditions, botanical sources, and seasonal variations (Karabagias et al. 2019). The highest content observed was registered in Oak (*Quercus* sp.) and hardy kiwi (*Actinidia arguta*) bee pollen samples from South Korea. The authors found that the concentration of potassium was 10267 mg/kg in oak bee pollen and 11670 mg/kg in hardy kiwi bee pollen, the concentration of phosphorus was 5174 mg/kg in oak bee pollen and 7666 mg/kg in hardy kiwi bee pollen, and the concentration of calcium was 1947 mg/kg in oak bee pollen and 5041 mg/kg in hardy kiwi bee pollen. The magnesium concentration was 1758 mg/kg in oak bee pollen and 2737 mg/kg in hardy kiwi bee pollen (Ghosh and Jung 2017). In comparison, the lowest concentrations of calcium and potassium were shown in Colombian bee pollen samples, with a range from 1.099 to 2.409 mg/kg, and from 3.058 to 7.616 mg/kg, respectively for calcium and potassium (Fuenmayor et al. 2014). Concerning magnesium, the lowest content (ranging from 68.73 to 793.35 mg/kg) was found in Moroccan bee pollen samples from the following botanical origins: *Coriandrum sativum*; *Ulex europaeus*; *Scorzonera cana*; *Trifolium pratense*; *Ulex europaeus*; *Reseda luteola*; *Spiraea salicifolia*; and *Lamium galeobdolon* (El Ghouizi et al. 2021).

Additionally, several microelements were found in bee pollen, such as Cu, Fe, Mn, Zn, and Se. Oak and hardy kiwi bee pollen samples from South Korea were found to have a high mineral content (Ghosh and Jung 2017). The findings showed that the contents of copper, iron, manganese, zinc, and selenium varied between 28 and 47 mg/kg, 322 and 345 mg/kg, 53 and 204 mg/kg, 96 and 102 mg/kg, and 0 and 9 mg/kg, respectively.

The lowest content of copper, iron, and manganese was registered in Brazilian and Chinese bee pollen samples (Silva et al. 2014; Yang et al. 2013). According to the quality criteria published by Campos et al. (2008), the following mineral content should be found in dried bee pollen: potassium (between 4000 and 20000 mg/kg), magnesium (between 200 and 3000 mg/kg), calcium (between 200 and 3000 mg/kg), phosphorus (between 800 and 6000 mg/kg), iron (between 11 and 170 mg/kg), zinc (between 30 and 250 mg/kg), copper (between 2 and 16 mg/kg), and manganese (between 20 and 110 mg/kg).

Several reports have shown the health benefits of minerals. For example, calcium can lower arterial hypertension and cholesterol levels, preventing osteoporosis (Cormick and Belizán 2019). Zinc and copper play a crucial role in the proper functioning of antioxidant enzymes such as superoxide dismutase (Lewandowski, Kepinska, and Milnerowicz 2019). The functioning of several vital enzymes, such as catalases, aconitases, and peroxidases, depends on iron (Lieu et al. 2001).

8.1.4 Sugar Profile

Approximately 94% of carbohydrates in bee pollen samples are represented by monosaccharides, especially fructose, glucose, and sucrose, with concentrations ranging from 3.44 to 25.65%, from 4.96 to 21.85%, and from 0.13 to 22.04%, respectively. There are also other minor sugars, such as turanose (ranging from 0.05 to 0.10%), maltose (ranging from 0.21 to 5.48%), trehalose (ranging from 0.37 to 1.93%), melizetose (ranging from 0.01 to 2.99%), and raffinose (ranging from 0.01 to 0.60%) (Oroian, Dranca, and Ursachi 2022; Gardana et al. 2018; Martins et al. 2011; Hsu et al. 2021; Prđun et al. 2021; Bakour et al. 2022; Fuenmayor B et al. 2014; Souza et al. 2018). The concentration of some individual sugars varied according to bee species. For instance, mannitol was not present in bee pollen produced by *Apis mellifera* (Qian et al. 2008), while high percentages of this sugar alcohol (between 20.8% and 34.9%) were reported in bee pollen produced by the stingless bee *Melipona subnitida* (da Silva et al.

2014; Belina-Aldemita et al. 2019). This is explained by the conversion of glucose and fructose into mannitol by *M. subnitida* or following the fermentation of bee pollen sugars by microorganisms in the hive (da Silva et al. 2014; Bhatt, Mohan, and Srivastava 2013).

It was revealed that the majority of sugars come from nectar mixed with the pollen grains of flowers during bee pollen formation (Nicolson 2011). In addition, significant differences in sugar composition were found between pollen grains collected by worker bees and bee pollen collected by humans. Bee pollen is richer in reducing sugars and lower in non-reducing sugars than the pollen isolated from flowers (Villanueva et al. 2002).

8.1.5 Fatty Acids

In addition to proteins and sugars, bee pollen contains a variable amount of lipids, 1–10% (w/w), in which fatty acids, sterols, and carotenoids figure as main components (Li et al. 2018). Due to their substantial role in reproduction, development, and bee nutrition, the identification and quantification of fatty acids in bee-collected pollen seem mandatory and are attracting the interest of many researchers worldwide.

Ghosh and Jung (2017) analysed two monofloral South Korean bee pollen, predominately *Quercus* sp. and *Actinidia arguta* pollen, using gas chromatography-flame ionization detection. The authors identified nine fatty acids, among which the highest concentrations were found in linolenic acid (1.4858%), palmitic acid (1.4378%), and linoleic acid (0.879%) for *Quercus* sp. bee pollen, and linolenic acid (1.3613%), palmitic acid (0.633%), and linoleic acid (0.1937%) for *Actinidia arguta* bee pollen. Moreover, Yang et al. (2013) examined the chemical composition of dried bee-collected pollen samples from different ecoregional areas of China and showed the presence of several fatty acids. Accordingly, elaidic acid (5.87 ± 0.38 to 41.24 ± 2.15%), palmitic acid (5.14 ± 0.25 to 35.89 ± 1.26%), linoleic acid (2.66 ± 0.34 to 24.38 ± 0.27%), and linolenic acid (4.11 ± 0.26 to 58.52 ± 4.89%) have been quantified as primary fatty acids. Other minor fatty acids, like stearic acid (0.39 ± 0.03 to 3.90 ± 0.35%), capric acid (0.38 ± 0.01 to 14.98 ± 0.76%), and lauric acid (0.13 ± 0.01 to 2.94 ± 0.32%), have also been quantified.

Dried bee pollen samples collected from Portugal showed the following fatty acid content: capric acid (2.98 to 8.47%), palmitic acid (7.83 to 30.05%), cis-9-octadecanoic acid (4.63 to 20.61%), linoleic acid (5.94 to 24.80%), α -linolenic acid (25.82 to 56.90%), arachidic acid (0.10 to 3.15%), and cis-11-eicosenoic acid (0.22 to 3.35%) (Feás et al. 2012). In a study conducted by Fuenmayor et al. (2014), the investigation of bee pollen samples harvested from different countries showed low levels of all fatty acids analysed (lauric acid, myristic acid, stearic acid, oleic acid, linoleic acid, *o*-linolenic acid, homo-γ -linolenic acid, arachidonic acid, and tricosanoic acid). The same findings were observed in another study published by Gardana et al. (2018) for Spanish, Italian, and Colombian bee pollen samples except for palmitic acid, which ranged from 20.6 ± 0.0% to 26.8 ± 1.5%, linoleic acid, which ranged from 10.6 ± 2.4% to 37.5 ± 4.3%, and α -linolenic acid, which ranged from18.4 ± 1.7% to 44.1 ± 5.4%.

Recently, scientific evidence has suggested that linolenic and linoleic acids improve the honey bee brood development and learning behaviour of adult bees (Arien et al. 2020; Arien, Dag, and Shafir 2018). Besides, and due to their substantial antimicrobial properties, fatty acids, particularly linoleic acid, linolenic acid, lauric acid, and myristic acid, play a critical role in colony hygiene (Dong et al. 2015).

Based on the literature, palmitic, myristic, and stearic acids represent the main saturated fatty acids in bee pollen and bee bread. However, the unsaturated fatty acids are represented by α -linolenic acid, linoleic acid, and oleic acid (Thakur and Nanda 2020a). Generally, from a qualitative standpoint, the lipid composition is almost the same in different bee-collected pollen samples. Still, their quantities and proportions vary and are influenced by several critical factors, such as botanical origin, pedoclimatic characteristics, and harvest season (Mayda et al. 2020). It is also important to note that the fatty acid composition may vary even within the same species and geographical origin (Thakur and Nanda 2020a).

8.1.6 Vitamins

Vitamins are among the organic molecules necessary for the normal functioning of several physiological processes, and they are involved in the regulation of cell differentiation and mineral metabolism. Additionally, some vitamins, such as ascorbic acid (vitamin C) and tocopherol (vitamin E), possess potent antioxidant properties, and others act as precursors of several enzymatic cofactors (Ferreira-Santos et al. 2021). Bee pollen contains many essential elements, including vitamins, which come essentially from airborne pollen grains. Generally, the bee pollen vitamin content varies between 0.02% and 0.7% (Farag and El-Rayes 2016). Till now, few studies have been conducted on the vitamin composition of bee pollen (Souza et al. 2018). A bee pollen sample with a predominance of *Rhododendron ponticum* pollen harvested from Turkey contained various vitamins, including vitamin A, B1, B2, B5, B6, B7, B12 along with vitamin C, E and vitamin K2. The authors showed that vitamins C, B2, E, and B6 are the most abundant, with concentrations of 16240 µg/100 g, 735 µg/100 g, 490 µg/100 g, and 455 µg/100 g, respectively. Moreover, Ecem Bayram (2021) analysed 28 bee-collected pollen samples from Brazil and revealed the presence of a vitamin B complex, namely, B1, B2, B3, and B6. In addition to the influence of botanic origin and pedoclimatic characteristics, the composition of bee pollen in vitamins is highly associated with the storage and conditioning processes.

8.1.7 Organic Acids

Based on literature data, different amounts and varieties of organic acids have already been detected in bee products, particularly bee pollen and bee bread. So far, limited research has targeted and documented the organic acid content of bee pollen. Recently, Çelik et al. (2022) investigated the organic acid profile of royal jelly, bee venom, bee pollen, and bee bread and showed the presence of gluconic acid, fumaric acid, suberic acid, tartaric acid, oxalic acid, formic acid, alpha-ketoglutaric acid, lactic acid, propionic acid, butyric acid, acetic acid, citric acid, succinic acid, and malic acid in bee pollen. Similarly, Oroian, Dranca, and Ursachi (2022) examined the chemical composition of fresh monofloral and multifloral Romanian bee pollen samples. Accordingly, gluconic acid (14.03 to 36.33 g/kg) and lactic acid (0.47 to 1.10 g/kg) have been quantified as abundant organic acids. Moreover, Kalaycıoğlu et al. (2017) studied four monofloral and six polyfloral bee-collected pollen samples from Turkey and stated that gluconic acid, lactic acid, and tartaric acid are the most quantified organic acids in all evaluated samples. Even though these molecules are quantified as minor compounds; their presence in bee-collected pollen provides a broad spectrum of biological activities, including antioxidant and antibacterial properties. Gluconic acid, a major organic acid of bee pollen, has been reported to have strong antibacterial activities against Gram-positive (*Staphylococcus aureus*) and Gram-negative bacteria (*Escherichia coli*). Most importantly, gluconic acid acts synergistically with polymyxin B, displaying high biofilm penetration and potent antibacterial action (Wu et al. 2021). In addition, tartaric acid, another bio-valuable organic acid present in bee pollen, has been reported to show an antibacterial effect against multidrug-resistant opportunistic pathogenic bacteria (Babayan et al. 2020). Thus, bee pollen extracts could be used as promising antimicrobials and may be incorporated as an active ingredient and food preservative agent.

8.2 PHYSICOCHEMICAL PROPERTIES

8.2.1 Water Activity, Moisture Content, and pH

Moisture, water activity, and the acidity (pH) of bee pollen are determining factors of the shelf life of this product. The research has shown a good correlation between these physicochemical parameters (Chirife, Zamora, and Motto 2006; Fernandez-Salguero, Gomez, and Carmona 1993). High moisture content in bee pollen caused later instability and increased the deteriorative reactions. In addition, high water activity and high pH values encourage the growth of microorganisms (bacteria, fungi, and yeasts) (Nogueira et al. 2012).

Moisture or water content is an essential criterion for marketing bee pollen. According to this criterion, bee pollen is classified into two types: fresh and dried. The first is kept in its original form and stored at a low temperature. The second is subjected to different drying processes, such as the conventional method at 100°C using the oven, lyophilization, Karl Fischer's method, and desiccation using sulfuric acid (Melo and Almeida-Muradian 2011; Campos et al. 2010). This parameter was studied in several bee pollen samples harvested from different countries. For instance, in South Korean bee pollen samples, it was 4.4 ± 0.19% in oak bee pollen and 13.8 ± 0.01% in hardy kiwi bee pollen (Ghosh and Jung 2017). Dried Brazilian bee pollen samples showed values ranging between 1% and 9.39% (Martins et al. 2011; de Arruda et al. 2017; Carpes et al. 2009; Souza et al. 2018), while in fresh bee pollen samples analysed by Rebelo, Ferreira, and Carvalho-Zilse (2016), the range of moisture was between 37.12% and 53.39%. In Morocco, only two studies investigated moisture parameters. In the first, the values ranged between 10.7% and 26.8% (El Ghouizi et al. 2021), and in the second, the sample revealed the following values: 3.34 ± 0.02% (Bakour et al. 2022). Following the criteria set for fresh bee pollen, the moisture content should range between 20% and 30%. In contrast, in dried bee pollen, the maximum limit varies depending on the established quality standard in each country. It should be less than 8% in Argentina, less than 4% in Brazil, less than 10% in Bulgaria, and less than 6% in Poland and Switzerland (Campos et al. 2010).

Regarding water activity, the published data showed that in a Moroccan bee pollen sample, the water activity was 0.30 ± 0.01 (Bakour et al. 2022); it ranged from 0.17 to 0.48 in Romanian bee pollen samples (Oroian, Dranca, and Ursachi 2022), and from 0.26 to 0.43 in bee pollen samples from Spain and Portugal (Nogueira et al. 2012). Brazilian bee pollen samples have a water activity ranging between 0.37 and 0.91 (Carpes et al. 2009; Rebelo, Ferreira, and Carvalho-Zilse 2016). It was revealed that dehydrated bee pollen is preserved against spoilage microorganisms because the deteriorative processes are reduced due to the reduction of water activity. Generally, bacteria grow when the water activity exceeds 0.91, while values higher than 0.61 are ideal for the growth of yeasts and moulds (Anjos et al. 2019b).

pH is an easy-to-assess parameter used as a quality indicator, and it is inversely correlated with the acidity of bee pollen (Fuenmayor B et al. 2014). It was shown that generally, this parameter ranged between 3.34 and 5.98. In Morocco, the values vary from 4.19 to 4.82 (El Ghouizi et al. 2021; Bakour et al. 2022), while Brazilian bee pollen samples have pH values varying from 3.34 to 5.9 (Duarte et al. 2018; Souza et al. 2018). In addition, the pH ranged from 3.40 to 5.98 in bee pollen samples harvested from Bulgaria, Hungary, Poland, China, Romania, Spain, Colombia, India, and Portugal (Oroian, Dranca, and Ursachi 2022; Fuenmayor et al. 2014; Nogueira et al. 2012; Gonçalves et al. 2018; Feás et al. 2012). According to Argentinian regulations, the ideal pH values must be between 4 and 6 (Coronel et al. 2004). It was shown that high microbial activity in bee pollen leads to decreased pH in bee pollen and increased acidity (Fuenmayor et al. 2014).

In sum, pH, water activity, and moisture are essential physicochemical parameters that give information about bee pollen's texture, quality, stability, safety, and shelf life (Nogueira et al. 2012). It was revealed that fresh frozen bee pollen has a nutritional value higher than dried bee pollen, while this latter is safer for the human diet than fresh bee pollen (high water content and water activity) due to the lower contamination with moulds, fungi, and bacteria (Anjos et al. 2019b).

8.2.2 Ash

Ash content is a parameter that gives information on the inorganic matter of bee pollen. It is affected by several factors, including the botanical source, pedo-geographical origin, seasonal variations, and climatic conditions (Negrao and Orsi 2018). According to Campos et al. (2008), the average ash content in bee pollen was between 2% and 6%, and it should not exceed 6%, while according to Brazilian and Argentinean specifications, it should not exceed 4%. This limit was sometimes exceeded due to impurities caused by poor pollen cleaning (Thakur and Nanda 2020a). The ash content in South Korean bee pollen samples was 5.3 ± 0.13% in oak bee pollen and 5.2 ± 0.04% in hardy kiwi bee pollen (Ghosh and Jung 2017). In Brazilian bee pollen samples, the ash values ranged between 1.05% and 4.61% (Martins et al. 2011; de Arruda et al. 2017; Rebelo, Ferreira, and Carvalho-Zilse 2016; Souza et al. 2018). The lowest ash content was registered in the samples analysed by Souza et al. (2018) from the following botanical origins: *Machaerium*; *Piptocarpha*; *Eucalyptus*; *Euterpe*; *Lithraea*; *Andira*; *Eupatorium*; *Brassica napus*; Caesalpiniaceae; *Mimosa scabrella*; Poaceae; Melastomataceae; *Ilex*. The ash content in Moroccan monofloral bee pollen samples (*Coriandrum sativum*; *Ulex europaeus*; *Scorzonera cana*; *Trifolium pretense*; *Reseda luteola*; *Spiraea salicifolia*; *Lamium galeobdolon*; and *Citrus aurantium*) ranged between 1.81% and 4.22% (El Ghouizi et al. 2021; Bakour et al. 2022).

8.2.3 Carbohydrates, Proteins, and Lipids

Macronutrients such as carbohydrates, proteins, and lipids are the main energetic ingredients present in high amounts in bee pollen (Bakour et al. 2022). The energetic value of bee pollen was calculated using Eq. (8.1) (Sánchez-Peña et al. 2017):

$$\begin{aligned}\text{Energy}\left(\text{kcal}\right) = 5.65\times\left(g\text{ of protein}\right)+4.1\times\left(gf\text{ of carbohydrate}\right)\\ +9.3\left(g\text{ of protein}\right)\end{aligned} \tag{8.1}$$

It was shown that bee pollen samples collected from Taiwan have high energetic values, ranging from 443.4 to 488.6 kcal/100g (Hsu et al. 2021), followed by samples collected from Portugal, with values ranging from 396.4 to 411.1 kcal/100g (Feás et al. 2012; Nogueira et al. 2012). However, the weakest energy was detected for samples collected from Ukraine, Moldova, Bulgaria, Serbia, and Hungary, with energy values ranging between 164.95 and 301.15 kcal/100 g (Spulber et al. 2018). It is worthy of note that this variability of energy of bee pollen samples could be due to their botanical origin.

Carbohydrates are the most common organic compounds in nature, as they can be found in plants, animal tissues, bacteria, and fungi. They are divided into mono-, di-, oligo-, and polysaccharides based on their chemical structure (Wang et al. 2022).

Carbohydrates play a crucial role in energy metabolism and homeostasis, since they are the primary energy source in the human diet. Aside from their nutritional value, carbohydrates are commonly used in the food industry as sweeteners, thickeners, and gelling agents (Niaz, Khan, and Shah 2020) and in developing new drugs (Jiang et al. 2021).

The carbohydrate content in South Korean bee pollen samples was 60.2 ± 0.92% in oak bee pollen and 60.6 ± 0.17% in hardy kiwi bee pollen (Ghosh and Jung 2017). In addition, the carbohydrate content in Brazilian bee pollen samples ranged from 15.5% to 71.64% in dried bee pollen (de Arruda et al. 2017; Carpes et al. 2009; Souza et al. 2018; Duarte et al. 2018) and from 25.66% to 44.27% in fresh bee pollen samples analysed by Rebelo, Ferreira, and Carvalho-Zilse (2016). It was found that total carbohydrates increased significantly with the drying temperature, as shown by Isik, Ozdemir, and Doymaz (2018), where the carbohydrate content in fresh bee pollen samples was 61.96 ± 0.59%, and when they were dried at 60°C, this content rose to 77.36 ± 0.29%. Bee pollen carbohydrate content may also vary according to its botanical origin (Kostić, Barać, et al. 2015).

Proteins are the most abundant and complex macromolecules in all biological systems, from unicellular and simple eukaryotes to vertebrates and higher mammals; they represent up to 50% of cells (Whitford 2013). Proteins are polymeric structures composed of various sequences of the 20 amino acids found in nature. Physiologically, proteins are involved in every task of cellular life, including cell structure maintenance, hormone and enzyme biosynthesis, biochemical catalysis, the body's formation blocks, cellular response, waste cleanup, and apoptosis initiation (LaPelusa and Kaushik 2021); this is why protein consumption is essential in the human diet. Both animals and plants can offer enough proteins; however, animal proteins are strongly linked to health risks, most notably cardiovascular disease and cancer (Bessada, Barreira, and Oliveira 2019).

Bee pollen is regarded as one of the most protein-rich natural products, containing up to 40% of protein available as bound-form and free amino acids (Campos et al. 2008). The pollen grain is the bee's primary and only source of protein; the protein content is considered a key criterion for increasing the longevity of bees (Quigley et al. 2018). Somerville and Nicol have classified the quality of pollen grains into three categories: category 1 is low quality, which contains less than 20% crude protein, category 2 medium quality, which includes a quantity of crude protein between 20% and 25%, and category 3 excellent quality, with more than 25% crude protein. (Somerville and Nicol 2006). Moroccan studies revealed protein content ranging from 19.86% to 30.32 % in eight monofloral bee pollen samples from the following botanical origins: *Coriandrum sativum*; *Ulex europaeus*; *Scorzonera cana*; *Trifolium pratense*; *Reseda luteola*; *Spiraea salicifolia*; *Lamium galeobdolon* (El Ghouizi et al. 2021), whereas in a monofloral *Citrus* bee pollen sample, the content of proteins was 27.53 ± 1.71% (Bakour et al. 2022). Brazilian bee pollen samples were analysed by Martins et al. (2011), de Arruda et al. (2017), Carpes et al. (2009), Souza et al. (2018), Rebelo, Ferreira, and Carvalho-Zilse (2016), and Duarte et al. (2018). The authors found a range between 4.5% and 37.63%.

The composition of bee pollen changes according to the botanical origin, processing, and storage conditions (Campos et al. 2008). Indeed, Isik, Ozdemir, and Doymaz (2018) observed that bee pollen's protein content can be affected by the drying air temperature, decreasing from 30.36 ± 0.63% in fresh samples to 13.28 ± 0.27% in the same samples dried at 60°C. Bee pollen can potentially be an environmentally benign and high-quality plant protein source for human consumption.

Lipids are the third most common constituent in bee pollen. In Brazilian bee pollen samples, the lipid content ranged from 0.37% to 14.00% (Martins et al. 2011; de Arruda et al. 2017; Carpes et al. 2009; Souza et al. 2018; Rebelo, Ferreira, and Carvalho-Zilse 2016; Duarte et al. 2018). Bee pollen samples from Bosnia and Herzegovina, India, China, Hungary, Poland, Romania, Spain, Bulgaria, Italy, Colombia, and Portugal revealed lipid content ranging from 2.52% to 6.78% (Aličić et al. 2020; Fuenmayor et al. 2014; Nogueira et al. 2012; Gardana et al. 2018; Gonçalves et al. 2018). In a study conducted by Prđun et al. (2021), Croatian bee pollen samples showed values ranging from 4.51% to 19.04%. The finding of Isik, Ozdemir, and Doymaz (2018) revealed that bee pollen lipids increased significantly during the process of drying; the content changed from 5.50 ± 0.04% in fresh bee pollen to 7.22 ± 0.02% in bee pollen dried at 60°C.

8.3 FUNCTIONAL PROPERTIES

8.3.1 Solubility

Typically, pollen solubility is influenced by its specific content of protein and carbohydrates as well as the proportions of soluble (e.g., honey, simple sugars, or nectar, salivary secretions, soluble vitamins, lower–molecular weight proteins [50–25 kDa], etc.) and insoluble (e.g., lipids, sporopollenin, lignin, cellulose, higher–molecular weight proteins [50 kDa], etc.) ingredients and resultant interlinkages (Kostić, Barać, et al. 2015; Thakur and Nanda 2020b). The study by Thakur and Nanda (2020a) showed values of bee pollen solubility between 84.91 and 87.56 g/100 g for coconut and rapeseed pollen, respectively, in accordance with those documented for beetroot orange juice powder (from 76.5 to 98.21 g/100 g).

Due to its significant impact on foaming, emulsification, and gelation, protein solubility has been considered one of the most important functional characteristics to be evaluated when investigating the techno-functional properties of organic products (Kostić, Barać et al. 2015). As documented in a previous investigation, the protein solubility of bee-collected pollens varies between samples. For instance, Kostić, Barać et al. (2015) examined 26 polyfloral samples and showed values ranging between 2.79 and 25.9 g/100 g. Likewise, a pollen sample containing predominantly *Helianthus annuus* L pollen grains presents a protein solubility of 3.64 g/100 g (Kostić et al. 2020).

Besides the amino acid content, the conformation/structure of proteins (e.g., hydrophobicity and surface), and processing conditions, the complex composition of bee-collected pollen directly impacts its solubility through the potential interactions between proteins and other bee pollen components (e.g., salts, lipids, phenolic compounds, and carbohydrates) (Thakur and Nanda 2020a). In addition to proteins, bee-collected pollen contains many simple and complex sugars, which may affect the sweetness and flavour of food products. Compared with protein solubility and due to the presence of nectar or honey (mostly made up of water-soluble sugars), starch and pectin bee pollen have a higher level of carbohydrate solubility. According to the results of Kostić, Barać et al. (2015), carbohydrate solubility varied significantly among samples, from 31.2 to 75.00 g/100 g. Besides the sugars contained in pollen flowers, a high amount of carbohydrate solubility is associated with the honey fraction in the bee pollen loads. Commonly, the sugar composition of bee pollen grains is mostly represented by water-soluble mono- and disaccharides such as glucose, fructose, and sucrose. Moreover, bee pollen also contains soluble sugars such as starch and pectin, which contribute largely to carbohydrate solubility. Naturally occurring polysaccharides in organic functional foods, including bee-collected pollen, exhibit distinct solubility due to their structural forms, monosaccharide composition, charge properties, molecular weight, configuration (α - or β -glycosidic), and linkage patterns, among others (Guo 2017). Also, insoluble polysaccharides like glucan, lignin, callose, sporopollenin, and cellulose are present in bee pollen loads (Krell 1996). Therefore, the ratio between soluble and insoluble carbohydrates and their interaction with other bee pollen active components, i.e., lipids, polyphenols, and proteins, could alter the carbohydrate solubility.

8.3.2 Water and Oil Absorption Capacity

Considering its effect on the quality and sensorial characteristics of food products, water absorption capacity (WAC) is an essential measurement to be carried out by the agri-food industries, especially for products where the humidity affects their physicochemical and microbiological quality. The texture and organoleptic properties of formulated foods are closely correlated with the WAC of their constituents. Based on previous studies, it has been shown that WAC varies between 0.92 and 2.25 g/g for 26 polyfloral bee pollens collected from different ecogeographical localities in Serbia (Kostić, Barać, et al. 2015) and ranged from 0.47 g/g (coriander bee pollen) to 0.72 g/g (coconut bee pollen) harvested from northwestern and northern regions of India, respectively (Thakur and Nanda 2020a). Moreover, Kostić et al. (2020) investigated the techno-functional and nutritional characteristics of monofloral sunflower bee pollen with a pollen predominance of *Helianthus annuus* L and showed a WAC equal to 0.87 g/g. The vast differences seen between samples could be associated with many specific factors, including the content of charged and uncharged sugars as well as their ratio, the structure and conformation of proteins and their specific composition of amino acids, polar mineral elements, and lipids that critically influence the water uptake process (Kinsella and Melachouris 1976; Kostić et al. 2020). It was suggested that the structure of lipid molecules, with polar and charged regions, improves the water-holding capacity of bee pollen (Ramli and Nora'aini Ali 2019). Thus, the proximate composition (lipids, carbohydrates, and proteins) may influence and reaffirm the significant differences in the WAC of the reviewed pollen samples. Additionally, it has been proven that protein–polysaccharide interactions directly impact the texture and stability of most food products (Dickinson 2008).

The oil absorption capacity (OAC) of many organic foods or their dietary components generally depends on their capacity to substantially entrap oil via an advanced capillary-attraction process. It is frequently dependent on insoluble compounds, especially their content of hydrophobic portions (Awuchi, Igwe, and Echeta 2019). Similar to the OAC values found in several functional products and/or their extracts, is the kidney bean (2.2–3.2 g/g), black gram (1.1–3.2 g/g), and quinoa protein isolate (0.94–3.19 g/g) (Wani et al. 2013a; Wani Idrees A, Sogi, and Gill 2013b; Y. Shen, Tang, and Li 2021), bee pollen presented different oil absorption capacities, with values ranging between 1.00 and 3.5 g/g and from 1.31 to 2.13 g/g for Serbian and Indian samples, respectively (Wani et al. 2013a;

Wani Idrees A, Sogi, and Gill 2013b). Besides the hydrophobic amino acids, the oil-holding ability of bee pollen has been attributed to the sporopollenin polymer complex, a highly cross-linked polymer and a major component of pollen exine. This complex polymer entangles and holds the oil within its matrix, consequently increasing bee-collected pollen's oil-holding capacity (Thakur and Nanda 2020a). The higher OAC in organic products acts as a flavour retainer and a healthier mouthfeel enhancer, which might enhance their applications in the food industries for multiple purposes.

8.3.3 Emulsifying Properties

Emulsions play crucial roles in food formulation/preparation and social acceptance. Consequently, foods made of emulsions are common (such as mayonnaise, coffee creamers, some yoghurt types, sauces, infant formulas, ice cream mixes, and several meat products). These products are recommended to be made in a particular way under specific conditions to ensure the emulsion stability and preserve the organoleptic characteristics of the final product over a long time (Dalgleish 1997).

Emulsifying stability (ESI) and emulsifying activity (EAI) are specific indices/indicators and key recommended parameters to evaluate numerous food products and their ingredients for their emulsifying properties. Limited reports have documented the emulsifying activities of bee-collected pollen. In this regard, Thakur and Nanda (2020a) reported that the ES and EA indexes ranged from 25.46 to 45.17 minutes and 21.62 to 80.54 g/m^3, respectively. In the same line, the emulsion stability index (ESI) and emulsion activity index (EAI) of multifloral Serbian samples were determined, and results showed values varying between 19.6 and 49.3 minutes for ESI and from 10.40 to 24.52 g/m^3 for EAI (Kostić et al. 2020). These parameters are influenced directly by many factors, including temperature, pH, mineral content, protein conformation, protein solubility (%), lipophilicity, and liquid droplets (size and distribution) (Avramenko, Low, and Nickerson 2013). Based on the study by Kostić and coworkers, the emulsion stability index was positively linked with protein solubility (%). However, the emulsion activity index was negatively correlated with the protein solubility, which reaffirms the impact of protein solubility proportion on the emulsion stability of food products (Kostić, Barać, et al. 2015). Accordingly, the protein molecular weight strongly influences the emulsifying characteristics of bee pollen. In fact, the authors highlighted that protein fractions with molecular weights lower than 50 kDa enhance the emulsion stability, while fractions with molecular weights higher than 50 kDa were negatively associated with ESI, signifying that smaller protein molecules might adsorb better at the oil/water interface and form stronger interfacial layers than higher molecules. In addition to proteins, and considering the oil/water emulsion interface, which comprises esters, monoglycerides, phospholipids, proteins, and/or their mixtures, other organic emulsifiers such as lipids, especially polar lipids, may play suitable emulsifying roles (Liang et al. 2013). Incorporating bee pollen, alone or combined with other beehive products, in food systems has attracted the attention of many food industries worldwide. In this line, previous studies have been carried out. The authors proved that bee pollen improved rheological, organoleptic, and sensory characteristics, increased the coagulation time and stability during cold storage, and supplemented the formulated yoghurts with nutritional components compared with the control ones (Atallah 2016; Atallah and Morsy 2017; Zlatev et al. 2018).

8.3.4 Foaming Properties

Typically, the texture and structure of many foods, including yoghurt, ice cream, cakes, baked bread, and whipped toppings, are strongly related to foams, systems created by trapping gas molecules beneath a thin layer of liquid or solid (Green et al. 2013). It was revealed that Serbian bee-collected pollen did not possess any foaming capacity (FC) or foam stability (FS). The Indian samples had FC and FS ranging from 6.21% to 8.69% and 17.50% to 20.00%, respectively (Thakur and Nanda 2020a). The authors suggested the incorporation of the investigated bee pollen samples in foam-derived formulations like yoghurt, ice cream, marshmallows, whipped cream, mousses, and cakes, while Kostić, Barać, et al. (2015) recommended the utilization of bee pollen as a promising foam depressant. Currently, researchers are targeting the possible integration of bee pollen in food systems, not only as a nutritional product but also as a functional food and bioactive ingredient to enhance and improve the quality characteristics of formulated products. In this context, Conte et al. (2018) explored the incorporation of polyfloral bee pollen as an organic and active complex matrix in the manufacture of gluten-free bread. The obtained results confirmed that pollen samples improved the techno-functional properties, lowered the staling rate, and enhanced the organoleptic acceptability of enriched bread without any change in the dough's appearance and physical properties. These investigations are a starting point for additional research into bee pollen as a value-added ingredient in other bakery products. In addition to its essential role as a techno-functional ingredient, and due to its high levels of phenolic compounds and other micro-antioxidant nutrients, several studies have recommended the integration of bee pollen at specific concentrations as a natural and alternative antioxidant substitute to prevent microbiological contamination and the lipid peroxidation of refrigerated pork sausages and black pudding (Almeida et al. 2017; Anjos et al. 2019a).

8.4 BIOLOGICAL ACTIVITIES OF BEE POLLEN

8.4.1 *In vitro* Studies

8.4.1.1 Antioxidant Effect

Oxidative stress orchestrates the development and progression of many diseases, including inflammation, diabetes mellitus, atherosclerosis, and so on (Forman and Zhang 2021). Reactive oxygen species (ROS) are the key compounds affecting the stability of proteins, lipids, and DNA that induce chronic inflammation (Pena et al. 2022). Scientific communities focus on natural resources for safety and undeniable efficacy (El-Saadony et al. 2021). The potential to attenuate the harmful effects of free radicals was the keystone of a long series of scientific experiments to confirm the different biological activities of bee pollen (Khalifa et al. 2021). The antioxidants provide bee pollen with excellent protective, nutraceutical, and therapeutic activities. This effect has been evaluated by numerous assays, including total antioxidant activity, free radical scavenging activity, iron chelating effect, radical cation decolourization, and ferric-reducing antioxidant power (Munteanu et Apetrei 2021). For instance, Fatrcová-Šramková et al. (2013) used two complementary tests to examine the antioxidant potential of Slovakian bee pollen of different origins. The authors found that the results differed according to species, with increasing effects in the order *Helianthus annuus* L < *Papaver somniferum* L < *Brassica napus* subsp. *napus* L. In the same context, LeBlanc et al. (2009) found that the mimosa pollen sample collected from the Sonoran desert displayed the highest antioxidant ability, with values of 90.45 ± 0.69% for DPPH and 4.15 ± 0.21 [Fe^{2+}] mM for FRAP.

8.4.1.2 Antidiabetic Effect

The management of glycaemia constitutes an effective strategy to affect the onset, development, and progression of diabetes. The first line of this strategy targets carbohydrate hydrolysing enzymes for managing diabetes complications and decreasing postprandial hyperglycaemia (El Adaouia Taleb et al. 2020). The carbohydrates are broken down by both pancreatic enzymes (α -amylase and α -glucosidase), with their inhibition resulting in reduced glucose uptake into the blood (Barik et al. 2020). Different drugs are used as inhibitors of these enzymes: acarbose, miglitol, and voglibose. However, many pharmacological studies show that some side effects appear after treatment with different enzyme inhibitors (Dong et al. 2022; Fisher 2022; Hisanaga et al. 2022). Combining these inhibitors with bioactive compounds recently showed promising results (Uuh Narvaez et Segura Campos 2022). Bee pollen exhibits a long series of actions because of the complexity of its composition. The study conducted by Khongkarat et al. (2022) on different partitioned extracts of bee pollen showed that the highest α -amylase inhibitory effect was provided by the dichloromethane fraction of bee pollen (*Xyris complanata*) with inhibitory activity of 54.8 ± 2.8% at a dosage of 2 mg/mL.

Additionally, Keskin et Özkök (2020) found that the IC_{50} of bee pollen against pancreatic α-amylase was 4.21 mg/mL. An on-line detection system was recently used for the first time to detect inhibitors of carbohydrate hydrolysing enzymes. The authors found that the chemical structure of bioactive compounds plays a pivotal role in the inhibition kinetics by interfering with enzyme structure and microenvironment polarity (Nan et al. 2022). As research progressed, it became clear that the bioactive compounds interact with enzymes synergistically to exhibit the highest enzyme inhibition effects (Uuh Narvaez and Segura Campos 2022; Perumal et al. 2022).

8.4.1.3 Anticancer Effect

Cancer is a global health issue that affected approximately 9.6 million patients in 2018 and was ranked as the second most common cause of death. Breast, colorectal, and cervical cancers are among the top ten cancers worldwide (Arung et al. 2021). Different chemical substances have been applied to treat cancer, but their adverse effects limit their use (Rouge-Bugat et al. 2015). Natural antioxidants extracted from beehive products have become a popular scientific trend for their pharmacological effects on different ailments, including cancer (Siddiqui et al. 2022). In fact, Arung et al. (2021) tested the effect of ethanolic extracts from different bee pollen samples on different cancer cell lines, including MCF-7, HeLa, and Caco-2. The authors found that *Homotrigona fimbriata* bee pollen exhibited higher cytotoxicity (82.2%, 55.5%, and 61.4% for MCF-7, HeLa, and Caco-2 cells, respectively) than bee pollen of *Tetragonula testaceitarsis*, *T. laeviceps*, *T. sarawakensis*, *H. bakeri*, *H. itama*, and *T. fuscobalteata*. Additionally, silver nanoparticles produced using bee pollen extract were assessed for cytotoxic activity against cancer. Applying these particles at a dose of 500 μL/mL showed less than 25% cell viability in two cancer cell lines (MCF-7 and HepG2). The authors found that the IC_{50} of silver nanoparticles produced using bee pollen extract on the cancer cell lines under study (HepG2 and MCF-7) was 47.6 ± 1.5 μg/mL and 96 ± 3.5 μg/mL, respectively (Al-Yousef et al. 2020). In the same context, Amalia, Diantini, and Subarnas (2020) found that the bee pollen exhibited a good antiproliferative effect in MCF-7 cells, with an IC_{50} value of 18.6 ± 0.03 mg/mL. Additionally, the authors noted many morphological changes in cells and signs of apoptosis. The apoptotic effect of water-soluble bee pollen could be due to the presence of caffeic acid phenethyl ester, and the major protein bands weighed between 50 and 75 kDa (Amalia, Diantini, and Subarnas 2020). In the study conducted by Wu and Lou on human prostate cancer, PC3 cells of different origins showed that the steroid fraction of bee pollen displayed strong cytotoxicity by enhancing apoptosis (Wu and Lou 2007). Flow-cytometric analysis and fluorescence microscopy revealed that treatment with the steroid fraction of bee pollen increased caspase-3 activity and decreased the expression of the anti-apoptotic protein Bcl-2 (Wu and Lou 2007). In the same regard, Kolesarova et al. (2013) confirmed that the expression of apoptosis markers, including Bcl-2, Bax, and caspase-3, was controlled by bee pollen.

8.4.1.4 Antibacterial Effect

The progression of the emergence of bacterial resistance constitutes a serious health concern worldwide. Given the limited number of effective antibiotics and their side effects due to their prolonged use, the search for new safe and effective agents is the main objective of many studies using bee-derived products (Giampieri et al. 2022; Sawicki et al. 2022; Dumitru et al. 2022; Suran 2022; Sonmez et al. 2022; Gülbandilar 2022). Bee pollen is a bee product effective against numerous bacteria and pathogenic yeasts (Çobanoğlu et al. 2022; Ilie et al. 2022). They reported that bee pollen has an antimicrobial effect against different pathogenic microbes, such as *Staphylococcus aureus*, *Escherichia coli*, *Pseudomonas aeruginosa*, *Enterococcus faecalis*, *Candida albicans*, and the clinically isolated strains *C. guillermondii*, *E. cloacae*, *C. famata*, *C. krusei*, *Salmonella* spp., *C. glabrata*, and *C. lusitaniae* (Ilie et al. 2022). On the contrary, Kahraman et al. (2022) reported that no detectable inhibition zones were produced by the bee pollen extract against *S. aureus*, *E. coli*, or *P. aeruginosa*. The antimicrobial effect of bee pollen is ascribed to its complex chemical composition, such as phenolic acids (chlorogenic acid, 4-hydroxybenzoic acid, ferulic acid, cinnamic acid, and gallic acid) and flavonoids (rutin, isorhamnetin, kaempferol, quercetin, chrysin, catechin, and epicatechin) (Ilie et al. 2022). It is worth noting that rutin exhibited an appreciable antibacterial effect against *P. aeruginosa* and methicillin-resistant *Staphylococcus aureus* (MRSA) urinary catheter biofilms, affecting *P. aeruginosa* protease, pyocyanin, rhamnolipid, and elastase production, and interfered with the membrane permeability of pathogenic bacteria (Ivanov et al. 2022). It has been found that rutin controls numerous genes implicated in the development of *P. aeruginosa* biofilms (Ivanov et al. 2022). Novel strategies are urgently required to fight against drug resistance using natural antimicrobial agents such as bioactive compounds. Mounting evidence has shown that the combination of rutin and quercetin with different antibiotics such as gentamicin and ceftriaxone exhibited remarkable antibacterial effects against *E. coli* (multidrug resistance), *Proteus mirabilis* (extensive drug resistance), and *Klebsiella pneumoniae* (pan drug resistance) (Alnour et al. 2022). Different bioactive compounds found in bee pollen in varying amounts are well known for their biological activities resulting from their synergetic interaction.

8.4.2 *In vivo* Studies

8.4.2.1 Antioxidative Stress

Antioxidant therapy is an effective approach to counteracting the adverse effects of oxidative stress. Many agents enhance the production of reactive oxygen species (ROS) that affect membrane lipids, proteins, and DNA (Juan et al. 2021). A study was planned to determine the protective effect of bee pollen against the testicular damage induced by methotrexate in adult rats (Saral et al. 2021). It was found that the administration of bee pollen alleviated the methotrexate-induced changes by decreasing malondialdehyde (MDA) and superoxide dismutase (SOD) activity, while the testosterone level was increased (Saral et al. 2021). In the same context, 100 mg/kg of bee pollen restored all male reproductive impairment in streptozotocin-intoxicated rats. The treatment showed that bee pollen significantly increased testis weight, testosterone, hormone (luteinizing hormone [LH] and follicle stimulating hormone [FSH]) levels, and quality of sperm by attenuating the harmful effects of ROS (Mohamed et al. 2018). Moreover, bee pollen attenuates the cytotoxic changes in the liver and kidney induced by sodium fluoride, restoring normal levels of antioxidant enzymes and serum electrolyte alterations (Al-Daihan and Bhat 2019). Interestingly, the study of the antioxidative mechanism that protects Caco-2 against dextran sulfate sodium DSS-induced cellular damage showed that pretreatment with bee pollen at a 100 µg/mL dose upregulated antioxidant mRNA expression (Li, Liang, et al. 2019). These genes are implicated in cellular defence against the deleterious effects of oxidative stress (Dinkova-Kostova et Talalay 2010; Nguyen, Nioi, and Pickett 2009; Kavian et al. 2018). The protective effect of bee pollen has been ascribed to its complex phenolic composition, which protects DNA from strand breakage (Bridi et al. 2022). Furthermore, bee pollen has been shown to be a promising natural agent against harmful changes in rat hearts induced by isoprenaline through lowering heart structural perturbations, increasing myocardial antioxidant enzyme activities (myocardial glutathione peroxidase, superoxide dismutase, and catalase), and elevating protein expression (Nrf-2 and HO-1) in heart tissues (Shen et al. 2019). Bee pollen contains a broad spectrum of bioactive compounds like phenolic acids and flavonoids, which act synergistically to attenuate the harmful effects of oxidative stress (Unuofin and Lebelo 2020).

8.4.2.2 Antidiabetic Effect

Diabetes mellitus is a metabolic disorder that induces a cluster of critical health conditions. The persistence of hyperglycaemia is highly related to the development of oxidative stress and stimulates the production of ROS (Unuofin and Lebelo 2020). Dietary approaches play a crucial role in the prevention and treatment of diabetes mellitus and in controlling glycaemic homeostasis (Blumenthal et al. 2010; Laaroussi, Bakour, Ousaaid, Aboulghazi, Ferreira-Santos, Genisheva, Teixeira, et al. 2020). The daily feeding of normal and diabetic rats for 16 weeks with two doses of bee pollen attenuates the perturbations induced by 10% D-glucose by preventing the initiation of type 2 diabetes, insulin resistance, dyslipidaemia, and hepatorenal injury (Laaroussi, Bakour, Ousaaid, Aboulghazi, Ferreira-Santos, Genisheva, Teixeira, et al. 2020). In the same context, the administration of bee pollen extract showed a significant decrease in fasting and post-prandial glycaemia in diabetic Wistar rats (Mohamed et al. 2018). The same findings are supported by the results obtained by Yamaguchi et al. (2007). In an attempt to determine the mechanism of action responsible for these beneficial properties of bee pollen, Zhang et al. (2022) suggested that bee pollen could affect the inflammation process in diabetic rats via the phosphatidylinositol 3-kinase/protein kinase B (PI3K/AKT) signalling pathway. Additionally, bee pollen modulates diabetes-induced gut microbiota perturbations, which present a close relationship with diabetes and could be the target of future experimental investigations (Zhang et al. 2022; Li et al. 2020). Other scientific reports showed that bee pollen improved insulin resistance in diabetic mice by controlling liver fat deposition and ameliorating β -cell proliferation, insulin synthesis, and insulin incretion (Yang et al. 2021). These beneficial effects are due to the presence of a wide spectrum of bioactive compounds acting synergistically, including naringin, apigenin, luteolin, resveratrol, and kaempferol, which are closely related to the plant origin (Laaroussi, Bakour, Ousaaid, Aboulghazi, Ferreira-Santos, Genisheva, Antonio Teixeira, et al. 2020; Zhang et al. 2022).

8.4.2.3 Anti-Inflammatory Effect

Bee pollen is a potential nutritious food with multifaceted properties, such as anti-inflammatory effects (Komosinska-Vassev et al. 2015). Since ancient times, bee products have been at the forefront for their effectiveness, safety, and fewer or no side effects. Experimental studies confirm the utility of these products in many animal models to attenuate inflammatory processes (Silva et al. 2012; Habryka, Kruczek, and Drygaś 2016; El-Seedi et al. 2022; El-Aidy et al. 2015; Othman et al. 2021). Bee pollen exhibits excellent ability to control inflammation by reducing the levels of inflammatory mediators (Li, Sun, et al. 2019). Inflammation is a normal physiological response against harmful infection and is considered the first point of the tissue recovery process (Q. Li, Sun, et al. 2019). However, its persistence is closely related to the pathogenesis of many diseases (Chen et al. 2018). Previously, it was proven that bee pollen ameliorates the perturbations induced by intrapouch injection carrageenan. Research experiments have demonstrated that bee pollen can decrease inflammatory exudates, leukocyte accumulation, nitric oxide production, vascular endothelial growth factor (VEGF), and tumour necrosis factor (TNF)-α production (Eteraf-Oskouei et al. 2020). Furthermore, the anti-inflammatory effect of bee pollen is due to its ability to inactivate the mitogen-activated protein kinase (MAPK) and nuclear factor-κ B (NF-κ B) signalling pathways (Li, Sun, et al. 2019). Bee pollen protects Caco-2 cells against inflammatory stimuli, preserving the cell's monolayer integrity and barrier function (Filannino et al. 2021). It was demonstrated that bee pollen extracts suppressed the effects of lipopolysaccharide-stimulated RAW264.7 macrophage cells by controlling the mRNA expression of key inflammation-associated mediator genes such as interleukin (IL)-10, cyclooxygenase-2, IL-6, and inducible nitric oxide synthase (Saisavoey et al. 2021; Li et al. 2017). On the contrary, Maruyama et al. (2010) found that water extract from bee pollen did not affect nitric oxide (NO) production or cyclooxygenase activities. The anti-inflammatory effects of bee pollen are attributed to its peptide composition, including phospholipids and unsaturated fatty acids (Li et al. 2017). Furthermore, bee pollen contains a broad spectrum of bioactive compounds, such as flavonoids, closely related to its therapeutic use in inflammatory disorders (Jannesar et al. 2017).

8.4.2.4 Immunostimulatory Effect of Bee Pollen

The immunostimulatory effect belongs to the long list of pharmacological properties of bee pollen. The search for immune system reinforcement agents constitutes a keystone in dealing with infections such as COVID-19. The immunostimulating effect of bee pollen was examined in Nile tilapias infected with *Aeromonas hydrophila*. The experimental tests revealed that the group treated with bee pollen showed a remarkable elevation in immunological potency through enhanced phagocytosis ability and increasing numbers of neutrophils and monocytes (El-Asely, Abbass, and Austin 2014). A bee pollen–supplemented diet for zebrafish improved the ability of the larvae to resist viral infections and increased neutrophil migration (Di Chiacchio et al. 2021). Bee pollen supplementation as a natural growth promoter in rabbits' diet promotes some immunological parameters such as immunoglobulin levels (IgG, IgM, and IgA) (Zeedan et al. 2017). In the same context, Al-Kahtani, Alaqil, and Abbas (2022) found that bee pollen supports broiler chicks' performance and immunological response by enhancing leukocyte cell viability, T and B lymphocyte proliferation, and immunoglobulin (IgM and IgA) production. Ovalbumin-sensitized mice treated with 200 mg/kg of bee pollen showed inhibited paw oedema, IgE and IgG1 production, and leukocyte migration (Medeiros et al. 2008). Shaldoum et al. (2021) have investigated the influence of bee pollen on doxorubicin-induced bone marrow/spleen immunosuppression in rats. The study proved the efficiency of bee pollen in decreasing the expression of pro-apoptotic genes (p-53, Bax, and caspase-3) and increasing anti-apoptotic gene expression (Bcl-2). The incubation of RAW 264.7 macrophages with bee pollen at a dose of 250 µg/mL stimulated MAPK and NF-κ B signalling pathways, which resulted in the production of proinflammatory factors. It has also been found that bee pollen augments the expression of cell surface molecules of macrophages, including the major histocompatibility complexes MHC class I/II, CD80/86, and a cluster of differentiation (Kim, Cho, and Byun 2018). Three types of heteropolysaccharides from bee pollen have been detected and are well known for their ability to enhance the production of NO, TNF-α , IL-1β , and IL-6 in RAW264.7 cells (Zhou et al. 2020).

8.4.2.5 Diuretic Effect

Bee pollen is described in Chinese medicine as an excellent diuretic agent (Ulbricht et al. 2009). Technological advancement permits the detection of a wide spectrum of active and safe compounds well known for their biological properties against different human ailments (Khalifa et al. 2021). This product's phytochemical complexity positively affects many physiological parameters, including kidney function markers (El Ghouizi et al. 2020). Data reflecting the level of available scientific evidence supporting the diuretic effect of bee pollen are of the order of two publications by El Ghouizi et al. (2020) and Elghouizi et al. (2022). In an attempt to assess the impact of bee pollen on diuresis, El Ghouizi et al. (2020) found that the administration of three different doses of bee pollen for 30 days increased urinary clearance of Na^+, Cl^-, and K^+ without affecting plasma concentrations of creatinine, urea, Na^+, Cl^-, and K^+. The diuretic activity of bee pollen is ascribed to its complex phytochemical composition, including luteolin, rutin, and kaempferol, all well known for their excellent diuretic effect (Boeing et al. 2017; Cechinel-Zanchett et al. 2020).

8.5 HEALTH BENEFITS OF BEE POLLEN

8.5.1 Bee Pollen for Breast Cancer

Cancer is the most challenging health issue threatening humankind throughout the world. Oxidative stress is fully linked to carcinogenesis; it is involved in gene mutation and increasing uncontrolled cell proliferation (Klaunig and Wang 2018). Inflammation is also important in the tumorigenesis process orchestrated by oxidative stress (Zhao et al. 2021). In this respect, the consumption of natural food products, notably vegetables, fruits, and whole grains, represents a promising factor in preventing many cancers. Plant-derived polyphenols have been demonstrated to play a key role in preventing carcinogenesis through their antioxidant and anti-inflammatory activities and triggering molecular signalling pathways (Guo, Kong, and Meydani 2009). Bee pollen is considered one of the richest functional food products due to its high amount of nutritional compounds, minerals, and polyphenols (Khalifa et al. 2021). Through this highly valued composition, bee pollen has shown promising *in vivo* and *in vitro* antitumoral activities against several cancer types (Denisow and Denisow-Pietrzyk 2016; Saraiva et al. 2018).

Regarding clinical applications of bee pollen, a prospective, randomized crossover trial was performed on 46 patients afflicted by postmenopausal disorders after breast cancer surgery and receiving antihormonal therapy. The study investigated whether bee pollen and honey are efficient in mitigating the symptoms of menopause in patients treated with tamoxifen and aromatase inhibitors/inactivators. The results showed that 70.9% of the patients receiving bee pollen and honey mixture revealed an improvement in postmenopausal complaints compared with 68.3% of the patients receiving honey; however, the post-analysis revealed that honey increased serum oestradiol levels, which may be a safety issue in breast cancer patients. This concern does not exist in the bee pollen and honey mixture group, which offers a safe alternative for treating postmenopausal symptoms in women after breast cancer surgery (Münstedt et al. 2015). It has been established that bee pollen positively impacts ovarian function. In a recent study performed on polycystic ovary syndrome, bee pollen was found to exert an oestradiol-like effect because of the presence of phytosterols such as β -sitosterol as well as reducing the expression of anti-apoptotic genes and increasing the expression of proapoptotic genes (Naseri, Khazaei, and Khazaei 2021; Rzepecka-Stojko et al. 2015).

Moreover, a randomized double-blinded clinical study was carried out on 300 women aged between 50 and 60 years old with and without breast cancer using a non-estrogenic pollen extract, PCC-100, to treat vasomotor symptoms during adjuvant hormonal treatment of breast cancer. The results showed a potential effect of the pollen extract on alleviating vasomotor symptoms in adjuvant hormone therapy, similarly to women with postmenopausal symptoms without breast cancer (Steven Weyers 2020).

8.5.2 Bee Pollen for Benign Prostatic Hyperplasia

Benign prostatic hyperplasia (BPH) is one of the most frequent benign tumours, referring to the non-cancerous proliferation of stromal and epithelial prostate cells, afflicting up to 90% of elderly men in their ninth decade. BHP places a significant burden on public health, since it may be related to other health outcomes, such as lower urinary tract symptoms (LUTS), that can negatively impact the life quality of ageing men (Egan 2016; Thorpe and Neal 2003; Speakman et al. 2015). Considering that dietary polyphenols have been detected in the prostate tissue, these phenolic compounds may affect the prostatic cells. In this respect, numerous *in vivo* experimental studies and clinical trials have investigated the beneficial effect of flavonoids and non-flavonoid compounds in treating BPH (Mitsunari et al. 2021; Sharma, Chadha, and Dhingra 2017; X. Yang et al. 2014).

As mentioned earlier, bee pollen is a functional bee product with high phenolic content. Consequently, recent *in vitro* and *in vivo* studies explored the effect of some monofloral bee pollen (*Brassica campestris* and *Nelumbo nucifera*) on different human prostate cancer cell lines and suggested that bee pollen induces cytotoxicity in the experimented cancer cells by triggering apoptosis and down-regulating oxidative stress, inflammation, and proliferation processes (Yang et al. 2014; Wu and Lou 2007; Tuoheti et al. 2020). Despite the pharmacological evidence of the effectiveness of bee pollen in the management of prostatic diseases, clinical trials in this regard remain scarce.

A double-blind, placebo-controlled clinical study was performed to explore the efficacy and safety of 3 months of intake of food supplemented with bee-collected pollen ethyl alcoholic extract on 47 male patients aged between 40 and 83 years suffering from symptomatic BPH.

The study was designed to test the efficiency of 320 mg of bee pollen–supplemented food as a high dose and 160 mg as a low dose. The results showed that the higher dose was more effective in improving BPH symptoms. Bee pollen extract significantly increased the urinary flow and decreased the residual urine volume in the group receiving the high dose. The subjective assessment results showed that both doses of bee pollen extract improved the life quality of the BPH symptomatic participants based on the International Prostate Symptom Score (IPSS) and the Quality of Life score (QOL). The results suggested that bee pollen–supplemented foods could be used to manage BPH symptoms (Murakami et al. 2008).

In contrast, most clinical trials have been designed with food supplements prepared with hand-collected flower pollen extracts. A recent pilot single-centre open-label clinical study was performed to assess the efficacy and safety of a food supplement enriched with pollen extract in managing LUTS in patients with BPH. Twenty-five male patients aged between 43 and 76 years with the clinical and medical diagnosis of LUTS related to BPH were recruited and treated with the combination for 3 months. The results showed that the food supplement improved the clinical symptoms without a statistical difference for age based on the IPSS score (Muraca et al. 2022).

Besides BPH, pollen extract preparations are becoming more beneficial for managing nonbacterial prostatitis. A multicentre, prospective, randomized, double-blind, placebo-controlled phase 3 study was conducted to evaluate the safety and efficacy of a standardized pollen extract (Cernilton) in 111 patients with chronic inflammatory prostatitis/chronic pelvic pain syndrome for 12 weeks. In this study, pollen extract was compared with a placebo, and the pollen group showed significant improvement in the total symptoms of pain based on the National Institutes for Health chronic prostatitis symptom index (NIH-CPSI) and QOL index Wagenlehner et al. (2009). Similar results were found by Iwamura et al. (2015) when pollen extract (Cernilton) was compared with the phytotherapeutic drug (Eviprostat) in a randomized, prospective study. The study was conducted on 100 patients who had chronic prostatitis/chronic pelvic pain syndrome. The results indicated that both treatments improved prostatitis symptoms using the QOL and NIH-CPSI scores. Another pollen extract (Prostat/Poltit) was used to explore its effect in treating chronic prostatitis/chronic pelvic pain syndrome in a randomized, double-blind, placebo-controlled study. In this study, 60 patients aged 20–55 years suffering from lower urinary tract symptoms were recruited and treated with Prostat/Poltit and a placebo for 6 months. Based on the findings of this study, pollen extract was shown to alleviate all prostate inflammation symptoms significantly.

The beneficial effect of pollen extract preparations is related to the active compounds found in bee pollen extract, such as phenolic compounds and phytosterols involved in the inhibition of cytokine liberation or inhibition of aromatase and 5α -reductase, responsible for testosterone conversion (Chambliss 2003; Bin Sayeed et al. 2016; Muraca et al. 2022).

8.6 FOOD SAFETY HAZARDS OF BEE POLLEN

8.6.1 Heavy Metals

Heavy metals are a group of toxic metals that have a density of at least 5 g/cm^3 of density and an atomic number higher than 20. Low concentrations of these elements can pose a threat to human health and the environment (Sharma and Agrawal 2005). Heavy metals are mainly found in the soil and marine ecosystems, while in the atmosphere, they are found in small quantities as particles or vapours. The concentration of these metals has increased in the biosphere due to anthropogenic factors such as industrialization, mining, industrial waste, urbanization, and natural factors (Sharma and Agrawal 2005; Morais, Costa, and Pereira 2012).

Exposure to heavy metals causes many health concerns due to the long-term accumulation of these metals in body organs such as bone, liver, and kidneys, which leads to damage and various disorders including renal dysfunction, Alzheimer's disease, osteoporosis, endocrine disorders, diabetes, and cancer (Alengebawy et al. 2021).

Debate continues about the quality criteria of bee pollen. For instance, the Technical Committee for Food Products of the International Organization for Standardization is working on standardizing bee products, including bee pollen (ISO/TC34/SC19/WG3). In addition, (Campos et al. 2008) have established a set of quality criteria to control and commercialize bee pollen of good quality. The recommended content limits in bee pollen for Cd (cadmium), Pb (lead), As (arsenic), and Mg (mercury) are not more than 0.03 mg/kg, 0.5 mg/kg, 0.5 mg/kg, and 0.01 mg/kg, respectively. In addition, by Polish standards, the recommended limits are 0.05 mg/kg for Cd, 0.20 mg/kg for As, and 0.50 mg/kg for Pb (Roman 2007; Roman et al. 2016). Moreover, it has been recommended to collect bee pollen from apiaries placed in areas at least 3 km from sources of pollution to avoid any possible contamination (Campos et al. 2008).

Published data have revealed that honey bees and bee products could be used as biological indicators of the pollution of a given region and its contamination by heavy metals. According to Perugini et al. (2011), honey bees can accumulate heavy metals (Hg, Pb, Cd, and Cr) in their bodies. The authors found that honey bees from urban areas, particularly those near airports, had the highest levels of lead and cadmium in comparison to those from natural reserves. Furthermore, the highest concentrations were found in samples analysed in July and September. Similarly, Al Naggar et al. (2013) found that the heavy metal content in Egyptian bee pollen, including lead (Pb), cadmium (Cd), copper (Cu), iron (Fe), and zinc (Zn), was higher than the recommended limits. The authors concluded that the contamination of bee pollen samples by these heavy metals significantly depends on the harvested season (higher in spring than summer), botanical origin (higher when honey bees collect pollen grains from clover only than when they collect from many floral sources, such as cotton, maize, eucalyptus, and pumpkins), and the degree of environmental pollution. In a study conducted by Kostić, Pešić, et al. (2015), Serbian bee pollen samples from six different regions were analysed to determine their mineral content. The results revealed that bee pollen samples of specific floral origin (*Helianthus*, *Sophora*, *Fraxinus*, and the Rosaceae family) were characterized by high levels of aluminium, iron, chromium, cobalt, and zinc. In contrast, the obtained data did not show that the soil type affected the mineral contents in the samples studied.

Concerning mercury, it was revealed that this heavy metal contaminated bee pollen from Slovakia, and a significant relationship was found between geographical origin and the content of mercury in bee pollen (Toth et al. 2016). In a study published by Zafeiraki et al. (2022), it was found that several bee pollen samples contained an amount of nickel (Ni), chromium (Cr), and arsenic (As). In addition, the calculation of carcinogenic risk, especially for Ni, showed relatively high-risk levels for adults.

According to Altunatmaz et al. (2017), the analysis of 20 bee pollen samples produced in Turkey, Canada, China, and Santiago revealed that cadmium (Cd) content ranged from 0.006 to 0.181 mg/kg, lead (Pb) content ranged from 0 to 0.479 mg/kg, and arsenic (As) content ranged from 0.006 to 1.035 mg/kg. In light of the criteria set by (Campos et al. 2008), 8 bee pollen samples exceeded the recommended limits for arsenic content, and 22 of 24 bee pollen samples analysed had a cadmium content that was higher than the limit. In comparison to the Polish standard, arsenic and cadmium values were higher than the recommended limits (As: 0.20 mg/kg and Cd: 0.05 mg/kg) in 13 samples for As and 15 samples for Cd.

8.6.2 Pesticide Contamination

Nowadays, acute pesticide poisoning has become a major health problem. The assessment of food contamination by pesticides is carried out compound by compound (Wesseling, Corriols, and Bravo 2005). Pesticides currently play an important role in people's continued food supply worldwide, which is why plant foods contain pesticide residues (Carvalho 2017).

The hive is directly exposed to pesticide contamination through pollen or nectar gathered from melliferous plants treated with these plant protection products or via water collection from an area impregnated with pesticide residues (Thompson 2010). Another source of pesticides in the hive is the conventional acaricides used to fight against the parasitic mite (*Varroa destructor*) (Tihelka 2018).

Recently, there has been an increasing amount of literature on pesticide content in bee pollen. In a review performed by Murcia-Morales et al. (2021), a critical evaluation was established for the literature data published over the 2011–2021 period in 19 countries around the world, from Africa, South America, Asia, and Europe, which analysed the content of pesticides in bee products. The results of these studies revealed the detection of 363 pesticide residues. These detected substances are represented mainly by insecticides, fungicides, metabolites, and herbicides: 41%, 33%, 14%, and

11%, respectively. The most frequently detected pesticides in these studies are chlorpyrifos (insecticide), coumaphos (acaricide), imidacloprid (insecticide), dimethoate (insecticide), tau-fluvalinate (acaricide), tebuconazole (fungicide), thiacloprid (insecticide), azoxystrobin (fungicide), carbendazim (fungicide), boscalid (fungicide), acetamiprid (insecticide), cypermethrin (insecticide), and thiamethoxam (insecticide). It was found that chlorpyrifos, carbendazim, and tebuconazole were most often detected in the 4114 bee pollen samples analysed.

According to Kasiotis et al. (2014), the analysis of bee pollen samples revealed the presence of 14 pesticides, including clothianidin, chlorpyrifos ethyl, imidacloprid, thiamethoxam, indoxacarb, thiacloprid, penconazole, and carbendazim, with a concentration range between 0.3 and 81.5 ng/g.

Daily bee pollen samples collected over 102 days from southern Germany were shown to contain 29 pesticides, including 12 insecticides, such as tebufenozide, spirodiclofen, fenoxycarb, chlorantraniliprole, diflubenzuron, and picaridin, 15 fungicides, including tebuconazole, fluopyram, trifloxystrobin, and myclobutanil, and only 2 herbicides (pendimethalin, pendimethalin) (Kasiotis et al. 2014). Additionally, in a study conducted by Fulton et al. (2019), the gas chromatography/mass spectrometry (GC-MS) analysis of 160 bee pollen samples collected from 49 apiaries in Virginia revealed the detection of 7 pesticides in 8% of the samples. The median concentrations of the pesticides detected are as follows: chlorothalonil (6.70 ng/g), chlorpyrifos (4.27 ng/g), bifenthrin (4.25 ng/g), cyhalothrin (28.9 ng/g), fluvalinate (6.06 ng/g), coralox (12.06 ng/g), and coumaphos (26.5 ng/g). It was shown that the likelihood of detecting these pesticide types was higher in pasture, agriculture, and urban landscapes than in forests.

It was concluded that bee pollen, which beekeepers collect at the entrance of a beehive for commercial purposes, was more contaminated by neonicotinoids, pyrethroids, and fungicides than bee bread (bee pollen stored in the hive). The prevalence of this contamination is more important during the spring season, which could be due to the degradation of some kinds of pesticides in bee bread (Tong et al. 2018). In addition, it was found that fluvalinate and coumaphos, acaricides used as mite treatments, were found at higher levels in bee bread than in bee pollen, and this was because the latter was not in contact with other acaricide-contaminated hive products such as wax and honey (Tong et al. 2018; Murcia-Morales et al. 2021).

Some pesticides found in bee pollen, such as thiacloprid, chlorpyrifos, and carbendazim, were recently banned and not renewed by the European Union (European Commission 2011, 2020). Unfortunately, in many countries, these pesticides are still used on a large scale. Due to the non-biodegradation and persistence of these chemicals, they accumulate in food, water, soil, and air. Several studies thus far have linked the widespread occurrence of pesticides to chronic human diseases such as reproductive disorders, neurodegenerative diseases, diabetes, cardiovascular diseases, chronic respiratory diseases, and chronic nephropathies (Mostafalou and Abdollahi 2013).

8.6.3 Bacteria, Mould, and Yeast

Analysing microbiological quality is an important step to ensure bee pollen's safety and good quality. Several parameters affect the microbial quality of bee pollen, such as hygiene procedures, the method of conservation, and geographical and botanical origins (Beev et al. 2018). According to the criteria set by (Campos et al. 2008) for microbiological quality requirements, bee pollen must be free from any contamination with *Salmonella*, *Staphylococcus aureus*, Enterobacteriaceae (max. 100/g), and *Escherichia coli*; the total aerobic plate count should be lower than 100000 per gram of bee pollen, while mould and yeast should be less than 50000 per gram of bee pollen.

Several researchers worldwide have investigated the microbiological quality criteria of bee pollen. For instance, Feás et al. (2012) examined 22 organic bee pollen samples collected from Portugal. They found that except for two samples, they were free from contamination by *E. coli*, sulfite-reducing clostridia, *Salmonella*, *S. aureus*, and faecal coliforms. In addition, most of the samples were contaminated with a range from <10 to 2450 for aerobic mesophiles and from <10 to 3560 for moulds and yeasts. Portuguese bee pollen was also analysed by Estevinho et al. (2012). The samples analysed were free from faecal coliforms, *E. coli*, sulfite-reducing clostridia, *Salmonella*, and *S. aureus*, whereas for the total number of aerobic mesophiles and moulds and yeasts, all examined bee pollen samples were positive in a range between <10 and 7.6 × 10^2 CFU/g and between <10 and 2.8 × 10^2 CFU/g, respectively. This study found a positive correlation between the growth of moulds and yeasts, the total number of aerobic mesophiles, and the water activity and pH, especially for values of pH higher than 4.96.

In a study conducted by Anjos et al. (2019a), the microbiological analysis of dried and frozen bee pollen samples showed that mesophilic bacteria ranged from 1.09 × 10^2 to 2.76 × 10^2 CFU/g in frozen bee pollen and between 1.01 × 10^3 and 8.97 × 10^1 in dried bee pollen; moulds and yeasts ranged between 1.51 × 10^3 and 7.57 × 10^2 CFU/g; *Bacillus cereus* enumeration revealed a content ranging between <10 and 4.63 × 10^2 CFU/g in frozen bee pollen, and between <10 and 6.26 × 10^2 CFU/g in dried bee pollen. It was concluded that the storage method affects the microbiological quality of bee pollen. Based on these results, the authors proposed that it is better to consume bee pollen frozen at −20°C than bee pollen dried in an electric oven. According to the quality criteria described by Campos et al. (2008), fresh bee pollen must be frozen for optimum quality to avoid its contamination by fungi and bacteria. The water content of dried bee pollen must not exceed 6%. In addition, it was recommended to pack bee pollen using food-grade materials.

Several attempts have been made to find a solution to bacteria and fungi contamination in bee pollen, such as using ozone treatments, irradiation, and fumigation using chemicals such as ethylene oxide (Yook, Lim, and Byun 1998; Hosny, Sabbah, and El-Bazza 2018). These suggestions are unsatisfactory because they may cause toxic residues.

8.6.4 Mycotoxins

One of the most important challenges of food safety is mycotoxin contamination. These substances are defined as secondary metabolite toxins produced by different species of microscopic filamentous fungi. They appear mainly in foods of plant origin and frequently occur in areas known for high humidity and temperature (Reverberi et al. 2010).

Mycotoxins are toxic for humans and animals, and exposure to these substances can occur via oral, inhalation, or dermal routes (Peraica et al. 1999). The most famous fungal genera that produce mycotoxins are *Fusarium*, *Penicillium*, *Aspergillus*, and *Alternaria* (Berthiller et al. 2013). Food and feed have been contaminated with various types of mycotoxins, including aflatoxins, deoxynivalenol, ochratoxins, patulin, fumonisins, zearalenone, T-2 toxin, and deoxynivalenol (Pitt 2000; Zheng, Richard, and Binder 2006).

Mycotoxins can cause different types of toxicity. Acute toxicity is characterized by liver and kidney impairment or interference of mycotoxins with protein synthesis, causing skin necrosis and in extreme cases, immune system deficiency. Acute mycotoxin toxicity also affects the nervous system and causes brain damage. The second type of mycotoxin poisoning is chronic toxicity induced by long-term ingestion of low doses of mycotoxins, which can affect the replication of genetic material and cause mutagenic and teratogenic damage (Pitt 2000; Fung and Clark 2004).

Bee pollen is a hive product that is easily contaminated with mycotoxins due to hygiene measures, inadequate treatment and storage methods, environmental conditions, and some physicochemical parameters such as moisture content, water activity, and pH value (Végh et al. 2021). The maximum tolerable levels in foodstuffs have been set by the Commission (2006) for many mycotoxins, such as patulin, aflatoxins, ochratoxin A, fumonisins, deoxynivalenol, zearalenone, and deoxynivalenol. Unfortunately, no such regulation has been established for bee products. To avoid mycotoxin contamination, Beev et al. (2018) recommended reducing the water activity of bee pollen to 0.60 and collecting it every 2 days.

In a study published by Kostić et al. (2017), 26 Serbian bee pollen samples were analysed for aflatoxin B1 content. The results revealed that all samples were contaminated with this aflatoxin at levels between 3.15 and 17.32 μg/kg. The authors suggested that analysis of bee pollen mycotoxins is necessary to ensure the good quality of bee pollen, because they found that aflatoxin B1 was detected even though some samples were not contaminated with fungi. In addition, in a study conducted by González et al. (2005), it was found that Spanish and Argentinian bee pollen samples were contaminated with ochratoxin A produced by *Aspergillus ochraceus*, *Aspergillus niger* aggregate, *Aspergillus carbonarius*, and *Penicillium verrucosum* with a range of 4.2 to 4.5 μg/ml, 1.4–1.8 μg/ml, 2.6–3.1 μg/ml, and 1.1–3.9 μg/ml, respectively. The bee pollen samples were also contaminated with aflatoxin B1 in a range of 3.5–9.3 μg/ml, and with aflatoxin B2 in a 1.2–3.1 μg/ml range.

Similarly, Kačániová et al. (2011) examined 45 Slovakian bee pollen samples for mycotoxin content. They found that the range of mycotoxins in frozen poppy pollen was as follows: total aflatoxin (15.80–16.20 μg/kg), deoxynivalenol (183.10–183.70 μg/kg), zearalenone (361.30–361.90 μg/kg), total ochratoxins (6.12–6.41 μg/kg), and T-2 toxin (299.00–299.60 μg/kg). In contrast, the content of mycotoxins in dried poppy pollen was 13.60–13.90 for total aflatoxin, 273.60–273.90 for deoxynivalenol, 351.50–351.70 for zearalenone, 10.89–10.98 for total ochratoxins, and 128.90-129.60 for T-2 toxin. Additionally, the evidence presented by the authors showed that UV treatment of poppy pollen for 1 hour significantly lowered the mycotoxin contamination compared with the dried and fresh bee pollen but did not fully protect the hygienic quality of the samples treated.

It is also worth noting that several preclinical studies have highlighted the toxicological effect of mycotoxins. For instance, Theumer et al. (2003) noted that subchronic administration (90 days) of aflatoxin B1 or a mixture of aflatoxin B1 and fumonisin B1 in rats caused a decrease in body weight, an increase in the mitosis of spleen mononuclear cells, and perturbation of immunobiological parameters. In addition, the haematotoxicity of mycotoxins, including deoxynivalenol, zearalenone, ochratoxin-B, neosolaniol, T-2, and nivalenol, was evaluated by Chattopadhyay et al. (2013) in Sprague-Dawley rats. The results revealed that 24 hours after the injection of 0.5 ml of the previously mentioned mycotoxins, a significant decrease in red and white blood cells was observed. This haematotoxicity was maximal with T-2 and minimal with deoxynivalenol.

A recent systematic review published on human epidemiological studies found a positive link between the consumption of aflatoxin-contaminated foods and the risk of liver cancer, and between the consumption of zearalenone-contaminated foods and breast cancer (Claeys et al. 2020).

8.7 CONCLUSION

In conclusion, bee pollen is a hive product rich in free amino acids, phenolic compounds, minerals, sugars, fatty acids, vitamins, and organic acids. A variety of factors influence this composition, including botanical origin, harvesting, processing, and packaging conditions. As a functional food, bee pollen can be used to protect against oxidative stress, diabetes, menopausal disorders, inflammation, cancer, and infectious diseases. However, to take advantage of the benefits of bee pollen, it is preferable to choose samples of high quality, free from any impurities such as heavy metals, pesticide contamination, mycotoxins, bacteria, moulds, and yeasts.

REFERENCES

Abbas, Munawar, Farhan Saeed, Faqir Muhammad Anjum, Muhammad Afzaal, Tabussam Tufail, Muhammad Shakeel Bashir, Adnan Ishtiaq, Shahzad Hussain, et Hafiz Ansar Rasul Suleria. 2017. "Natural polyphenols: An overview". *International Journal of Food Properties* 20(8). Taylor & Francis: 1689–99.

Agrawal, AD. 2011. "Pharmacological activities of flavonoids: A review". *International Journal of Pharmaceutical Sciences and Nanotechnology* 4(2): 1394–98.

Akram, M, HM Asif, M Uzair, Naveed Akhtar, Asadullah Madni, SM Ali Shah, ZU Hasan, et Asmat Ullah. 2011. "Amino acids: A review article". *Journal of Medicinal Plants Research* 5(17): 3997–4000.

Al-Daihan, Sooad, et Ramesa Shafi Bhat. 2019. "Protective effect of bee pollen against sodium fluoride induced hepatonephrotoxicity and serum electrolyte changes in rats". *Fluoride* 52(1). International Society for Fluoride Research, Inc.: 9–17.

Alengebawy, Ahmed, Sara Taha Abdelkhalek, Sundas Rana Qureshi, et Man-Qun Wang. 2021. "Heavy metals and pesticides toxicity in agricultural soil and plants: Ecological risks and human health implications". *Toxics* 9(3). MDPI: 42.

Aličić, Damir, Ivana Flanjak, Đurđica Ačkar, Midhat Jašić, Jurislav Babić, et Drago Šubarić. 2020. "Physicochemical properties and antioxidant capacity of bee pollen collected in Tuzla Canton (B&H)". *Journal of Central European Agriculture* 21(1). Agronomski fakultet Zagreb: 42–50.

Al-Kahtani, Saad N, Abdulaziz A Alaqil, et Ahmed O Abbas. 2022. "Modulation of antioxidant defense, immune response, and growth performance by inclusion of propolis and bee pollen into broiler diets". *Animals* 12(13). MDPI: 1658.

Al-Kahtani, Saad N, El-Kazafy Taha, Khalid Ali Khan, Mohammad Javed Ansari, Soha A Farag, Dalia MB Shawer, et El-Said Mohamed Elnabawy. 2020. "Effect of harvest season on the nutritional value of bee pollen protein". *PLOS One* 15(12). Public Library of Science San Francisco, CA: e0241393.

Almeida, Jacqueline de Florio, Daiane Pereira, Mirelli Bianchin, Cristiane de Moura, Manuel Vicente Plata-Oviedo, Charles Windson Isidoro Haminiuk, et Ingridy Simone Ribeiro. 2017. "Lyophilized bee pollen extract: A natural antioxidant source to prevent Lipid oxidation in refrigerated sausages". *Food Science and Technology*, 7, 299-305.

Al Naggar, Yahya A, El-Saied A Naiem, Amal I Seif, et Mohamed H Mona. 2013. "Honey bees and their products as a bio-in-dicator of environmental pollution with heavy metals". *Mellifera* 13: 1–20.

Alnour, Tarig MS, Eltayib H Ahmed-Abakur, Elmutuz H Elssaig, Faisel M Abuduhier, et Mohammad Fahad Ullah. 2022. "Antimicrobial synergistic effects of dietary flavonoids rutin and quercetin in combination with antibiotics gentamicin and ceftriaxone against E. coli (MDR) and P. mirabilis (XDR) strains isolated from human infections: Implications for food–medicine interactions". *Italian Journal of Food Science* 34(2): 34–42.

Altunatmaz, Sema Sandikci, Duygu Tarhan, Filiz Aksu, Umıt Bora Barutçu, et Mehmet Erman Or. 2017. "Mineral element and heavy metal (cadmium, lead and arsenic) levels of bee pollen in Turkey". *Food Science and Technology* 37. SciELO: 136–41.

Al-Yousef, Hanan M, Musarat Amina, Ali S Alqahtani, Mohammed S Alqahtani, Abdul Malik, Mohammad Rafe Hatshan, Mohammed Rafiq H Siddiqui, Mujeeb Khan, Mohammed Rafi Shaik, et Mohammad Shamsul Ola 2020. "Pollen bee aqueous extract-based synthesis of silver nanoparticles and evaluation of their anti-cancer and anti-bacterial activities". *Processes* 8(5). MDPI: 524.

Amalia, Eri, Ajeng Diantini, et Anas Subarnas. 2020. "Watersoluble propolis and bee pollen of Trigona spp. from south Sulawesi Indonesia induce apoptosis in the human breast cancer MCF7 cell line". *Oncology Letters* 20(5). Spandidos Publications: 1–.

Anjos, Ofélia, Rodrigo Fernandes, Susana M Cardoso, Teresa Delgado, Nelson Farinha, Vanessa Paula, Letícia M Estevinho, et Solange T Carpes. 2019a. "Bee pollen as a natural antioxidant source to prevent lipid oxidation in black pudding". *LWT* 111. Elsevier: 869–75.

Anjos, Ofelia, Vanessa B Paula, Teresa Delgado, et Leticia M Estevinho. 2019b. "Influence of the storage conditions on the quality of bee pollen" *Zemdirbyste-Agriculture*, 106(1): 87–94.

Anklam, Elke. 1998. "A review of the analytical methods to determine the geographical and botanical origin of honey". *Food Chemistry* 63(4): 549–62. doi:10.1016/S0308-8146(98)00057-0.

Arien, Yael, Arnon Dag, et Sharoni Shafir. 2018. "Omega-6: 3 ratio more than absolute lipid level in diet affects associative learning in honey bees". *Frontiers in Psychology* 9. Frontiers Media SA: 1001.

Arien, Yael, Arnon Dag, Shiran Yona, Zipora Tietel, Taly Lapidot Cohen, et Sharoni Shafir. 2020. "Effect of diet lipids and omega-6: 3 ratio on honey bee brood development, adult survival and body composition". *Journal of Insect Physiology* 124. Elsevier: 104074.

Arruda, Vanilda Aparecida Soares de, Alexandre Vieria dos Santos, Davi Figueiredo Sampaio, Elias da Silva Araújo, André Luís de Castro Peixoto, Maria Leticia Fernandes Estevinho, et Ligia Bicudo de Almeida-Muradian. 2017. "Microbiological quality and physicochemical characterization of Brazilian bee pollen". *Journal of Apicultural Research* 56(3). Taylor & Francis: 231–38.

Arung, Enos Tangke, Rico Ramadhan, Binti Khairunnisa, Yhiya Amen, Masako Matsumoto, Maki Nagata, Irawan Wijaya Kusuma, Swandari Paramita, Nataniel Tandirogang, et Naomichi Takemoto 2021. "Cytotoxicity effect of honey, bee pollen, and propolis from seven stingless bees in some cancer cell lines". *Saudi Journal of Biological Sciences* 28(12). Elsevier: 7182–89.

Atallah, AA. 2016. "The production of bio-yoghurt with probiotic bacteria, royal jelly and bee pollen grains". *Journal of Nutrition and Food Sciences* 6: 510.

Atallah, AA, et KM Morsy. 2017. "Effect of incorporating royal jelly and bee pollen grains on texture and microstructure profile of probiotic yoghurt". *Journal of Food Processing and Technology* 8(9). OMICS International: 1–4.

Avramenko, Nicole A, Nicholas H Low, et Michael T Nickerson. 2013. "The effects of limited enzymatic hydrolysis on the physicochemical and emulsifying properties of a lentil protein isolate". *Food Research International* 51(1). Elsevier: 162–69.

Awuchi, Chiniza-Godswill, Victory Somtochukwu Igwe, et Chinelo Kate Echeta. 2019. "The functional properties of foods and flours". *International Journal of Advanced Academic Research* 5(11): 139–60.

Azad, Sabina. 2018. "Amino acids: Its types and uses". *International Journal of Clinical and Diagnostic Pathology* 1(1): 13–6.

Babayan, Bella G, Aram R Mikaelyan, Nona L Asatryan, Samvel A Bagdasaryan, et Marina A Melkumyan. 2020. "The effect of tartaric acid new derivatives against the multidrug resistant opportunistic pathogenic soil strains of pseudomonas fluorescens". *Test Engineering and Management* 83(5–6). Mattingley Publishing Co., Inc.: 8516–21.

Bakour, Meryem, Hassan Laaroussi, Pedro Ferreira-Santos, Zlatina Genisheva, Driss Ousaaid, José Antonio Teixeira, et Badiaa Lyoussi. 2022. "Exploring the palynological, chemical, and bioactive properties of non-studied bee pollen and honey from morocco". *Molecules* 27(18). Multidisciplinary Digital Publishing Institute: 5777.

Barik, Sisir Kumar, Wendy R Russell, Kim M Moar, Morven Cruickshank, Lorraine Scobbie, Gary Duncan, et Nigel Hoggard. 2020. "The anthocyanins in black currants regulate postprandial hyperglycaemia primarily by inhibiting α-glucosidase while other phenolics modulate salivary α-amylase, glucose uptake and sugar transporters". *The Journal of Nutritional Biochemistry* 78. Elsevier: 108325.

Beev, G, D Stratev, I Vashin, D Pavlov, et D Dinkov. 2018. "Quality assessment of bee pollen: A cross sectional survey in Bulgaria". *Journal of Food Quality and Hazards Control* 5(1): 11–6.

Belina-Aldemita, Ma Desiree, Christine Opper, Matthias Schreiner, et Stefano D'Amico. 2019. "Nutritional composition of pot-pollen produced by stingless bees (Tetragonula biroi Friese) from the Philippines". *Journal of Food Composition and Analysis* 82. Elsevier: 103215.

Berthiller, Franz, Colin Crews, Chiara Dall'Asta, Sarah De Saeger, Geert Haesaert, Petr Karlovsky, Isabelle P Oswald, Walburga Seefelder, Gerrit Speijers, et Joerg Stroka. 2013. "Masked mycotoxins: A review". *Molecular Nutrition and Food Research* 57(1). Wiley Online Library: 165–86.

Bessada, Sílvia MF, João CM Barreira, M Beatriz, and PP Oliveira. 2019. "Pulses and food security: Dietary protein, digestibility, bioactive and functional properties". *Trends in Food Science and Technology* 93. Elsevier: 53–68.

Bhatt, Sheelendra Mangal, Anand Mohan, et Suresh Kumar Srivastava. 2013. "Challenges in enzymatic route of mannitol production". *International Scholarly Research Notices*, doi: 10.5402/2013/91418.

Blumenthal, James A, Michael A Babyak, Andrew Sherwood, Linda Craighead, Pao-Hwa Lin, Julie Johnson, Lana L Watkins, Jenny T Wang, Cynthia Kuhn, et Mark Feinglos 2010. "Effects of the dietary approaches to stop hypertension diet alone and in combination with exercise and caloric restriction on insulin sensitivity and lipids". *Hypertension* 55(5). American Heart Association: 1199–205.

Boeing, Thaise, Luisa Mota da Silva, Mariha Mariott, Sergio Faloni de Andrade, et Priscila de Souza. 2017. "Diuretic and natriuretic effect of luteolin in normotensive and hypertensive rats: Role of muscarinic acetylcholine receptors". *Pharmacological Reports : PR* 69(6): 1121–24. doi:10.1016/j.pharep.2017.05.010.

Bridi, Raquel, Javier Echeverría, Arturo Larena, Paula Nuñez Pizarro, Elias Atala, Adriano Costa De Camargo, Won Young Oh, Fereidoon Shahidi, Olga Garcia, et Kong S Ah-Hen. 2022. "Honeybee pollen from Southern Chile: Phenolic profile, antioxidant capacity, bioaccessibility, and inhibition of DNA damage". *Frontiers in Pharmacology* 13. Frontiers Media SA: 775219.

Brodschneider, Robert, et Karl Crailsheim. 2010. "Nutrition and health in honey bees". *Apidologie* 41(3): 278–94. doi:10.1051/apido/2010012.

Bruni, Renato, et Gianni Sacchetti. 2009. "Factors affecting polyphenol biosynthesis in wild and field grown St. John's Wort (Hypericum perforatum L. Hypericaceae/Guttiferae)". *Molecules* 14(2). Molecular Diversity Preservation International: 682–725.

Campos, Maria, Kenneth R Markham, Kevin A Mitchell, et A Proenca da Cunha. 1997. "An approach to the characterization of bee pollens via their flavonoid/phenolic profiles". *Phytochemical Analysis: An International Journal of Plant Chemical and Biochemical Techniques* 8(4). Wiley Online Library: 181–85.

Campos, Maria GR, Rosemary F Webby, Kenneth R Markham, Kevin A Mitchell, et Antonio P Da Cunha. 2003. "Age-induced diminution of free radical scavenging capacity in bee pollens and the contribution of constituent flavonoids". *Journal of Agricultural and Food Chemistry* 51(3). ACS Publications: 742–45.

Campos, Maria GR, Stefan Bogdanov, Ligia Bicudo de Almeida-Muradian, Teresa Szczesna, Yanina Mancebo, Christian Frigerio, et Francisco Ferreira. 2008. "Pollen composition and standardisation of analytical methods". *Journal of Apicultural Research* 47(2). Taylor & Francis: 154–61. doi:10.1080/00218839.2008.11101443.

Campos, Maria GR, Christian Frigerio, Joana Lopes, et Stefan Bogdanov. 2010. "What is the future of Bee-Pollen?" *Journal of Apiproduct and Apimedical Science* 2(4). International Bee Research Association: 131–44.

Canale, Angelo, Giovanni Benelli, Antonella Castagna, Cristina Sgherri, Piera Poli, Andrea Serra, Marcello Mele, Annamaria Ranieri, Francesca Signorini, et Matteo Bientinesi. 2016. "Microwave-assisted drying for the conservation of honeybee pollen". *Materials* 9(5). MDPI: 363.

Carpes, Solange T, Ingridy SR Cabral, Cynthia Fernandez P Luz, Jailson P Capeletti, Severino Matias Alencar, et Maria Lúcia Masson. 2009. "Palynological and physicochemical characterization of Apis mellifera L. bee pollen in the Southern region of Brazil". *Journal of Food, Agriculture and Environment* 7: 667–73.

Carvalho, Fernando P. 2017. "Pesticides, environment, and food safety". *Food and Energy Security* 6(2). Wiley Online Library: 48–60.

Castagna, Antonella, Giovanni Benelli, Giuseppe Conte, Cristina Sgherri, Francesca Signorini, Cristiano Nicolella, Annamaria Ranieri, et Angelo Canale. 2020. "Drying techniques and storage: Do they affect the nutritional value of bee-collected pollen?" *Molecules* 25(21). MDPI: 4925.

Castillo, Julian, Obdulio Benavente, et José A del Río. 1992. "Naringin and neohesperidin levels during development of leaves, flower buds, and fruits of Citrus aurantium". *Plant Physiology* 99(1). American Society of Plant Biologists: 67–73.

Cavazza, Antonella, Claudio Corradini, Marilena Musci, et Paola Salvadeo. 2013. "High-performance liquid chromatographic phenolic compound fingerprint for authenticity assessment of honey". *Journal of the Science of Food and Agriculture* 93(5). Wiley Online Library: 1169–75.

Cechinel-Zanchett, Camile Cecconi, Luísa Nathália Bolda Mariano, Thaise Boeing, Juliana de Carvalho da Costa, Luísa Mota Da Silva, Jairo K Bastos, Valdir Cechinel-Filho, et Priscila de Souza. 2020. "Diuretic and renal protective effect of kaempferol 3-O-alpha-l-rhamnoside (afzelin) in normotensive and hypertensive rats". *Journal of Natural Products* 83(6). ACS Publications: 1980–89.

Çelik, Saffet, Yusuf Can Gerçek, Aslı Özkök, et Nesrin Ecem Bayram. 2022. "Organic acids and their derivatives: Minor components of bee pollen, bee bread, royal jelly and bee venom". *European Food Research and Technology* 248. Springer: 1–21.

Chambliss, Walter G. 2003. "A critical review of Graminex flower pollen extract for symptomatic relief of lower urinary tract symptoms (LUTS) in men". *Flower Pollen Extract and its Effect for Prostate Health* 4: 1–21.

Chattopadhyay, Pronobesh, Aadesh Upadhyay, Amit Agnihotri, Sanjeev Karmakar, Danswerang Ghoyary, et Vijay Veer. 2013. "Comparative hematoxicity of fusirium mycotoxin in experimental Sprague-Dawley rats". *Toxicology International* 20(1). Wolters Kluwer--Medknow Publications: 25.

Chen, Linlin, Huidan Deng, Hengmin Cui, Jing Fang, Zhicai Zuo, Junliang Deng, Yinglun Li, Xun Wang, et Ling Zhao. 2018. "Inflammatory responses and inflammation-associated diseases in organs". *Oncotarget* 9(6). Impact Journals, LLC: 7204.

Chirife, Jorge, María Clara Zamora, et Aldo Motto. 2006. "The correlation between water activity and% moisture in honey: Fundamental aspects and application to Argentine honeys". *Journal of Food Engineering* 72(3). Elsevier: 287–92.

Claeys, Liesel, Chiara Romano, Karl De Ruyck, Hayley Wilson, Beatrice Fervers, Michael Korenjak, Jiri Zavadil, Marc J Gunter, Sarah De Saeger, et Marthe De Boevre 2020. "Mycotoxin exposure and human cancer risk: A systematic review of epidemiological studies". *Comprehensive Reviews in Food Science and Food Safety* 19(4). Wiley Online Library: 1449–64.

Çobanoğlu, Duygu Nur, İlginç Kizilpinar Temizer, Esra Deniz Candan, Ufuk Yolcu, et Aytaç Güder. 2022. "Evaluation of the nutritional value of bee pollen by palynological, antioxidant, antimicrobial, and elemental characteristics". *European Food Research and Technology*. Springer, 1–19.

The Commission of the European Communities. 2006. "Commission Regulation (EC) no 1881/2006 of 19 December 2006 setting maximum levels for certain contaminants in foodstuff". *Official Journal of the European Union:* 5–24.

Conte, Paola, Alessandra Del Caro, Federica Balestra, Antonio Piga, et Costantino Fadda. 2018. "Bee pollen as a functional ingredient in gluten-free bread: A physical-chemical, technological and sensory approach". *LWT* 90. Elsevier: 1–7.

Cormick, Gabriela, et Jose M Belizán. 2019. "Calcium intake and health". *Nutrients* 11(7). MDPI: 1606.

Coronel, Bertha Baldi, Diego Grasso, Silvia Chaves Pereira, et Gonzalo Fernández. 2004. "Caracterización bromatológica del polen apícola argentino". *Ciencia, Docencia y Tecnología* 15(29). Universidad Nacional de Entre Ríos: 145–81.

Dalgleish, Douglas G. 1997. "Adsorption of protein and the stability of emulsions". *Trends in Food Science and Technology* 8(1). Elsevier: 1–6.

De-Melo, Adriane Alexandre Machado, et Ligia Bicudo de Almeida-Muradian. 2017. "Chemical composition of bee pollen". In *Bee Products-Chemical and Biological Properties*, José M. Alvarez-Suarez (ed). Springer, Switzerland, 221–59.

Denisow, Bożena, et Marta Denisow-Pietrzyk. 2016. "Biological and therapeutic properties of bee pollen: A review". *Journal of the Science of Food and Agriculture* 96(13). Wiley Online Library: 4303–9.

Di Chiacchio, Isabela M, Isadora M Paiva, Danilo JM de Abreu, Elisângela EN Carvalho, Pedro J Martínez, Stephan M Carvalho, Victoriano Mulero, et Luis David S Murgas. 2021. "Bee pollen as a dietary supplement for fish: Effect on the reproductive performance of zebrafish and the immunological response of their offspring". *Fish and Shellfish Immunology* 119. Elsevier: 300–7.

Dibos, Chloé. 2010. "Interactions Plante-Pollinisateur: Caractérisation de la Qualité du Pollen de Deux Cucurbitacées Durant Son Ontogenèse, Sa Présentation et Son Transport sur le Corps de l'Abeille Domestique". PhD Thesis. Université d'Avignon.

Dickinson, Eric. 2008. "Interfacial structure and stability of food emulsions as affected by protein–polysaccharide interactions". *Soft Matter* 4(5). Royal Society of Chemistry: 932–42.

Dong, Jie, Yiting Yang, Xue Wang, et Hongcheng Zhang. 2015. "Fatty acid profiles of 20 species of monofloral bee pollen from China". *Journal of Apicultural Research* 54(5). Taylor & Francis: 503–11.

Dong, Yuesheng, Liping Sui, Fan Yang, Xinxiu Ren, Yan Xing, et Zhilong Xiu. 2022. "Reducing the intestinal side effects of acarbose by baicalein through the regulation of gut microbiota: An in vitro study". *Food Chemistry* 394. Elsevier: 133561.

Duarte, Alysson Wagner Fernandes, Maria Raphaella dos Santos Vasconcelos, Melissa Oda-Souza, Favízia Freitas de Oliveira, et Ana Maria Queijeiro López. 2018. "Honey and bee pollen produced by Meliponini (Apidae) in Alagoas, Brazil: multivariate analysis of physicochemical and antioxidant profiles". *Food Science and Technology* 38. SciELO Brasil: 493–503.

Dumitru, Corina Dana, Ionela Andreea Neacsu, Alexandru Mihai Grumezescu, et Ecaterina Andronescu. 2022. "Bee-derived products: Chemical composition and applications in skin tissue engineering". *Pharmaceutics* 14(4). MDPI: 750.

Ecem Bayram, Nesrin. 2021. "Vitamin, mineral, polyphenol, amino acid profile of bee pollen from Rhododendron ponticum (source of 'mad honey'): Nutritional and palynological approach". *Journal of Food Measurement and Characterization* 15(3). Springer: 2659–66.

Egan, Kathryn Brigham. 2016. "The epidemiology of benign prostatic hyperplasia associated with lower urinary tract symptoms: Prevalence and incident rates". *Urologic Clinics* 43(3). Elsevier: 289–97.

El Adaouia Taleb, Rabia, Noureddine Djebli, Hadjer Chenini, Huseyin Sahin, et Sevgi Kolayli. 2020. "In vivo and in vitro anti-diabetic activity of ethanolic propolis extract". *Journal of Food Biochemistry* 44(7). Wiley Online Library: e13267.

El-Aidy, Waleed K, Ahmad A Ebeid, Abd El-Raouf, M Sallam, Ibrahim E Muhammad, Ayman T Abbas, MA Kamal, et Sayed Sartaj Sohrab. 2015. "Evaluation of propolis, honey, and royal jelly in amelioration of peripheral blood leukocytes and lung inflammation in mouse conalbumin-induced asthma model". *Saudi Journal of Biological Sciences* 22(6). Elsevier: 780–88.

El-Asely, Amel M, Amany A Abbass, et Brian Austin. 2014. "Honey bee pollen improves growth, immunity and protection of Nile tilapia (Oreochromis niloticus) against infection with Aeromonas hydrophila". *Fish and Shellfish Immunology* 40(2). Elsevier: 500–6.

El Ghouizi, Asmae, Nawal El Menyiy, Soraia I Falcão, Miguel Vilas-Boas, et Badiaa Lyoussi. 2020. "Chemical composition, antioxidant activity, and diuretic effect of Moroccan fresh bee pollen in rats". *Veterinary World* 13(7). Veterinary World: 1251.

El Ghouizi, Asmae, Nawal El Menyiy, Meryem Bakour, et Badiaa Lyoussi. 2021. "Moroccan monofloral bee pollen: Botanical origin, physicochemical characterization, and antioxidant activities". Édité par Vera Lavelli. *Journal of Food Quality* (janvier): 1–10. doi:10.1155/2021/8877266.

El Ghouizi, Asmae, Noori Al-Waili, Nawal El Menyiy, Salma Elfetri, Abderrazak Aboulghazi, Ahmed Al-Waili, et Badiaa Lyoussi. 2022. "Protective effect of bee pollen in acute kidney injury, proteinuria, and crystalluria induced by ethylene glycol ingestion in rats". *Scientific Reports* 12(1). Nature Publishing Group: 1–11.

El-Saadony, Mohamed T, Nidal M Zabermawi, Nehal M Zabermawi, Maryam A Burollus, Manal E Shafi, Mahmoud Alagawany, Nahed Yehia, Ahmed M Askar, Sara A Alsafy, et Ahmed E Noreldin. 2021. "Nutritional aspects and health benefits of bioactive plant compounds against infectious diseases: A review". *Food Reviews International*. Taylor & Francis: 1–23.

El-Seedi, Hesham R, Nehal Eid, Abd El-Wahed, A Aida, Mostafa E Rateb, Hanan S Afifi, Ahmed F Algethami, Chao Zhao, Yahya Al Naggar, et Sultan M Alsharif. 2022. "Honey bee products: Preclinical and clinical studies of their anti-inflammatory and immunomodulatory properties". *Frontiers in Nutrition* 8: 1109.

Estevinho, Leticia M, Sandra Rodrigues, Ana P Pereira, et Xesús Feás. 2012. "Portuguese bee pollen: Palynological study, nutritional and microbiological evaluation". *International Journal of Food Science and Technology* 47(2). Wiley Online Library: 429–35.

Eteraf-Oskouei, Tahereh, Ayda Shafiee-Khamneh, Fariba Heshmati-Afshar, et Abbas Delazar. 2020. "Anti-inflammatory and anti-angiogenesis effect of bee pollen methanolic extract using air pouch model of inflammation". *Research in Pharmaceutical Sciences* 15(1). Medknow Publications: 66.

European Commission. 2011. "Commission Implementing Regulation (EU) 542/2011 of 1 June 2011 amending Implementing Regulation (EU) no 540/2011 implementing Regulation (EC) no 1107/2009 of the European Parliament and of the Council as regards the list of approved active substances to take into account Directive 2011/58/EU amending Council Directive 91/414/EEC to renew the inclusion of carbendazim as active substance". *Official Journal of the European Union* 153: 189–91.

———. 2020. "Commission Implementing Regulation (EU) 2020/18 of 10 January 2020 concerning the non-renewal of the approval of the active substance chlorpyrifos, in accordance with Regulation (EC) no 1107/2009 of the European Parliament and of the Council concerning the placing of plant protection products on the market, and amending the Annex to Commission Implementing Regulation (EU) no 540/2011". *Official Journal of the European Union* 7: 14–6.

Farag, Soha A, et T El-Rayes. 2016. "Research article effect of bee-pollen supplementation on performance, carcass traits and blood parameters of broiler chickens". *Asian Journal of Animal and Veterinary Advances* 11(3): 168–77.

Fatrcová-Šramková, Katarína, Janka Nôžková, Miroslava Kačániová, Magda Máriássyová, Katarína Rovná, et Michal Stričík. 2013. "Antioxidant and antimicrobial properties of monofloral bee pollen". *Journal of Environmental Science and Health, Part B* 48(2). Taylor & Francis: 133–38.

Feás, Xesús, M Pilar Vázquez-Tato, Leticia Estevinho, Julio A Seijas, et Antonio Iglesias. 2012. "Organic bee pollen: Botanical origin, nutritional value, bioactive compounds, antioxidant activity and microbiological quality". *Molecules* 17(7). MDPI: 8359–77.

Fernandez-Salguero, J, R Gomez, et MA Carmona. 1993. "Water activity in selected high-moisture foods". *Journal of Food Composition and Analysis* 6(4). Elsevier: 364–69.

Ferreira-Santos, Pedro, Zlatina Genisheva, Claudia Botelho, Cristina Rocha, et José António Teixeira. 2021. "Valorization of natural antioxidants for nutritional and health applications". In V. Waisundara (Ed.), *Antioxidants*- Benefits, Sources, Mechanisms of Action. IntechOpen, pp. 1–33.

Ferreres, Federico, Christina García-Viguera, Francisco Tomás-Lorente, et Francisco A Tomás-Barberán. 1993. "Hesperetin: A marker of the floral origin of citrus honey". *Journal of the Science of Food and Agriculture* 61(1). Wiley Online Library: 121–23.

Filannino, Pasquale, Raffaella Di Cagno, Olimpia Vincentini, Daniela Pinto, Andrea Polo, Francesca Maialetti, Annalisa Porrelli, et Marco Gobbetti. 2021. "Nutrients bioaccessibility and anti-inflammatory features of fermented bee pollen: A comprehensive investigation". *Frontiers in Microbiology* 12. Frontiers Media SA: 622091.

Fisher, Miles. 2022. "Acarbose and alpha glucosidase inhibitors". *Diabetes Drug Notes*. Wiley Online Library, 229–38.

Forman, Henry Jay, et Hongqiao Zhang. 2021. "Targeting oxidative stress in disease: Promise and limitations of antioxidant therapy". *Nature Reviews Drug Discovery* 20(9). Nature Publishing Group: 689–709.

Fuenmayor, Carlos, Carlos Zuluaga, Consuelo Díaz M, Quicazán de C Marta, María Cosio, et Saverio Mannino. 2014. "Evaluation of the physicochemical and functional properties of Colombian bee pollen". *Revista MVZ Córdoba* 19(1). Universidad de Córdoba, Montería, Colombia: 4003–14.

Fulton, Corie A, Kara E Huff Hartz, Richard D Fell, Carlyle C Brewster, John D Reeve, et Michael J Lydy. 2019. "An assessment of pesticide exposures and land use of honey bees in Virginia". *Chemosphere* 222(mai). England: 489–93. doi:10.1016/j.chemosphere.2019.01.156.

Fung, Frederick, et Richard F Clark. 2004. "Health effects of mycotoxins: A toxicological overview". *Journal of Toxicology: Clinical Toxicology* 42(2). Taylor & Francis: 217–34.

Gardana, Claudio, Cristian Del Bo, Marta C Quicazán, Ana Ruby Corrrea, et Paolo Simonetti. 2018. "Nutrients, phytochemicals and botanical origin of commercial bee pollen from different geographical areas". *Journal of Food Composition and Analysis* 73. Elsevier: 29–38.

Ghosh, Sampat, et Chuleui Jung. 2017. "Nutritional value of bee-collected pollens of hardy kiwi, Actinidia arguta (Actinidiaceae) and oak, Quercus sp.(Fagaceae)". *Journal of Asia-Pacific Entomology* 20(1). Elsevier: 245–51.

Giampieri, Francesca, Jose Luis Quiles, Danila Cianciosi, Tamara Yuliett Forbes-Hernández, Francisco Josè Orantes-Bermejo, José Miguel Alvarez-Suarez, et Maurizio Battino. 2022. "Bee products: An emblematic example of underutilized sources of bioactive compounds". *Journal of Agricultural and Food Chemistry* 70(23). ACS Publications: 6833–48.

Gonçalves, Paulo JS, Letícia M Estevinho, Ana Paula Pereira, João MC Sousa, et Ofélia Anjos. 2018. “Computational intelligence applied to discriminate bee pollen quality and botanical origin”. *Food Chemistry* 267. Elsevier: 36–42.

González, G, MJ Hinojo, R Mateo, A Medina, et M Jiménez. 2005. “Occurrence of mycotoxin producing fungi in bee pollen”. *International Journal of Food Microbiology* 105(1). Elsevier: 1–9.

Green, Ali J, Karen A Littlejohn, Paul Hooley, et Philip W Cox. 2013. “Formation and stability of food foams and aerated emulsions: Hydrophobins as novel functional ingredients”. *Current Opinion in Colloid and Interface Science* 18(4). Elsevier: 292–301.

Groot, Antonius Petrus de. 1952. “Protein and amino acid requirements of the honeybee (Apis mellifica L.)”. *Experientia* 8(5), 192–194.

Gülbandilar, Aysel. 2022. “Antimicrobial activities of propolis samples collected from different provinces of Turkey”. *Mas Journal of Applied Sciences* 7(2): 433–42.

Guo, Mark Q. 2017. “Polysaccharides: Structure and solubility”. In *Solubility of Polysaccharides*, édité par Xinzhong Hu, traduit par Lianzhong Ai, ED - Zhenbo Xu, Ch. 2. IntechOpen, Rijeka. doi:10.5772/intechopen.71570.

Guo, Weimin, EunHee Kong, et Mohsen Meydani. 2009. “Dietary polyphenols, inflammation, and cancer”. *Nutrition and Cancer* 61(6). Taylor & Francis: 807–10.

Habryka, Celina, Marek Kruczek, et Barbara Drygaś. 2016. “Bee products used in apitherapy”. *World Scientific News* 48: 254–58.

Hisanaga, Etsuko, Takaaki Sano, Yuji Kumakura, Yozo Yokoyama, Isao Nakajima, et Hitoshi Takagi. 2022. “Case of pneumatosis cystoides intestinalis with intra-abdominal free air developed during treatment with voglibose”. *JGH Open* 6(9). Wiley Online Library: 643–46.

Hollman, Peter CH. 2004. “Absorption, bioavailability, and metabolism of flavonoids”. *Pharmaceutical Biology* 42(supl). Taylor & Francis: 74–83.

Hosny, AS, FM Sabbah, et ZE El-Bazza. 2018. “Studies on the microbial decontamination of Egyptian bee pollen by γ radiation”. *Egyptian Pharmaceutical Journal* 17(3). Medknow Publications: 190.

Hsu, Pei-Shou, Tzu-Hsien Wu, Meng-Yuan Huang, Dun-Yan Wang, et Ming-Cheng Wu. 2021. “Nutritive value of 11 bee pollen samples from major floral sources in Taiwan”. *Foods* 10(9). MDPI: 2229.

Ikeda, K. 1908. “A new flavor enhancer”. *J. Tokyo Chem. Soc.* 30: 820–36.

Ilie, Cornelia-Ioana, Eliza Oprea, Elisabeta-Irina Geana, Angela Spoiala, Mihaela Buleandra, Gratiela Gradisteanu Pircalabioru, Irinel Adriana Badea, Denisa Ficai, Ecaterina Andronescu, et Anton Ficai 2022. “Bee pollen extracts: Chemical composition, antioxidant properties, and effect on the growth of selected probiotic and pathogenic bacteria”. *Antioxidants* 11(5). MDPI: 959.

Isik, Ayla, Murat Ozdemir, et Ibrahim Doymaz. 2018. “Effect of hot air drying on quality characteristics and physicochemical properties of bee pollen”. *Food Science and Technology* 39(1). SciELO, Brasil: 224–31.

Ivanov, Marija, Katarina Novović, Milka Malešević, Miroslav Dinić, Dejan Stojković, Branko Jovčić, et Marina Soković. 2022. “Polyphenols as inhibitors of antibiotic resistant bacteria—Mechanisms underlying Rutin interference with bacterial virulence”. *Pharmaceuticals* 15(3). MDPI: 385.

Iwamura, Hiromichi, Takuya Koie, Osamu Soma, Teppei Matsumoto, Atsushi Imai, Shingo Hatakeyama, Takahiro Yoneyama, Yasuhiro Hashimoto, et Chikara Ohyama. 2015. “Eviprostat has an identical effect compared to pollen extract (cernilton) in patients with chronic prostatitis/chronic pelvic pain syndrome: A randomized, prospective study”. *BMC Urology* 15(1). BioMed Central: 1–6.

Jannesar, Masoomeh, Maryam Sharif Shoushtari, Ahmad Majd, et Zahra Pourpak. 2017. “Bee pollen flavonoids as a therapeutic agent in allergic and immunological disorders”. *Iranian Journal of Allergy, Asthma, and Immunology.* 16(3): 171–182.

Jiang, Hongfei, Xiaofei Qin, Qi Wang, Qi Xu, Jie Wang, Yudong Wu, Wujun Chen, Chao Wang, Tingting Zhang, et Dongming Xing 2021. “Application of carbohydrates in approved small molecule drugs: A review”. *European Journal of Medicinal Chemistry* 223. Elsevier: 113633.

Juan, Celia Andrés, José Manuel Pérez de la Lastra, Francisco J Plou, et Eduardo Pérez-Lebeña. 2021. “The chemistry of reactive oxygen species (ROS) revisited: Outlining their role in biological macromolecules (DNA, lipids and proteins) and induced pathologies”. *International Journal of Molecular Sciences* 22(9). MDPI: 4642.

Kačániová, Miroslava, Miroslav Juráček, Róbert Chlebo, Vladimíra Kňazovická, Miriam Kadasi-Horáková, Simona Kunová, Jadža Lejková, Peter Haščík, Ján Mareček, et Milan Šimko. 2011. “Mycobiota and mycotoxins in bee pollen collected from different areas of Slovakia”. *Journal of Environmental Science and Health, Part B* 46(7). Taylor & Francis: 623–29.

Kahraman, Hatice Ahu, Hidayet Tutun, Muhammet Mükerrem Kaya, Melike Sultan Usluer, Soner Tutun, Ceren Yaman, Sedat Sevin, et Erhan Keyvan. 2022. "Ethanolic extract of Turkish bee pollen and propolis: Phenolic composition, antiradical, antiproliferative and antibacterial activities". *Biotechnology and Biotechnological Equipment* 36(1). Taylor & Francis: 44–55.

Kalaycıoğlu, Zeynep, Hakan Kaygusuz, Serhat Döker, Sevgi Kolaylı, et F Bedia Erim. 2017. "Characterization of Turkish honeybee pollens by principal component analysis based on their individual organic acids, sugars, minerals, and antioxidant activities". *LWT* 84. Elsevier: 402–8.

Karabagias, Ioannis K, Artemis P Louppis, Anastasia Badeka, Chara Papastephanou, et Michael G Kontominas. 2019. "Nutritional aspects and botanical origin recognition of Mediterranean honeys based on the "mineral imprint"with the application of supervised and non-supervised statistical techniques". *European Food Research and Technology* 245(9). Springer: 1939–49.

Kasiotis, Konstantinos M, Chris Anagnostopoulos, Pelagia Anastasiadou, et Kyriaki Machera. 2014. "Pesticide residues in honeybees, honey and bee pollen by LC-MS/MS screening: Reported death incidents in honeybees". *The Science of the Total Environment* 485–486(juillet). Netherlands: 633–42. doi:10.1016/j.scitotenv.2014.03.042.

Kavian, Niloufar, Souad Mehlal, Mohamed Jeljeli, Nathaniel Edward Bennett Saidu, Carole Nicco, Olivier Cerles, Sandrine Chouzenoux, Anne Cauvet, Claire Camus, et Mehdi Ait-Djoudi 2018. "The Nrf2-antioxidant response element signaling pathway controls fibrosis and autoimmunity in scleroderma". *Frontiers in Immunology* 9. Frontiers Media SA: 1896.

Keskin, Merve, et Asli Özkök. 2020. "α-amylase Inhibition Properties of Bee Pollen and Bee Bread (Perga)". *Hacettepe Journal of Biology and Chemistry*, août. doi:10.15671/hjbc.774901.

Khalifa, Shaden AM, Mohamed H Elashal, Nermeen Yosri, Ming Du, Syed G Musharraf, Lutfun Nahar, Satyajit D Sarker, Zhiming Guo, Wei Cao, et Xiaobo Zou 2021. "Bee pollen: Current status and therapeutic potential". *Nutrients* 13(6). Multidisciplinary Digital Publishing Institute: 1876.

Khongkarat, Phanthiwa, Prapun Traiyasut, Preecha Phuwapraisirisan, et Chanpen Chanchao. 2022. "First report of fatty acids in Mimosadiplotricha bee pollen with in vitro lipase inhibitory activity". *PeerJ* 10. PeerJ Inc: e12722.

Kim, Yi-Eun, Eun-Ji Cho, et Eui-Hong Byun. 2018. "Immunomodulatory effect of bee pollen extract in macrophage cells". *Korean Journal of Food Science and Technology* 50(4). Korean Society of Food Science and Technology: 437–43.

Kinsella, John E, et Nicholas Melachouris. 1976. "Functional properties of proteins in foods: A survey". *Critical Reviews in Food Science and Nutrition* 7(3). Taylor & Francis: 219–80.

Klaunig, James E, et Zemin Wang. 2018. "Oxidative stress in carcinogenesis". *Current Opinion in Toxicology* 7. Elsevier: 116–21.

Kolesarova, A, Z Bakova, M Capcarova, B Galik, M Juracek, M Simko, R Toman, et AV Sirotkin. 2013. "Consumption of bee pollen affects rat ovarian functions". *Journal of Animal Physiology and Animal Nutrition* 97(6). Wiley Online Library: 1059–65.

Komosinska-Vassev, Katarzyna, Pawel Olczyk, Justyna Kaźmierczak, Lukasz Mencner, et Krystyna Olczyk. 2015. "Bee pollen: Chemical composition and therapeutic application". *Evidence-Based Complementary and Alternative Medicine* 2015: 1–6. doi:10.1155/2015/297425.

Kostić, Aleksandar Ž, Mirjana B Pešić, Mirjana D Mosić, Biljana P Dojčinović, Maja M Natić, et Jelena Đ Trifković. 2015. "Mineral content of bee pollen from Serbia". *Arhiv Za Higijenu Rada i Toksikologiju* 66(4). Croatia: 251–58. doi:10.1515/aiht-2015-66-2630.

Kostić, Aleksandar Ž, Miroljub B Barać, Sladjana P Stanojević, Dušanka M Milojković-Opsenica, Živoslav Lj Tešić, Branko Šikoparija, Predrag Radišić, Marija Prentović, et Mirjana B Pešić. 2015. "Physicochemical composition and techno-functional properties of bee pollen collected in Serbia". *LWT – Food Science and Technology* 62(1). Elsevier: 301–9.

Kostić, Aleksandar Ž, Danijel D Milinčić, Bojana D Špirović Trifunović, Sladjana P Stanojević, Steva Lević, Nebojša Nedić, Viktor Nedović, Živoslav Lj Tešić, et Mirjana B Pešić. 2020. "Nutritional and techno-functional properties of monofloral bee-collected sunflower (Helianthus annuus L.) pollen". *Emirates Journal of Food and Agriculture* 32: 768–77.

Kostić, Aleksandar Ž, Tanja S Petrović, Vesna S Krnjaja, Nebojša M Nedić, Živoslav Lj Tešić, Dušanka M Milojković-Opsenica, Miroljub B Barać, Slađana P Stanojević, et Mirjana B Pešić. 2017. "Mold/aflatoxin contamination of honey bee collected pollen from different Serbian regions". *Journal of Apicultural Research* 56(1). Taylor & Francis: 13–20.

Krell, Rainer. 1996. *Value-Added Products from Beekeeping*. Food & Agriculture Organization.

Laaroussi, Hassan, Meryem Bakour, Driss Ousaaid, Abderrazak Aboulghazi, Pedro Ferreira-Santos, Zlatina Genisheva, José Antonio Teixeira, et Badiaa Lyoussi. 2020. "Effect of antioxidant-rich propolis and bee pollen extracts against D-glucose induced Type 2 diabetes in rats". *Food Research International*, octobre: 109802. doi:10.1016/j.foodres.2020.109802.

———. 2020. "Effect of antioxidant-rich propolis and bee pollen extracts against D-glucose induced type 2 diabetes in rats". *Food Research International* 138(décembre): 109802. doi:10.1016/j.foodres.2020.109802.

LaPelusa, Andrew, et Ravi Kaushik. 2021. "Physiology, proteins". In *StatPearls [Internet]*. StatPearls Publishing.

Lapert, Flore. 2010. "L'Association Centrale d'entraide Vétérinaire: historique, missions, avenir". PhD Thesis, These pour le Doctorat Veterinaire. École Nationale Veterinaire d'Alfort, Année. http://theses. vet-alfort. fr/telecharger. php.

LeBlanc, Blaise W, Owen K Davis, Stephen Boue, Anthony DeLucca, et Thomas Deeby. 2009. "Antioxidant activity of Sonoran Desert bee pollen". *Food Chemistry* 115(4). Elsevier: 1299–305.

Lewandowski, Łukasz, Marta Kepinska, et Halina Milnerowicz. 2019. "The copper-zinc superoxide dismutase activity in selected diseases". *European Journal of Clinical Investigation* 49(1). Wiley Online Library: e13036.

Li, Qiangqiang, Xinwen Liang, Liang Zhao, Zhongyin Zhang, Xiaofeng Xue, Kai Wang, et Liming Wu. 2017. "UPLC-Q-exactive Orbitrap/MS-based lipidomics approach to characterize lipid extracts from bee pollen and their in vitro anti-inflammatory properties". *Journal of Agricultural and Food Chemistry* 65(32). ACS Publications: 6848–60.

Li, Qiang-Qiang, Kai Wang, Maria Cristina Marcucci, Alexandra Christine Helena Frankland Sawaya, Lin Hu, Xiao-Feng Xue, Li-Ming Wu, et Fu-Liang Hu. 2018. "Nutrient-rich bee pollen: A treasure trove of active natural metabolites". *Journal of Functional Foods* 49. Elsevier: 472–84.

Li, Qiangqiang, Xinwen Liang, Nana Guo, Lin Hu, Yajun Wu, Xiaofeng Xue, Liming Wu, Kai Wang. 2019. "Protective effects of bee pollen extract on the Caco-2 intestinal barrier dysfunctions induced by dextran sulfate sodium". *Biomedicine and Pharmacotherapy* 117. Elsevier: 109200.

Li, Qiangqiang, Minghui Sun, Zhengrui Wan, Junshi Liang, Mirko Betti, Yuliya Hrynets, Xiaofeng Xue, Liming Wu, et Kai Wang. 2019. "Bee pollen extracts modulate serum metabolism in lipopolysaccharide-induced acute lung injury mice with anti-inflammatory effects". *Journal of Agricultural and Food Chemistry* 67(28). ACS Publications: 7855–68.

Li, Wei-Zheng, Kyle Stirling, Jun-Jie Yang, et Lei Zhang. 2020. "Gut microbiota and diabetes: From correlation to causality and mechanism". *World Journal of Diabetes* 11(7). Baishideng Publishing Group Inc: 293.

Liang, Miao, Peng Zhang, Xi Shu, Changgeng Liu, et Jinian Shu. 2013. "Characterization of pollen by MALDI-TOF lipid profiling". *International Journal of Mass Spectrometry* 334. Elsevier: 13–8.

Lieu, Pauline T, Marja Heiskala, Per A Peterson, et Young Yang. 2001. "The roles of iron in health and disease". *Molecular Aspects of Medicine* 22(1–2). Elsevier: 1–87.

Martins, Marcia CT, Marcelo A Morgano, Eduardo Vicente, Sueli R Baggio, et Delia B Rodriguez-Amaya. 2011. "Physicochemical composition of bee pollen from eleven Brazilian states". *Journal of Apicultural Science* 55(2): 107–16.

Maruyama, Hiroe, Takashi Sakamoto, Yoko Araki, et Hideaki Hara. 2010. "Anti-inflammatory effect of bee pollen ethanol extract from Cistus sp. of Spanish on carrageenan-induced rat hind paw edema". *BMC Complementary and Alternative Medicine* 10(1). BioMed Central: 1–11.

Mayda, Nazlı, Asli Özkök, Nesrin Ecem Bayram, Yusuf Gercek, et Kadriye Sorkun. 2020. "Bee bread and bee pollen of different plant sources: Determination of phenolic content, antioxidant activity, fatty acid and element profiles". *Journal of Food Measurement and Characterization*, mars. doi:10.1007/s11694-020-00427-y.

Medeiros, KCP, CAV Figueiredo, TB Figueredo, KRL Freire, FAR Santos, Neuza Maria Alcântara-Neves, TMS Silva, et MR Piuvezam. 2008. "Anti-allergic effect of bee pollen phenolic extract and myricetin in ovalbumin-sensitized mice". *Journal of Ethnopharmacology* 119(1). Elsevier: 41–6.

Melin, E. 2011. *Botanique apicole*. Ecol e d'Apiculture de la Région wallonne & Institut de Botanique. Université de Liège, Belgique.

Melo, Illana Louise Pereira de, et Ligia Bicudo de Almeida-Muradian. 2011. "Comparison of methodologies for moisture determination on dried bee pollen samples". *Food Science and Technology* 31(1): 194–97.

Mitsunari, Kensuke, Yasuyoshi Miyata, Tomohiro Matsuo, Yuta Mukae, Asato Otsubo, Junki Harada, Tsubasa Kondo, Tsuyoshi Matsuda, Kojiro Ohba, et Hideki Sakai. 2021. "Pharmacological effects and potential clinical usefulness of polyphenols in benign prostatic hyperplasia". *Molecules* 26(2). MDPI: 450.

Mohamed, Nema A, Osama M Ahmed, Walaa G Hozayen, et Marium A Ahmed. 2018. "Ameliorative effects of bee pollen and date palm pollen on the glycemic state and male sexual dysfunctions in streptozotocin-Induced diabetic Wistar rats". *Biomedicine and Pharmacotherapy* 97. Elsevier: 9–18.

Morais, Simone, F Garcia Costa, et Mde L Pereira. 2012. "Heavy metals and human health". *Environmental Health–Emerging Issues and Practice* 10(1). InTech Rijeka: 227–45.

Mostafalou, Sara, et Mohammad Abdollahi. 2013. "Pesticides and human chronic diseases: Evidences, mechanisms, and perspectives". *Toxicology and Applied Pharmacology* 268(2). Elsevier: 157–77.

Münstedt, Karsten, Benjamin Voss, Uwe Kullmer, Ursula Schneider, et Jutta Hübner. 2015. "Bee pollen and honey for the alleviation of hot flushes and other menopausal symptoms in breast cancer patients". *Molecular and Clinical Oncology* 3(4). Spandidos Publications: 869–74.

Munteanu, Irina Georgiana, et Constantin Apetrei. 2021. "Analytical methods used in determining antioxidant activity: A review". *International Journal of Molecular Sciences* 22(7). MDPI: 3380.

Muraca, Lucia, Antonio Scuteri, Elisabetta Burdino, Gianmarco Marcianò, Vincenzo Rania, Luca Catarisano, Alessandro Casarella, Erika Cione, Caterina Palleria, et Manuela Colosimo. 2022. "Effectiveness and safety of a new nutrient fixed combination containing pollen extract plus Teupolioside, in the management of LUTS in patients with benign prostatic hypertrophy: A pilot study". *Life* 12(7). MDPI: 965.

Murakami, Maki, Osamu Tsukada, Kiyoshi Okihara, Ken Hashimoto, Hideo Yamada, et Hideyo Yamaguchi. 2008. "Beneficial effect of honeybee-collected pollen lump extract on benign prostatic hyperplasia (BPH)—A double-blind, placebo-controlled clinical trial". *Food Science and Technology Research* 14(3). Japanese Society for Food Science and Technology: 306–10.

Murcia-Morales, María, Horacio Heinzen, Piedad Parrilla-Vázquez, María del Mar Gómez-Ramos, et Amadeo R Fernández-Alba. 2022. "Presence and distribution of pesticides in apicultural products: A critical appraisal". *TrAC Trends in Analytical Chemistry* 146. Elsevier: 116506.

NAD, AT, and Dinkova-Kostova P Talalay. 2010. "H: Quinone acceptor oxidoreductase 1 (NQO1), a multifunctional antioxidant enzyme and exceptionally versatile cytoprotector". *Archives of Biochemistry and Biophysics* 501: 116–23.

Nan, Xiaoke, Weijuan Jia, Yuankuan Zhang, Hong Wang, Zongtao Lin, et Shizhong Chen. 2022. "An on-line detection system for screening small molecule inhibitors of α-amylase and α-glucosidase in Prunus mume". *Journal of Chromatography A* 1663. Elsevier: 462754.

Narvaez, Uuh, Jonatan Jafet, et Maira Rubí Segura Campos. 2022. "Combination therapy of bioactive compounds with acarbose: A proposal to control hyperglycemia in type 2 diabetes". *Journal of Food Biochemistry*. Wiley Online Library: e14268.

Naseri, Leila, Mohammad Rasoul Khazaei, et Mozafar Khazaei. 2021. "Potential therapeutic effect of bee pollen and metformin combination on testosterone and estradiol levels, apoptotic markers and total antioxidant capacity in a rat model of polycystic ovary syndrome". *International Journal of Fertility and Sterility* 15(2). Royan Institute: 101.

Negrao, Adriana F, et Ricardo O Orsi. 2018. "Harvesting season and botanical origin interferes in production and nutritional composition of bee pollen". *Anais da Academia Brasileira de Ciências* 90(1). SciELO, Brasil: 325–32.

Nguyen, Truyen, Paul Nioi, et Cecil B Pickett. 2009. "The Nrf2-antioxidant response element signaling pathway and its activation by oxidative stress". *Journal of Biological Chemistry* 284(20). ASBMB: 13291–95.

Niaz, Kamal, Fazlullah Khan, et Muhammad Ajmal Shah. 2020. "Analysis of carbohydrates (monosaccharides, polysaccharides)". In Seyed Mohammad Nabavi, Mina Saeedi, Seyed Fazel Nabavi, and et Ana Sanches Silva (Eds), *Recent Advances in Natural Products Analysis*. Elsevier, 621–33.

Nicolson, Susan W. 2011. "Bee food: The chemistry and nutritional value of nectar, pollen and mixtures of the two". *African Zoology* 46(2). Taylor & Francis: 197–204.

Nicolson, Susan W, Susana Da Silva Das Neves, Hannelie Human, et Christian WW Pirk. 2018. "Digestibility and nutritional value of fresh and stored pollen for honey bees (Apis Mellifera Scutellata)". *Journal of Insect Physiology* 107(mai): 302–8. doi:10.1016/j.jinsphys.2017.12.008.

Nogueira, Carla, Antonio Iglesias, Xesus Feás, et Leticia M Estevinho. 2012. "Commercial bee pollen with different geographical origins: A comprehensive approach". *International Journal of Molecular Sciences* 13(9). Molecular Diversity Preservation International (MDPI): 11173–87.

Oroian, Mircea, Florina Dranca, et Florin Ursachi. 2022. "Characterization of Romanian bee pollen—An important nutritional source". *Foods* 11(17). MDPI: 2633.

Othman, Zaidatul Akmal, Zaida Zakaria, Joseph Bagi Suleiman, Victor Udo Nna, Aminah Che Romli, Wan Syaheedah Wan Ghazali, et Mahaneem Mohamed. 2021. "Bee bread ameliorates vascular inflammation and impaired vasorelaxation in obesity-induced vascular damage rat model: The role of eNOS/NO/cGMP-signaling pathway". *International Journal of Molecular Sciences* 22(8). Multidisciplinary Digital Publishing Institute: 4225.

Pena, Eduardo, Samia El Alam, Patricia Siques, et Julio Brito. 2022. "Oxidative stress and diseases associated with high-altitude exposure". *Antioxidants* 11(2). MDPI: 267.

Peraica, Maja, B Radić, A Lucić, et M Pavlović. 1999. "Toxic effects of mycotoxins in humans". *Bulletin of the World Health Organization* 77(9). World Health Organization: 754.

Perugini, Monia, Maurizio Manera, Lisa Grotta, Maria Cesarina Abete, Renata Tarasco, et Michele Amorena. 2011. "Heavy metal (Hg, Cr, Cd, and Pb) contamination in urban areas and wildlife reserves: Honeybees as bioindicators". *Biological Trace Element Research* 140(2). Springer: 170–76.

Perumal, Nithiyaa, Meenakshii Nallappan, Shamarina Shohaimi, Nur Kartinee Kassim, Thiam Tsui Tee, et Yew Hoong Cheah. 2022. "Synergistic antidiabetic activity of Taraxacum officinale (L.) Weber ex FH Wigg and Momordica charantia L. polyherbal combination". *Biomedicine and Pharmacotherapy* 145. Elsevier: 112401.

Pitt, JI. 2000. "Toxigenic fungi and mycotoxins". *British Medical Bulletin* 56(1). Oxford University Press: 184–92.

Prđun, Saša, Lidija Svečnjak, Mato Valentić, Zvonimir Marijanović, et Igor Jerković. 2021. "Characterization of bee pollen: Physico-chemical properties, headspace composition and FTIR spectral profiles". *Foods* 10(9). MDPI: 2103.

Qian, Wei Liang, Zeeshan Khan, David G Watson, et James Fearnley. 2008. "Analysis of sugars in bee pollen and propolis by ligand exchange chromatography in combination with pulsed amperometric detection and mass spectrometry". *Journal of Food Composition and Analysis* 21(1). Elsevier: 78–83.

Quigley, P Tyler, Gro V. Amdam, et Olav Rueppell. 2018. "Honeybee workers as models of aging". In *Conn's Handbook of Models for Human Aging*, 533–47. Elsevier. doi:10.1016/B978-0-12-811353-0.00040-3.

Ramli, Nur Amalina, et Sofiah Hamzah Nora'aini Ali. 2019. "Indo-Malayan stingless bees' propolis extract-loaded Chitosomes: Characterization and storage stability evaluation". *International Journal of Recent Technology and Engineering* 7(6): 560–67.

Rasouli, Hassan, Mohammad Hosein Farzaei, et Reza Khodarahmi. 2017. "Polyphenols and their benefits: A review". *International Journal of Food Properties* 20(Sup2). Taylor & Francis: 1700–41.

Rebelo, Kemilla Sarmento, Antônio Gilberto Ferreira, et Gislene Almeida Carvalho-Zilse. 2016. "Physicochemical characteristics of pollen collected by Amazonian stingless bees". *Ciência Rural* 46(5). SciELO: 927–32.

Reverberi, Massimo, Alessandra Ricelli, Slaven Zjalic, Anna A Fabbri, et Corrado Fanelli. 2010. "Natural functions of mycotoxins and control of their biosynthesis in fungi". *Applied Microbiology and Biotechnology* 87(3). Springer: 899–911.

Roman, Adam. 2007. "Content of some trace elements in fresh honeybee pollen". *Polish Journal of Food and Nutrition Sciences* 57(4C): 475–78.

Roman, Adam, Ewa Popiela-Pleban, Paweł Migdał, et Wojciech Kruszyński. 2016. "As, Cr, Cd, and Pb in bee products from a Polish industrialized region". *Open Chemistry* 14(1). De Gruyter Open Access: 33–36.

Rouge-Bugat, Marie-Eve, Donia Lassoued, Joy Bacrie, Nathalie Boussier, Jean-Pierre Delord, Stéphane Oustric, Eric Bauvin, Maryse Lapeyre-Mestre, François Bertucci, et Pascale Grosclaude. 2015. "Guideline sheets on the side effects of anticancer drugs are useful for general practitioners". *Supportive Care in Cancer* 23(12). Springer: 3473–80.

Rzepecka-Stojko, Anna, Jerzy Stojko, Anna Kurek-Górecka, Michał Górecki, Agata Kabała-Dzik, Robert Kubina, Aleksandra Moździerz, et Ewa Buszman. 2015. "Polyphenols from bee pollen: Structure, absorption, metabolism and biological activity". *Molecules* 20(12). Multidisciplinary Digital Publishing Institute: 21732–49.

Saisavoey, Tanatorn, Papassara Sangtanoo, Chanpen Chanchao, Onrapak Reamtong, et Aphichart Karnchanatat. 2021. "Identification of novel anti-inflammatory peptides from bee pollen (Apis mellifera) hydrolysate in lipopolysaccharide-stimulated RAW264. 7 macrophages". *Journal of Apicultural Research* 60(2). Taylor & Francis: 280–89.

Sánchez-Peña, M Judith, Fabiola Márquez-Sandoval, Ana C Ramírez-Anguiano, Sandra F Velasco-Ramírez, Gabriela Macedo-Ojeda, et Luis J González-Ortiz. 2017. "Calculating the metabolizable energy of macronutrients: A critical review of Atwater's results". *Nutrition Reviews* 75(1). United States: 37–48. doi:10.1093/nutrit/nuw044.

Saraiva, Louise CF, Francisco VM Cunha, Daniel ROD Léllis, et Lívio CC Nunes. 2018. "Composition, biological activity and toxicity of bee pollen: State of the art". *Boletin Latinoamericano y del Caribe de Plantas Medicinales y Aromaticas* 17(5): 426–440.

Saral, Ozlem, Eda Dokumacioglu, Sinan Saral, Aysegul Sumer, Ozgur Bulmus, Seyma Ozer Kaya, et Sinan Canpolat. 2021. "The effect of bee pollen on reproductive and biochemical parameters in methotrexate-induced testicular damage in adult rats". *Journal of Basic and Clinical Physiology and Pharmacology* 32(5). De Gruyter: 1001–6.

Sawicki, Tomasz, Małgorzata Starowicz, Lucyna Kłębukowska, et Paweł Hanus. 2022. "The profile of polyphenolic compounds, contents of total phenolics and flavonoids, and antioxidant and antimicrobial properties of bee products". *Molecules* 27(4). MDPI: 1301.

Sayeed, Bin, Muhammad Shahdaat, Selim Muhammad Rezaul Karim, Tasnuva Sharmin, et Mohammed Monzur Morshed. 2016. "Critical analysis on characterization, systemic effect, and therapeutic potential of beta-sitosterol: A plant-derived orphan phytosterol". *Medicines* 3(4). MDPI: 29.

Shaldoum, Fayez, Attalla F El-kott, Marwa Mohamed Ahmed Ouda, et Eman M Abd-Ella. 2021. "Immunomodulatory effects of bee pollen on doxorubicin-induced bone marrow/spleen immunosuppression in rat". *Journal of Food Biochemistry* 45(6). Wiley Online Library: e13747.

Sharma, Rajesh Kumar, et Madhoolika Agrawal. 2005. "Biological effects of heavy metals: An overview". *Journal of Environmental Biology* 26(2): 301–13.

Sharma, Meenakshi, Renu Chadha, et Neelima Dhingra. 2017. "Phytotherapeutic agents for benign prostatic hyperplasia: An overview". *Mini-Reviews in Medicinal Chemistry* 17(14). Bentham Science Publishers: 1346–63.

Shen, Yanting, Xiao Tang, et Yonghui Li. 2021. "Drying methods affect physicochemical and functional properties of quinoa protein isolate". *Food Chemistry* 339. Elsevier: 127823.

Shen, Zhenhuang, Qianqian Geng, Haibo Huang, Hong Yao, Tianyu Du, Lifu Chen, Zhenhong Wu, Xiaoqing Miao, et Peiying Shi. 2019. "Antioxidative and cardioprotective effects of Schisandra chinensis bee pollen extract on isoprenaline-induced myocardial infarction in rats". *Molecules* 24(6). MDPI: 1090.

Siddiqui, Arif Jamal, Sadaf Jahan, Ritu Singh, Juhi Saxena, Syed Amir Ashraf, Andleeb Khan, Ranjay Kumar Choudhary, Santhanaraj Balakrishnan, Riadh Badraoui, et Fevzi Bardakci. 2022. "Plants in anticancer drug discovery: from molecular mechanism to chemoprevention". *BioMed Research International* 2022. Hindawi, p. 18.

Silva, Girliane Regina da, Tamires Botelho da Natividade, Celso Amorim Camara, Eva Monica Sarmento da Silva, Francisco de Assis Ribeiro dos Santos, et Tania Maria Sarmento Silva. 2014. "Identification of sugar, amino acids and minerals from the pollen of Jandaíra stingless bees (Melipona subnitida)". *Food and Nutrition Sciences* 5. Scientific Research Publishing, pp. 1015–1021.

Silva, João Carlos, Sandra Rodrigues, Xesús Feás, et Leticia M Estevinho. 2012. "Antimicrobial activity, phenolic profile and role in the inflammation of propolis". *Food and Chemical Toxicology* 50(5). Elsevier: 1790–95.

Somerville, DC, and HI Nicol. 2006. "Crude protein and amino acid composition of honey bee-collected pollen pellets from south-east Australia and a note on laboratory disparity". *Australian Journal of Experimental Agriculture* 46(1): 141–9.

Sonmez, Emine, Meral Kekecoglu, Arif Bozdeveci, et Sengul Alpay Karaoglu. 2022. "Chemical profiling and antimicrobial effect of Anatolian honey bee venom". *Toxicon* 213. Elsevier: 1–6.

Souza, Bianca de Rodrigues, José Augusto Gasparotto Sattler, Adriane Alexandre Machado de Melo, Daniel Granato, Ortrud Monika Barth, et Aroni Sattler. 2018. "Physicochemical parameters and content of B-complex vitamins: An exploratory study of bee pollen from Southern Brazilian states" *Revista Chilena de Nutricion* 45: 232–242.

Speakman, Mark, Roger Kirby, Scott Doyle, et Chris Ioannou. 2015. "Burden of male lower urinary tract symptoms (LUTS) suggestive of benign prostatic hyperplasia (BPH)–focus on the UK". *BJU International* 115(4). Wiley Online Library: 508–19.

Spulber, Roxana, Muhsin Dogaroglu, Narcisa Babeanu, et Ovidiu Popa. 2018. "Physicochemical characteristics of fresh bee pollen from different botanical origins". *Romanian Biotechnology Letters* 23: 13357–65.

Suran, Jelena. 2022. "Beehives as a natural source of novel antimicrobials". In Mahendra Rai and et Ivan Kosalec (Eds), *Promising Antimicrobials from Natural Products*. Springer, 373–95.

Taha, El-Kazafy A, Saad Al-Kahtani, et Reda Taha. 2019. "Protein content and amino acids composition of bee-pollens from major floral sources in Al-Ahsa, eastern Saudi Arabia". *Saudi Journal of Biological Sciences* 26(2): 232–37. doi:10.1016/j.sjbs.2017.06.003.

Thakur, Mamta, et Vikas Nanda. 2020a. "Composition and functionality of bee pollen: A review". *Trends in Food Science and Technology* 98. Elsevier: 82–106.

———. 2020b. "Exploring the physical, functional, thermal, and textural properties of bee pollen from different botanical origins of India". *Journal of Food Process Engineering* 43(1). Wiley Online Library: e12935.

Themelis, Thomas, Roberto Gotti, Serena Orlandini, et Rita Gatti. 2019. "Quantitative amino acids profile of monofloral bee pollens by microwave hydrolysis and fluorimetric high performance liquid chromatography". *Journal of Pharmaceutical and Biomedical Analysis* 173(septembre): 144–53. doi:10.1016/j.jpba.2019.05.031.

Theumer, MG, AG Lopez, DT Masih, SN Chulze, et HR Rubinstein. 2003. "Immunobiological effects of AFB1 and AFB1–FB1 mixture in experimental subchronic mycotoxicoses in rats". *Toxicology* 186(1–2). Elsevier: 159–70.

Thompson, Helen M. 2010. "Risk assessment for honey bees and pesticides—Recent developments and 'new issues'". *Pest Management Science* 66(11). Wiley Online Library: 1157–62.

Thorpe, A, et D Neal. 2003. "Benign prostatic hyperplasia". *The Lancet* 361(9366). Elsevier: 1359–67.

Tihelka, Erik. 2018. "Effects of synthetic and organic acaricides on honey bee health: A review". *Slovenian Veterinary Research* 55(3). Veterinarska Fakulteta, Univerza v. Ljubljani: 114–40.

Tong, Zhou, Jinsheng Duan, Yancan Wu, Qiongqiong Liu, Qibao He, Yanhong Shi, Yu Linsheng, et Haiqun Cao. 2018. "A survey of multiple pesticide residues in pollen and beebread collected in China". *Science of the Total Environment* 640. Elsevier: 1578–86.

Toth, Tomas, Miriama Kopernicka, Rastislav Sabo, et Tatiana Kopernicka. 2016. "The evaluation of mercury in honey bees and their products from Eastern Slovakia". *[Scientific Papers]: Animal Science & Biotechnologies/Lucrari Stiintifice: Zootehnie si Biotehnologii* 49(1): 257–260.

Tungmunnithum, Duangjai, Areeya Thongboonyou, Apinan Pholboon, et Aujana Yangsabai. 2018. "Flavonoids and other phenolic compounds from medicinal plants for pharmaceutical and medical aspects: An overview". *Medicines* 5(3). MDPI: 93.

Tuoheti, Tuhanguli, Hafiz Abdul Rasheed, Ling Meng, et Ming sheng Dong. 2020. "High hydrostatic pressure enhances the anti-proliferative properties of lotus bee pollen on the human prostate cancer PC-3 cells via increased metabolites". *Journal of Ethnopharmacology* 261. Elsevier: 113057.

Ulbricht, Catherine, Julie Conquer, Nicole Giese, Karta Purkh Singh Khalsa, Joshua Sklar, Wendy Weissner, et Jen Woods. 2009. "An evidence-based systematic review of bee pollen by the Natural Standard Research Collaboration". *Journal of Dietary Supplements* 6(3). Taylor & Francis: 290–312.

Unuofin, Jeremiah Oshiomame, et Sogolo Lucky Lebelo. 2020. "Antioxidant effects and mechanisms of medicinal plants and their bioactive compounds for the prevention and treatment of type 2 diabetes: An updated review". *Oxidative Medicine and Cellular Longevity* 2020. Hindawi, p. 36.

Végh, Rita, Mariann Csóka, Csilla Sörös, et László Sipos. 2021. "Food safety hazards of bee pollen–A review". *Trends in Food Science and Technology* 114. Elsevier: 490–509.

Villanueva, MT, A Orzáez, Díaz Marquina, R Bravo Serrano, et G Blazquez Abellán. 2002. "The importance of bee-collected pollen in the diet: A study of its composition". *International Journal of Food Sciences and Nutrition* 53(3). Taylor & Francis: 217–24.

Wagenlehner, Florian ME, Henning Schneider, Martin Ludwig, Jörg Schnitker, Elmar Brähler, et Wolfgang Weidner. 2009. "A pollen extract (cernilton) in patients with inflammatory chronic prostatitis–chronic pelvic pain syndrome: A multicentre, randomised, prospective, double-blind, placebo-controlled phase 3 study". *European Urology* 56(3). Elsevier: 544–51.

Wang, Junqiao, Jing Zhao, Shaoping Nie, Mingyong Xie, et Shaoping Li. 2022. "MALDI mass spectrometry in food carbohydrates analysis: A review of recent researches". *Food Chemistry*. Elsevier: 133968.

Wani, Idrees A, Dalbir S Sogi, et Balmeet S Gill. 2013a. "Physicochemical and functional properties of flours from three black gram (*Phaseolus mungo* L.) cultivars". *International Journal of Food Science and Technology* 48(4). Wiley Online Library: 771–77.

Wani, Idrees Ahmed, Dalbir Singh Sogi, Ali Abas Wani, et Balmeet Singh Gill. 2013b. "Physico-chemical and functional properties of flours from Indian kidney bean (*Phaseolus vulgaris* L.) cultivars". *LWT – Food Science and Technology* 53(1). Elsevier: 278–84.

Wesseling, Catharina, Marianela Corriols, et Viria Bravo. 2005. "Acute pesticide poisoning and pesticide registration in Central America". *Toxicology and Applied Pharmacology* 207(2). Elsevier: 697–705.

Weyers, Steven. 2020. "The effect of the non-estrogenic pollen extract PCC-100 on hot flushes during adjuvant hormonal treatment for breast cancer in postmenopausal women and in menopausal women without a history of breast cancer". clinicaltrials.gov. Clinical Trial Registration NCT03684096, Mar. 2020. Accessed: Oct. 16, 2022 [Online]. Available: https://clinicaltrials.gov/ct2/show/NCT03684096.

Whitford, David. 2013. *Proteins: Structure and Function*. John Wiley & Sons.

Wu, Yao-Dong, et Yi-Jia Lou. 2007. "A steroid fraction of chloroform extract from bee pollen of Brassica campestris induces apoptosis in human prostate cancer PC-3 cells". *Phytotherapy Research: An International Journal Devoted to Pharmacological and Toxicological Evaluation of Natural Product Derivatives* 21(11). Wiley Online Library: 1087–91.

Wu, Shuangmei, Chen Xu, Yiwen Zhu, Liang Zheng, Ludan Zhang, Yang Hu, Yu Bingran, Yuguang Wang, et Fu-Jian Xu. 2021. "Biofilm-sensitive photodynamic nanoparticles for enhanced penetration and antibacterial efficiency". *Advanced Functional Materials* 31(33). Wiley Online Library: 2103591.

Yamaguchi, Masayoshi, Reiko Hamamoto, Satoshi Uchiyama, Kaori Ishiyama, et Ken Hashimoto. 2007. "Preventive effects of bee pollen Cistus ladaniferus extract on bone loss in streptozotocin-diabetic rats in vivo". *Journal of Health Science* 53(2). The Pharmaceutical Society of Japan: 190–95.

Yang, Kai, Dan Wu, Ye Xingqian, Donghong Liu, Jianchu Chen, et Peilong Sun. 2013. "Characterization of chemical composition of bee pollen in China". *Journal of Agricultural and Food Chemistry* 61(3). ACS Publications: 708–18.

Yang, Xian, Liuliu Yuan, Chaomei Xiong, Chunping Yin, et Jinlan Ruan. 2014. "Abacopteris penangiana exerts testosterone-induced benign prostatic hyperplasia protective effect through regulating inflammatory responses, reducing oxidative stress and anti-proliferative". *Journal of Ethnopharmacology* 157. Elsevier: 105–13.

Yang, Siwen, Yunhe Qu, Jiyu Chen, Si Chen, Lin Sun, Yifa Zhou, et Yuying Fan. 2021. "Bee pollen polysaccharide from Rosa rugosa Thunb.(Rosaceae) promotes pancreatic β-cell proliferation and insulin secretion". *Frontiers in Pharmacology* 12. Frontiers Media SA, p. 688073.

Yook, Hong-Sun, Seong-Il Lim, et Myung-Woo Byun. 1998. "Changes in microbiological and physicochemical properties of bee pollen by application of gamma irradiation and ozone treatment". *Journal of Food Protection* 61(2). International Association for Food Protection: 217–20.

Zafeiraki, Effrosyni, Konstantinos M Kasiotis, Paul Nisianakis, Electra Manea-Karga, et Kyriaki Machera. 2022. "Occurrence and human health risk assessment of mineral elements and pesticides residues in bee pollen". *Food and Chemical Toxicology* 161. Elsevier: 112826.

Zeedan, A Battaa Elneney, Ahmed Abuoghaba, et Khaled El-Kholy. 2017. "Effect of bee pollen at different levels as natural additives on immunity and productive performance in rabbit males". *Egypt. Poultry Science* 37(janvier): 213–31.

Zhang, Jinjin, Wei Cao, Haoan Zhao, Sen Guo, Qian Wang, Ni Cheng, et Naisheng Bai. 2022. "Protective mechanism of Fagopyrum esculentum Moench. Bee pollen EtOH extract against Type II diabetes in a high-fat diet/streptozocin-induced C57BL/6J mice". *Frontiers in Nutrition* 9: 925351.

Zhao, Huakan, Lei Wu, Guifang Yan, Yu Chen, Mingyue Zhou, Yongzhong Wu, et Yongsheng Li. 2021. "Inflammation and tumor progression: Signaling pathways and targeted intervention". *Signal Transduction and Targeted Therapy* 6(1). Nature Publishing Group: 1–46.

Zheng, Michael Z, John L Richard, et Johann Binder. 2006. "A review of rapid methods for the analysis of mycotoxins". *Mycopathologia* 161(5). Springer: 261–73.

Zhou, Wangting, Ya Zhao, Yamei Yan, Jia Mi, Lu Lu, Qing Luo, Xiaoying Li, Xiaoxiong Zeng, et Youlong Cao. 2020. "Antioxidant and immunomodulatory activities in vitro of polysaccharides from bee collected pollen of Chinese wolfberry". *International Journal of Biological Macromolecules* 163. Elsevier: 190–99.

Zlatev, Zlatin, Ira Taneva, Stanka Baycheva, et Mitko Petev. 2018. "A comparative analysis of physicochemical indicators and sensory characteristics of yogurt with added honey and bee pollen". *Bulgarian Journal of Agricultural Science* 24: 132–44.

Zuluaga-Domínguez, Carlos, Juan Serrato-Bermudez, et Marta Quicazán. 2018. "Influence of drying-related operations on microbiological, structural and physicochemical aspects for processing of bee-pollen". *Engineering in Agriculture, Environment and Food* 11(2). Elsevier: 57–64.

9 Propolis

Syeda Tasmia Asma, Ulas Acaroz, Otilia Bobiş, Syed Rizwan Ali Shah, Damla Arslan-Acaroz, Narimane Segueni, Anna Kurek-Górecka, Paweł Olczyk, Vikas Nanda and Gulzar Ahmad Nayik

9.1 INTRODUCTION

The name "propolis" originated from two Greek terms in combination, "pro" (means in defense of) and "polis" (means city), meaning defender of the city, providing a defensive effect to hives (Simone-Finstrom et al., 2017). Propolis (commonly known as bee glue) is a sticky, balsamic, and resinous substance released from different sources of plants (including flowers, bud exudates, and plant leaves). Propolis is prepared by the bees (*Apis mellifera*) by merging the plant-based sap with beeswax and salivary secretions, resulting in the formation of darker green, adhesive material used by honey bees to cover and protect the hive against microbes and predators (Bankova, 2005; Martinotti & Ranzato, 2015). Propolis has been effectively used as a nutrition-rich and therapeutic product for the well-being of humans since ancient times (Wagh, 2013).

Propolis shows numerous biological properties due to its bioactive compounds, including antibacterial (Elkhateeb et al., 2022), antiviral (Asma et al., 2022; Kontogiannis et al., 2022; Silva-Beltrán et al., 2022), antifungal (Afata et al., 2022; Gülbandılar, 2022), free radical scavenging (Chien et al., 2022), antioxidant (Al-Juhaimi et al., 2022), antitumor (Oršolić & Jazvinšćak Jembrek, 2022), antiparasitic (Elmahallawy et al., 2022), hepato-protective (Hasan et al., 2022), immunomodulatory (AbdullKadem et al., 2022), skin-regeneration (Elkhateeb et al., 2022), and cytotoxic (de Oliveira Cardoso et al., 2022) activities have been reported in several studies. The antibacterial, antifungal, and antioxidant effects of propolis extracts are crucial in food industries because they can delay the process of lipid peroxidation and increase the shelf life of several food items (Namae, 2022; Oliveira et al., 2022; Tavares et al., 2022).

9.2 COMPOSITION OF PROPOLIS

Propolis has been characterized into different classes depending on the plant sources and the geographic locations. The most commonly known propolis types with their geographic locations include red propolis (originating from Mexico, Cuba, and Brazil), poplar propolis (originating from Asia, Europe, New Zealand, and North America), Brazilian propolis (originating from Brazil), Pacific propolis (originating from Indonesia, Taiwan, and Okinawa), green propolis (originating from Alecrim), birch propolis (originating from Russia), Mediterranean propolis (originating from Malta, Greece, Sicily, and Crete), and Chinese propolis (originating from China). Those from Brazil and New Zealand are the most eminent propolis types, which grabbed the attention of researchers and commercialists (Bhargava et al., 2021). In former times, propolis quality was evaluated based on botanical sources, species of honey bees, and weather conditions. Several analytical techniques. such as mass spectrometry (MS), gas chromatography (GC), gas chromatography-mass spectrometry (GC-MS), thin layer chromatography (TLC), high-performance liquid chromatography (HPLC), and nuclear magnetic resonance (NMR), have been employed to conduct evaluations of propolis chemotypes. Raw propolis comprises plant resins, different essential oils, pollens, waxes, and some organic substances in varying amounts (Fokt et al., 2010), as represented in Table 9.1 and Figure 9.1.

DOI: 10.1201/9781003354116-9

TABLE 9.1
Some Newly Identified Compounds in Propolis

Geographical origin	Type of compound	Chemical name	Reference
Cameroon	Fatty acid	Hexatriacontanoic acid	(Mahamat et al., 2020)
	Isoflavonol	2′,3′-dihydroxypropyltetraeicosanoate 2-hydroxy-8-prenylbinochanin A Triacontyl p-coumarate	(Tamfu et al., 2020)
Colombia	Vitamin E like δ-tocotrienol derivative	5-hydroxy-8b-oxo-garcinoic acid (trans-13′carboxy-5-hydroxy-8b-oxo-d- tocotrienol) Garcinoic acid (trans-13′carboxy-d-tocotrienol) 5,6-dioxo-garcinoic acid (trans-13′carboxy-5,6dioxo-d-tocotrienol)	(Popova et al., 2020)
Brazil	Cinnamic acid derivatives	5-isoprenyl caffeic acid, p-coumaric-acid ester	(Xu et al., 2020)
		(E)-3-[4-hydroxy-3-((E)-3-formyl-2-butenyl) phenyl]-2-propenoic acid	(Tani et al., 2019)
	Flavanoid dimers	Propolol A Propolones (A–D) Propolonones (A–C)	(Banzato et al., 2020)
Indonesia	–	Sulawesin-A Sulawesin-B 4-(4′-hydroxy-3′-methoxyphenyl)-3,5,7-tri-hydroxycoumarin	(Miyata et al., 2019)
Libya	Flavanonol derivative	Taxifolin-3-acetyl-4′-methyl ether	(Siheri et al., 2019)
Midwest Brazil	Prenylated aromatic compound	(E)-3-[4-hydroxy-3-(2-hydroxy-3-methylbut-3-en-1-yl)-5-(3-methylbut-2-en-1-yl) phenyl] propenoic acid	(Costa Jr et al., 2020)
Morocco	Alkaloids	Trigonelline	(Belmehdi et al., 2021)
Pitcairn Island	Cycloartane triterpenes	3-oxo-cycloart-24E-en-21,26-diol-21-acetate 3-oxo-cycloart-24E-en-21,26-diol-26-acetate 3-oxo-cycloart-24E-en-21,26-diol Diacetate of 3-oxo-9,19-cycloartane triterpenoid	(Georgieva et al., 2018)
Spain	Igstroside derivative	p-HPEA-elenolic acid monoaldehyde	(Fernández-Calderón et al., 2020)
	Lignane	1-Acetoxypinoresinol	
	Oleuperin derivative	3,4-DHPEA-elenolic acid dialdehyde	
	Phenolic acid	Vanillic acid	
Vietnam	Homoisoflavanes	10,11-dihydroxydracaenone C (3R)-7,4′-dihydroxyhomoisoflavane	(Georgieva et al., 2019)
	Flavane	(2R)-7,4′dihydroxy-5-methoxy-8methylflavane	
	Xanthones	7-geranyloxy-1,3-dihydroxyx-anthone A mixture of 3-geranyloxy-1,7-dihydroxy-xanthone (cochinchinone G)	
Zambia and Tanzania	Flavanones	5-Hydroxy-4″,4″-dimethyl-5″methyl-5″-H-dihydro-furanol [2″,3″,6,7] flavanone 6-(1,1-dimethylallyl) pinocembrin	(Alenezi et al., 2020)

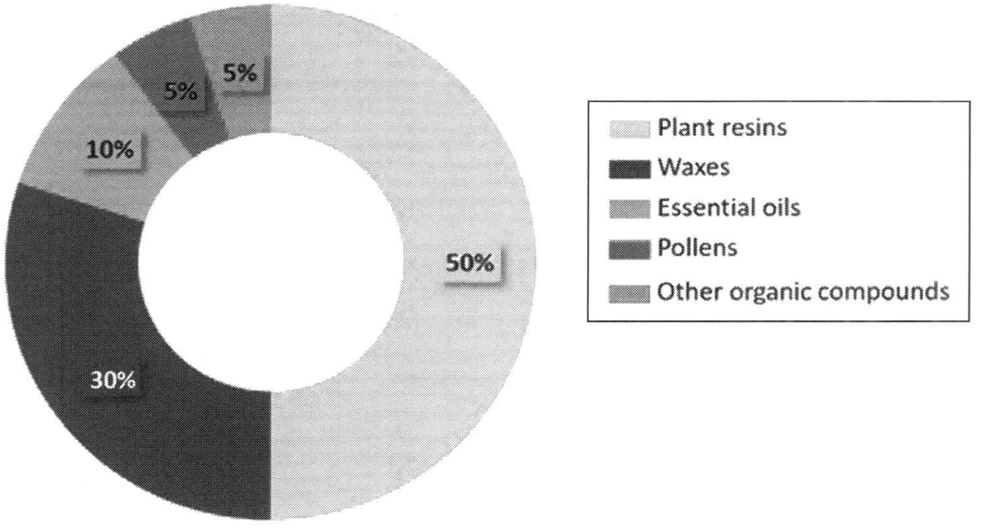

FIGURE 9.1 Chemical composition of propolis.

9.2.1 Phenolic Compounds

Several investigations have revealed that the phenolic components in propolis are responsible for its pharmacological activities due to the significant antiradical impacts of phenolic acids, flavonoid compounds, and their esters (Galeotti et al., 2018; Martinello & Mutinelli, 2021; Woźniak et al., 2020). Touzani et al. evaluated the phenolic components from different samples of propolis obtained from Asian and African origins, showing high antioxidant and antimicrobial activities interrelated with the high content of phenolic components (Touzani et al., 2021). Many investigators have documented that the total phenolic and flavonoid contents are crucial due to their potential biological activities, such as antimicrobial activities (Gajger et al., 2017; Hudz et al., 2020).

9.2.2 Phenolic Acids

Mezgabu et al. identified coumaric acid, caffeic acid, chlorogenic acid, and gallic acid as phenolic acids in an Algerian propolis sample (Mezgabu et al., 2016). Rufatto et al. reported that Brazilian red propolis is rich in major phenolic acids like ferulic acid, caffeic acid, isoferulic acid, and coumaric acid (Rufatto et al., 2018).

9.2.3 Flavonoids

Flavonoids have been studied as key components of propolis, possessing several pharmacological activities. The flavonoid content is a standard to determine propolis quality (Cui-ping et al., 2014). Flavonoids have been documented to exhibit several biological properties, including antimicrobial (antiviral, antifungal, and antibacterial) and anti-inflammatory activities (Boulechfar, Zellagui et al., 2022; Cerqueira et al., 2022; Kustiawan et al., 2022). Flavonoids identified in propolis have been categorized based on their chemical configurations, such as flavanols, flavonols, flavanones, and flavones (Asma et al., 2022).

9.2.4 Terpenoids

Terpenes have been identified as possessing anti-inflammatory, antimicrobial, and antitumor effects (Afrouzan et al., 2018). These terpenoids can be categorized into three categories: mono- and sesquiterpenoids, diterpenoids, and triterpenoids. Sesquiterpenes (volatiles) are observed in propolis at about 1% concentration. The biological and aromatic characteristics of volatile compounds make them essential for propolis characterization (Bankova et al., 2014). Santos et al. identified six different diterpenes: cis-communic acid, trans-communic acid, abietic acid, epi-cupressic acid, iso-cupressic acid, and epi-13-torulosol from Brazilian brown propolis (Santos et al., 2021). The structures of some mono- and sesquiterpenes, diterpenes, and triterpenes are presented in Figure 9.2.

9.3 PROPOLIS PROCESSING

Nowadays, propolis is gaining popularity all over the world for nutritional and therapeutic purposes. The crude propolis can be extracted by applying different methods. The method selection for extract preparation is crucial for the propolis product's high quality and economic yield. The most widely used conventional techniques include maceration and Soxhlet extraction (Socha et al., 2015). Generally, conventional extraction techniques need large-volume extraction solvents and are time-consuming. To overcome the shortcomings of conventional methods, researchers have modified

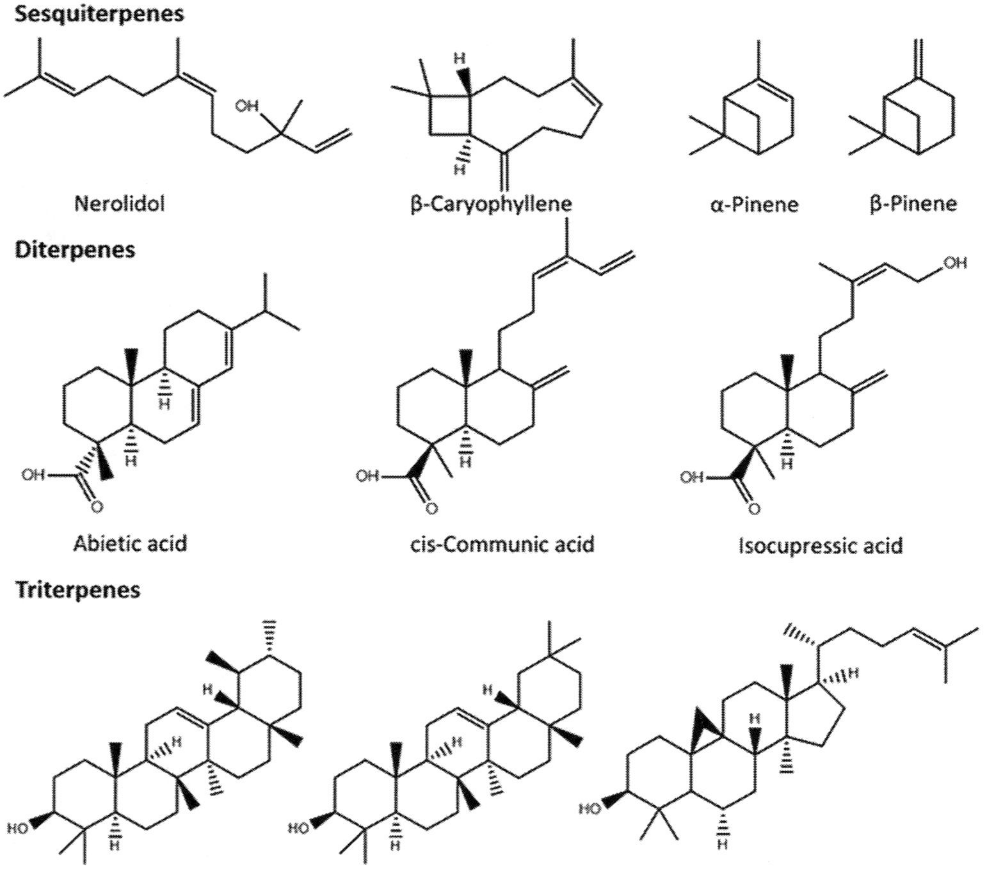

FIGURE 9.2 Structures of some mono- and sesquiterpenes, diterpenes, and triterpenes.

the conventional techniques and designed modern extraction methods. Current methods include microwave-based extraction, ultrasound-based extraction, solid-phase extraction, supercritical fluid extraction, and high-pressure extraction.

9.3.1 Processing of Powder Propolis

The processing of propolis extract in a powder state mostly begins with conventionally prepared liquid extracts using volatile extraction solvents like methanol (MeOH), ethanol, water, acetone, 2-propanol (2-PrOH), dimethylsulfoxide (DMSO), glacial acetic acid, etc. (Sosnowski, 1983). Later, the obtained liquid extracts can be evaporated to prepare dried or powder extracts. Furthermore, the liquid extracts can also be dealt with by a spray-drying technique or by fluidized bed technology with a suitable solid carrier and subsequently with or without adding a small amount of colloidal silicon dioxide (SiO_2), magnesium stearate ($Mg(C_{18}H_{35}O_2)_2$), or any related anticaking agent.

On the other hand, propolis extract in a powder state can also be directly processed by supercritical fluid extraction (SFE). Over the last two decades, SFE has been significantly employed to extract bioactive components from different food materials (Gomez-Lopez et al., 2022). This technology overcomes the shortcomings of conventional extraction methods by providing various benefits such as lower energy consumption, controlled pressure and temperature, ease of extraction process, and solvent-free extraction (West, 2019; Yousefi et al., 2019).

Spray-drying extraction can also be performed to prepare a solid extract, which is encapsulated/carried by pharma- or food-based materials. This type of extraction technology is named encapsulation. Encapsulation technology offers enhanced physical and chemical stability against humidification causing caking (Busch et al., 2017; Sato et al., 2020; Šturm et al., 2019).

9.4 PROPERTIES AND HEALTH BENEFITS OF PROPOLIS

A type of bee product, propolis is identified as the most powerful natural antibiotic. It protects against internal diseases caused by fungi, yeast, or bacteria, and also protects the beehive from outside dangers or insect predators (Bankova, 2005). The raw material is represented by the vegetal material of resinous origin from different parts of the plants and wax added by the bee (Anjum et al., 2019). The resinous materials are found as exudates on leaves and buds, and resins and gums on tree branches or trunks of the trees. It may have different colors depending on its botanical and geographical origin (Figure 9.3).

9.4.1 Propolis Properties

Propolis components have three distinctive primary sources: vegetal sources (plant exudates and resins from different tree buds – poplar, birch, pine, chestnut, tropical plants – and lipophilic substances from tree wounds); animal sources: bee wax and saliva; and finally, accidental materials from the "production chain" – honey, propolis, or pollen (Greenaway et al., 1990). Determination of the physical parameters of propolis is based mainly on direct observation and includes color, appearance, consistency, taste, and odor (Bogdanov & Bankova, 2016; López et al., 2003). The color of propolis varies considerably with the botanical plant source and geographical origin from yellow to red, brown, or green (Dezmirean et al., 2020). The major botanical sources of brown propolis (poplar-type, characteristic of the temperate zones of Europe, North America, Russia, New Zealand, and Australia) are chestnut (*Aesculus hippocastanum*), beech (*Fagus* sp.), oak (*Quercus* sp.), silver birch buds (*Betula verrucosa*), *Populus* species buds, pine (*Pinus* sp.), and willow (*Salix* sp.) (Bankova et al., 2002; Garcia-Viguera et al., 1993; Greenaway et al., 1988; Isidorov et al., 2014; Markham et al., 1996; Popova et al., 2013).

Green propolis mainly originates from Brazil and is the main botanical source of resins via apical buds and fresh leaves of *Baccharis dracunculifolia* (Berretta et al., 2017; Moise & Bobiş, 2020).

FIGURE 9.3 Ground propolis samples according to different geographical origins. (Original images courtesy of APHIS Laboratory, University of Agricultural Sciences and Veterinary Medicine Cluj-Napoca, Romania. Otilia Bobiş.)

Red propolis mainly originates from Brazil and Cuba but has different botanical sources. While the bees produce Cuban propolis from *Clusia rosea* resin (Hossain et al., 2022), Brazilian red propolis has as the key botanical source the red resinous exudates from the branches of *Dalbergia ecastophyllum* L. (Moise & Bobiş, 2020; Salatino et al., 2021). *Apis mellifera* bees produce the above-described propolis types, but in tropical South America, another variety of propolis is formed by stingless bee species (*Meliponine*), called geopropolis. Meliponines gather resinous materials from different plants and mix them with beeswax and soil or clay; hence the name of the product (Pereira et al., 2021; Lavinas et al., 2019).

9.4.2 Selected Health Benefits of Propolis

Propolis originates from diverse plant origins and has been applied in traditional medicines for ages. This product has different biological applications, including anti-inflammatory, antioxidant, anticancer, antimicrobial, antiviral, antidepressant, analgesic, and anxiolytic, as well as immunomodulatory activities (Anjum et al., 2019; Asma et al., 2022; Berretta et al., 2017; Bhargava et al., 2021; Hossain et al., 2022; Moise & Bobiş, 2020; Pereira et al., 2021; Ripari et al., 2021; Rivera-Yañez et al., 2020). It is also used in treating other medical disorders (metabolic diseases), improving the healing or management of wounds, and relieving many other related discomforts (Martinotti & Ranzato, 2015; Oryan et al., 2018; Balica et al., 2021).

9.4.2.1 Antioxidant Activity

All bee products, especially propolis, are considered important sources of natural compounds with antioxidant potential (Balica et al., 2021; Bankova et al., 2002; Dezmirean et al., 2020; Hossain et al., 2022; Moise & Bobiş, 2020). The antioxidant activity of propolis has been studied by employing different *in vitro* methods, including 2,2-diphenyl-1-picryl-hydrazine (DPPH), ferric reducing antioxidant power (FRAP), Trolox equivalent antioxidant capacity (TEAC), cupric ions reducing antioxidant power (CUPRAC), H_2O_2 scavenging activity, superoxide anion radical scavenging ability, β-carotene bleaching, etc. (Gülçin et al., 2010; Kumazawa et al., 2004; Miguel et al., 2010; Nichitoi et al., 2021; Özkök et al., 2021; Russo et al., 2002). Miguel et al. revealed that the seasonal collection of propolis might influence the content of bioactive compounds, which is theoretically to be expected, as different plants or vegetal sources are available as raw materials for the production of propolis during different seasons. The antioxidant effect of propolis was mainly attributed to polyphenols, their increased redox potential making them reductive agents, oxygen quenchers, and hydrogen donors (Miguel et al., 2010).

9.4.2.2 Antibacterial Activity

Honeybees utilize propolis in the hive to cover cracks and holes, embalm different invaders, and clean the combs after a cycle of metamorphosis: in conclusion, to maintain cleanliness and a bacteria-free environment (Erler & Moritz, 2016; Spivak et al., 2019). Generally, the antibacterial potential of propolis extracts (bactericidal or bacteriostatic) is determined in two different ways: direct microbial inhibition and indirect action against pathogenic microbes by immune system stimulation (Bhargava et al., 2021). The hypothesized mechanism includes inhibiting bacterial division, lowering bacterial mobility, and direct action upon the bacterial membrane by increasing the permeability of bacterial cells toward the bioactive compound (Hossain et al., 2022; Ristivojević et al., 2018). Most of the scientific investigations on the antibacterial action of propolis have demonstrated its effectiveness against Gram-positive bacterial species and showed restricted activity against Gram-negative ones (Dezmirean et al., 2017; Hossain et al., 2022), depending on the region of collection, the species of bee, and the solvent used (Dezmirean et al., 2020; Moise & Bobiş, 2020).

9.4.2.3 Biomedical Activity against Chronic Diseases

Cancer, diabetes mellitus, and obesity are the most prevalent diseases in the modern era. Fighting against these diseases has been a major object of contemporary medicine, and finding appropriate treatments has been a concern of doctors for several decades now. Preclinical studies *in vitro* and *in vivo*, followed by clinical studies, have been performed to investigate the effect of propolis and propolis extracts on the above-mentioned metabolic diseases (Balica et al., 2021; Hossain et al., 2022; Rivera-Yañez et al., 2020). Diabetes mellitus is categorized by an elevated level of blood sugar due to deficiency in insulin secretion and dyslipidemia (generally, diabetic patients have low-density lipoprotein cholesterol and high levels of triglycerides), inducing oxidative stress and causing cellular dysfunction (American Diabetes Association, 2014). Obesity is also a substantial health threat linked with increased type 2 diabetes, cardiovascular disorders, dyslipidemia, and cancer (Bentham et al., 2017). Several literature studies are available on propolis in weight control and its antitumoral effects; a few are presented in Table 9.2.

9.5 CYTOTOXIC EFFECTS OF PROPOLIS

Kaewmuangmoon et al. examined the cytotoxic effect of two types of propolis extracts against human submandibular salivary gland cancer (HSG) cell lines. Propolis extracts were prepared using hexane (HEX) and dichloromethane (DCM). HEX extracts (at 50, 100, and 200 μg/mL) and DCM extracts (at 25, 50, 100, and 200 μg/mL) exhibited a notable increase in cytotoxicity and a considerable decrease in cell viability. The HEX extracts showed an IC_{50} in the lactate dehydrogenase (LDH) (42.32 ± 1.00) and MTT assays (61.3 ± 5.4), while the DCM extracts displayed an IC_{50} in the LDH (39.94 ± 0.22) and MTT (42.93 ± 2.7) assays (Kaewmuangmoon et al., 2022). Cytotoxic effects of Brazilian (red and green) and Mexican (brown) propolis extracts and some phenolic compounds (such as quercetin, caffeic acid, and rutin) have been determined using the MTT assay. The toxic concentrations (TC_{50}) were measured. The outcomes revealed that the brown propolis extract showed significant antiviral activity against corona virus strain (HCoV-229E) in human fibro-blast lung cells (MRC-5) at lower concentrations than the green and red propolis extract. The cytotoxicity was observed at higher concentrations >30 μg/mL in MRC-5 cells (Silva-Beltrán et al., 2022).

The cytotoxicity of ethanolic extract of Peruvian propolis was investigated in peripheral-blood mononuclear cells (PBMCs), Henrietta Lacks cells (HeLa), and human gingival fibroblasts (HGF-1). The PBMC, HeLa, and HGF-1 cells were cultured and treated with propolis extract. The cytotoxicity was investigated by measuring cell viability using the MTT assay. The outcomes suggested that

TABLE 9.2
Latest Experimental Study Models of Propolis Extracts on Metabolic Diseases

Experimental model	Effects	Reference
Anti-diabetic and anti-hyperlipidemic activities		
D-glucose-induced diabetes in rats	↓ in blood glucose level in fasting, insulin resistance, and body weight	(Laaroussi et al., 2020)
Streptozotocin-induced (STZ) diabetes in rats	Decrease in blood sugar levels; enhanced pancreatic, renal, and hepatic tissue levels, histological observation	(El Adaouia Taleb et al., 2020)
STZ diabetes in rats	↓ in blood glucose in fasting	(Nna et al., 2019)
STZ diabetes in rats	↓ in blood glucose, urea, and serum creatinine	(El Menyiy et al., 2019)
High-fat diet mouse	↓ in cholesterol and triglyceride levels, and atherogenic plasma index	(Oršolić et al., 2019)
HFD-STZ-induced type 2 diabetes in rats	Reduction of β-cell failure, attenuation of inflammation, ↓ in MDA content, ↑ in GPx, SOD, and total antioxidant effect in the kidney, promotion of hepatic genes	(Chen et al., 2018)
STZ type 2 diabetes in rats	↓ blood glucose in fasting and serum insulin level, ↑ in β-cells number, ↓ MDA content and body weight	(Nna et al., 2018)
STZ diabetes in rats	↓ in blood glucose	(Rivera-Yañez et al., 2018)
STZ type 1 diabetes in rats	↓ in fasting blood glucose level	(Usman et al., 2018)
Anti-obesity effect		
Obese HLD-supplemented rats	↓ in body weight and abdominal fat accumulation after 4-week HERP treatment	(Prata et al., 2022)
HF diet-fed male mice	↓ in HF-induced adiposity index, body weight, and adipocyte hypertrophy. Also improving fasting glucose levels	(Cardinault et al., 2020)
Obese C57BBL/6J mice	Thermogenesis escalation in white adipose tissue, and activation of metabolism of creatine	(Nishikawa et al., 2020)
HFD-induced obese Wistar rats	↓ weight gain of perirenal, mesenteric, and total white adipose tissues, and ↓ in level of cholesterol and triglyceride in liver and plasma	(Zheng et al., 2020)
HFD-fed mice	↓ in weight gain, insulin resistance proinflammatory cytokines, and fat accumulation in liver	(Cai et al., 2020)
Anticancer effect		
Human promyelocytic leukemia (HL60), murine fibroblast (L929), human monocytic leukemia (THP-1), human osteosarcoma (MG63), human mesenchymal hMSCs cells	Mildly cytotoxic toward cancer cells and accelerated apoptosis caspase 3/7 activity	(Campoccia et al., 2021)
Human tumor cell lines: glioma (U-251), ovarian cancer expressing multiple drug resistance phenotype (NCI-ADR/RES), prostate cancer (PC-3), and breast cancer (MCF7)	Different outcomes obtained for propolones separated from Brazilian propolis; displayed antiproliferative activity against U251, MCF7, NCI-ADR/ RES, and PC-3 cancer cells after 48 h of treatment	(Banzato et al., 2020)
C6 glioma cells	Apoptotic and cytotoxic effects on C6 glioma cancer cells and protective effect in the antioxidant defense system	(Coskun et al., 2020)

(Continued)

TABLE 9.2 (CONTINUED)
Latest Experimental Study Models of Propolis Extracts on Metabolic Diseases

Experimental model	Effects	Reference
Colorectal adenocarcinoma (HT-29) (ATCC® HTB-38™) and monocytic THP-1 (ATCC® TIB-202™) cells	Cuban propolis and its major component nemorosome impact the crosstalk among colon cancer cells by causing inhibition of M2-like macrophage viability	(Frión-Herrera et al., 2020)
C6 and cervical cancer cell lines (CasKi, HeLa, and SiHa)	EEP exhibited an anti-proliferative activity on glioma cells better than temozolomide and also reduced proliferation and cell viability	(Rivero-Cruz et al., 2020)
Human breast carcinoma (MCF7, T47D, MDA-MB-231, and -453), colon carcinoma (COLO-320, HCT116, DLD-1, and HT-29), cervical carcinoma (HeLa, ME-180, CaSki, and SKG-II), fibrosarcoma (HT1080), liver carcinoma (Huh6 and Huh7), lung carcinoma (A549 and H1299), normal human lung fibroblast (MRC5), osteosarcoma (U2OS and Saos-2), and ovarian carcinoma (SKOV-3)	CAPE exhibited anticancer activity by mediating cell cycle arrest and apoptosis; CAPE-MotAb nanoparticles showed higher potency of the mentioned processes	(Wang et al., 2020)
Human squamous cell carcinoma presenting in the tongue (SCC-25) and human gingival fibroblasts (HGFs)	Anticancer effect in SCC-25 cells and cytotoxicity toward murine macrophages. HEEP and EEP did not reveal any cytotoxic influence in the normal gingival fibroblasts	(Wezgowiec et al., 2020)
Fibroblasts as normal and cancer cell lines (A431 and KB)	Dose-dependent cytotoxic effect on cancer cell lines and no cytotoxic influence on fibroblast cells	(Asgharpour et al., 2019)
Four control and four experimental rat groups (two groups fed with two different concentrations of propolis)	Red propolis reduced preneoplastic lesions in rats	(Braga et al., 2019)

STZ: streptozotocin; HFD: high-fat diet; HLD: hyperlipid dietary; HERP: hydroalcoholic extract of red propolis; EEP: ethanolic extract of propolis; HEEP: hexane–ethanol extract of propolis
Source: Adapted from Balica et al., 2021; Bhargava et al., 2021; Rivera-Yañez et al., 2020).

Peruvian propolis can safely be used on the HGF-1 cell lines (Millones-Gómez et al., 2022). Boke Sarıkahya et al. determined the chemical and total phenolic content of different Turkish propolis samples and compared their antioxidant, cytotoxic, and antiviral potential. The cytotoxicity was analyzed against normal human embryonic kidney (HEK-293), adenocarcinoma human alveolar (A-549), MDA-MB-231, HeLa, and PC-3 cell lines using the MTT assay. A significant cytotoxic effect was observed in A-549 cell lines, with 3.32 ± 0.21 μg/mL IC_{50} (Boke Sarikahya et al., 2022). Many investigations have reported the cytotoxic activities of propolis against breast cancer (MCF-7), colon adenocarcinoma (CaCo-2), osteosarcoma (MG-63), human pancreatic cancer (PANC-1), promyelocytic leukemia (HL-60), A-549, HeLa, Hep-2, and MDA-MB-231 cell lines (El-aziz et al., 2021; Forma & Bryś, 2021; Sarıkahya et al., 2021; Surek et al., 2021).

The anticancer, antioxidant, and cytotoxic effects of phenolic-rich Algerian propolis extracts collected from two different origins, Oum El Bouaghi and Collo, named PREPO and PREPC, respectively, were studied. The cytotoxicity was evaluated via cell counting kit-8 assay (CCK-8) against human hepatocellular carcinoma (Hep-G2) cell lines. The outcomes exhibited a dose-dependent cytotoxic effect with PREPO and PREPC treatment against Hep-G2. Moreover, PREPC exhibited considerably higher cytotoxicity against Hep-G2 (IC_{50}, 12.22 ± 0.05 μg/mL) than PREPO

(IC_{50}, 18.7 ± 0.33 μg/mL) (Boulechfar et al., 2022). Several investigations have reported significant cytotoxicity of propolis samples from Algeria against several cancer cell lines, such as PANC-1, MDA-MB-231, PC3, MCF-7, A-549, and myelogenous leukemia (K-562) cell lines (Bakchiche et al., 2020; Brihoum et al., 2018; Rouibah et al., 2018). Propolis was obtained from seven geographic eras of Cyprus, and their cytotoxicity was investigated using an MTT assay against A-549, CaCo-2, HeLa, MCF, MDA, PANC-1, PC-3, and SH-SY5Y cell lines. All the collected propolis extracts demonstrated cytotoxic effects in a dose-dependent way against all tested cell lines (Nalbantsoy et al., 2022).

9.6 FOOD APPLICATIONS OF PROPOLIS

Food industries have been using chemicals to inhibit microorganism growth, responsible for food deterioration, for a long time (Davidson et al., 2012). Numerous research studies have been conducted to find a natural alternative. In this regard, bio-preservatives have attracted scientists' attention in the food industry. Bio-preservatives are natural substances derived from plants, animals, and microorganisms incorporated into food to enhance shelf life (Batiha et al., 2021; Pisoschi et al., 2018). Propolis utilization as a natural food additive is of great interest for several reasons. In general, propolis has a broad spectrum of biological properties, and its use as a diet supplement has been proven to enhance health and prevent diseases (Azemin et al., 2017; Pobiega et al., 2019; Hossain et al., 2022; Wieczorek et al., 2022).

9.6.1 Propolis as Bio-Preservative

In the food industry, propolis is used in extracted form. The most frequently used extracts are ethanol and water extracts. Generally, the extracts are polyphenol rich, with the highest antioxidant and antimicrobial activities (Zabaiou et al., 2017). Propolis extracts are used to reduce pathogens in fresh and transformed foods such as ready-to-eat foods. Propolis extracts can be applied superficially or by immersing foods directly. Another alternative method is adding propolis to coatings based on polymers (Pobiega et al., 2019). Propolis extracts were reported to be effective in inhibiting or preventing the growth of foodborne pathogens in different meat products such as Carpio meat (Payandan et al., 2017), sausages (Viera et al., 2016), fresh oriental sausages (Ali et al., 2010), beef patties (Vargas-Sánchez et al., 2014), etc.

Propolis aqueous extract (0, 4, 8, and 12% [v/w]) tested on the sensory, physicochemical, and microbial characteristics of chicken breast marinated during 12 days of storage at 5°C was reported to decrease microbial population growth in a dose-dependent approach. Propolis extract might also control the rise in pH and total volatile nitrogen (TVN). The sensory characteristics of ready-to-cook kebab chickens were improved by applying propolis aqueous extract. Adding up to 8% (v/w) propolis extract was reported to have no discernible detrimental effects on the color and general acceptance of the tested chickens (Mahdavi-Roshan et al., 2022). Regarding fruits, propolis research is more abundant, and several fruits have been tested, such as orange (Yang et al., 2010), mandarins (Matny, 2015), papaya (Cheong & Zahid, 2014), dragon fruits (Zahid et al., 2013), etc. Water and ethanol extracts of propolis were reported to prevent decay in cherries stored at 0°C for four weeks. The ethanol extract was the most effective (Candir et al., 2009). Propolis was also determined to be a promising approach to the preservation of fish (Spinelli et al., 2015), eggs (Akpinar et al., 2015), and dairy products (Cottica et al., 2015).

9.6.2 Propolis in Active Packaging

Natural products with their dual effects (antioxidant and antimicrobial) can be utilized in active packaging. Using propolis or other natural product active substances in active packaging is challenging. The substances can be incorporated in edible coating biopolymers or active packaging films. Active substances included in edible coatings or films are released to the food surface in a controlled manner (Batiha et al., 2021). Active packaging films containing propolis extracts can be prepared using different methods. The simplest method is solvent casting. Another method, extrusion, is a dry procedure involving high temperature and shear forces. Commercially available films coated with propolis extracts are another alternative and straightforward approach. However, incorporating propolis extracts into biopolymer-based films greatly alters their physical properties and structural characterization (Yong & Liu, 2021). Edible coatings are in a liquid form but are prepared using the same formulation of active packaging films. Propolis coating is prepared by mixing propolis extracts with biopolymers, emulsifiers, plasticizers, and other active compounds such as nanoparticles or essential oils (Moreno et al., 2020). Food surfaces can be coated with edible solutions by spraying, dipping, or brushing edible solutions over food surfaces. The coated foods are then dried at room temperature. Finally, they are appropriately packaged and stored (Suhag et al., 2020).

Cheese, salami, and salmon slices have also been studied. The results indicated that coating commercial polyethylene (PE) films using propolis ethanolic extract was effective against contamination with *Listeria monocytogenes*, depending on the tested food (Skowron et al., 2019). In a recent study, pullulan/chitosan-based multifunctional edible composite films were developed. The synthesized film was characterized using physicochemical methods and tested for its applicability to pork packaging. The incorporation of ZnO NPs associated with propolis was found to improve the property of blocking UV. This property did not cause any loss in transparency. In addition, the mechanical strength was enhanced and reinforced by approximately 25%. Hydrophobicity and the barrier property of water vapor increased slightly in the tested film. The results indicated that the tested film exhibited good antioxidant and antibacterial activity. The total count of aerobic microbes and peroxide value decreased in pork belly and meat packaged with the developed edible film (Roy et al., 2021). Edible coatings enriched with polyphenol propolis extract were effective on fruits, vegetables, and fresh-cut products. Experiments demonstrated good antibacterial and antifungal action against foodborne pathogens (*Colletotrichum*, *Penicillium* species, *E. coli*, and *L. monocytogenes*) (Ali et al., 2014, 2015; Feás et al., 2014; Moreno et al., 2020; Zahid et al., 2013).

Edible coatings with propolis possessed several advantages and were found to display a significant role in sensory quality, such as appearance and color. The visual and sensory qualities were also improved (Pobiega et al., 2020). The edible coating containing propolis was documented to suppress weight loss during storage. In addition, it was also reported to have a positive impact on firmness, Total Soluble Solid (TSS), and Titratable Acidity (TA) in several experiments on different kinds of vegetables and fruits, like papaya (*Carica papaya* L.) (Barrera Bello et al., 2012; Cheong & Zahid, 2014), chili (*Capsicum annuum* L.) (Ali et al., 2014, 2015), mango (*Mangifera indica* L.) (Al-Qurashi & Awad, 2018), and banana (*Musa* spp.) during storage (Awad & Al-Qurashi, 2019). A new coating was synthesized by carboxymethyl cellulose (CMC) containing ethanolic propolis extract and evaluated for the ability to maintain stored chicken breast meat quality and enhance the product's shelf life. The new coating demonstrated a higher capability to inhibit the growth of microorganisms, prevent lipid oxidation, and maintain the overall acceptability of the studied chicken breast meat (El Sheikha et al., 2022). The effect on sensory acceptance, physicochemical, and microbial properties of *Colossoma macropomum* fillets coated with green propolis extract incorporated into a sodium alginate bilayer was recently investigated. The tested coating was reported to reduce microbial deterioration and extend shelf life by 11 days. It also showed high sensory acceptance (Cruz et al., 2021).

9.7 NON-FOOD APPLICATIONS OF PROPOLIS

Propolis contains polyphenols, which exhibit massive free radical–scavenging activity. The latter property may bring beneficial effects on skin health, with particular emphasis on preventing and treating sunburn. Topical application of propolis is beneficial as a sunscreen and regenerative agent suitable as a component of cosmetic formulations and a constituent of medicines in dermal treatment, such as ointments and dressings for wounds and burns (Kurek-Górecka et al., 2022). Among the polyphenolic components of propolis, caffeic acid, p-coumaric acid, ferulic acid, quercetin, and kaempferol are essential ingredients of commercial photoprotective cosmetics such as sunscreen creams, lotions, sticks, and lipsticks. Many studies confirm that propolis exhibits protective and preventive properties from UV radiation (Batista et al., 2018; Bolfa et al., 2013). It may be applied as a natural filter with strong antioxidative, anti-inflammatory, and regenerative activities. Propolis absorbs radiation in the UV light range and plays a significant role in the photoprotective mechanism against UVA and UVB light (Butnariu & Giuchici, 2011; Saewan & Jimtaisong, 2015).

Exposure to UV radiation leads to erythema, hyperpigmentation, and premature skin aging, and may lead to different forms of cancer. UV radiation initiates an oxidative mechanism in the biological environment. It is responsible for forming oxygen free radicals that enhance oxidative stress in skin exposure to solar radiation. Hence, the disturbed balance between antioxidants and reactive oxygen generative systems promotes pathological implications. As a source of antioxidants, propolis may protect against UV radiation by absorbing UV photons and help reduce skin sunburn. Several studies conducted by Olczyk et al. confirmed the regenerative effect of propolis on an experimental animal model. Biochemical analyses demonstrate that propolis is an effective apitherapeutic agent that brings a satisfactory outcome in repairing tissue damage (Olczyk et al., 2012, 2013a, 2013b, 2014).

9.7.1 Photoprotective Effect of Propolis

Numerous studies indicate that propolis exhibits promising bioactivities, including photoprotective, antioxidant, anti-inflammatory, and anti-aging properties. Therefore, it might be a potential sunscreen for cosmetics (Batista et al., 2018; Bolfa et al., 2013; Fu et al., 2022). Polyphenolic components are largely responsible for the multidirectional properties of propolis. The mechanism of propolis's photoprotective effect is linked closely with its compounds' anti-inflammatory and antioxidant action. The mechanism of the antioxidative capacity of polyphenols depends on enzymatic activity inhibition; they inhibit reactive oxygen species (ROS). Moreover, polyphenols chelate metal ions. They participate in free radical formation and scavenge ROS, blocking the cascade of reactions associated with lipid peroxidation. Furthermore, they demonstrate synergistic activity with other antioxidants. The anti-inflammatory activity of polyphenols is a result of their antioxidative capacity. The anti-inflammatory action of polyphenols is correlated with the hydroxyl groups (C-3' and C-4') in the B ring. Caffeic acid phenethyl ester (CAPE) exhibits strong inhibitory activity on the arachidonic acid metabolic tract enzymes. Chrysin, kaempferol, quercetin, and galangin modulate mRNA expression and inhibit the synthesis of IL1β, a cytokine that performs a vital function in the appearance and spread of inflammatory reactions (Batista et al., 2018; Kurek-Górecka et al., 2013).

Among the compounds determining the photoprotective effect of propolis are caffeic acid, p-ferulic acid, coumaric acid, kaempferol, quercetin, and CAPE. Moreover, López et al. determined that red propolis extract from Brazil contains benzophenone, used as an organic filter (López et al., 2014). The possibility of using propolis in a sunscreen formulation to prevent skin damage with effective photostability was studied using green propolis. Lyophilized green propolis extracts from Brazil have been combined with Permulen TR-1 gel. This combined gel exhibited sun protection at all tested concentrations, from 0.02, 0.03, 0.05, 0.07, to 0.1 mg/mL. The best results during the measurement of solar protection factor according to the Mansur method demonstrated the formulation with the highest concentration of hydroalcoholic green propolis extract (0.1 mg/mL). Therefore, the mentioned study indicated that green propolis extract might be used in sunscreen formulation and showed good photostability, which had been tested utilizing a light chamber with UV radiation at 365 nm. Following the HET-CAM test, Permulen TR-1 gel with green propolis extract has been classified as safe for direct application with lower skin irritation potential. Test results demonstrated slight hyperemia and hemorrhage after 300 s in contact with the studied formulation (Almeida et al., 2020). Despite the undoubted benefits of using various photoprotective dosage forms based on propolis, scientists have indicated disadvantages limiting the local use of sunscreen formulations containing the aforementioned apitherapeutic agent. These limitations include instability, photocytotoxicity, skin permeability, and poor water solubility of the main bioactive compounds (Fu et al., 2022). Overall, based on the conducted studies, it is highlighted that propolis, as a source of polyphenols, may protect from UV radiation and sunburn. Moreover, it provides sun protection and can be incorporated into different formulations of sunscreen.

9.7.2 Regenerative Effect of Propolis

Propolis is well recognized for its regenerative, immunomodulatory, and protective properties. Due to the potent influence of propolis on the prooxidative/antioxidative balance, modulation can be used in managing tissue damage of various types and severity. The regenerative effect of propolis on damaged tissues results – among other things – from the content of compounds that enhance systemic and cellular metabolism by NADPH-2 tetrazolium reductase and ATPase activation (Staniczek et al., 2021). These increases in the mitotic index value may be accompanied by accelerated tissue regeneration and shortened healing time. Numerous studies indicate that propolis exhibits promising actions in the healing of wounds. Typically, the wound healing or management process consists of interrelated stages, including (i) hemostasis, (ii) inflammation, (iii) proliferation, and (iv) remodeling. Several studies have shown that propolis is a promising apitherapeutic for burn tissue treatment, enhancing fibroblast proliferation, activation, and growth capacity (Abreu et al., 2012; Sehn et al., 2009; Pessolato et al., 2011). The interaction induces the healing process among the extracellular matrix (ECM) and different cell types. The components of ECM include glycosaminoglycans, proteoglycans, elastin, and collagen. Studies performed by Olczyk et al. demonstrated that propolis ointment accelerates the repair of burned tissue, stimulating the accumulation of total glycosaminoglycans, dermatan/chondroitin sulfates, and hyaluronic acid during the subsequent phases of wound healing. The content of total glycosaminoglycans increased up to the 15th day. Then, the amount of glycosaminoglycans was reduced from the 15th to the 21st day of the experiment (Olczyk et al., 2013a). Moreover, applying propolis ointment to burn wounds inflicted on pigs stimulates two types of collagen accumulation (types I and III are primary collagen types) in the burn wound matrix (Olczyk et al., 2013a, 2013b).

Moreover, during wound repair, propolis may exhibit an inhibitory effect on reconstituting fibronectin, which determines collagen fibrogenesis. Increased fibronectin accumulation is connected with intensified creation and secretion of collagen in fibromatous tissues. Considering this aspect, propolis seems to reduce the risk of keloid development (Olczyk et al., 2014). In addition, other glycoproteins, including laminin and vitronectin, are involved in the cellular and molecular mechanisms engaged in the regenerative tissue process. Laminin is needed for the migration and further proliferation of epidermal cells, while vitronectin induces the migration and proliferation of keratinocytes and stimulates the synthesis of proteins. During experimental burn wounds treated with propolis ointment, increased content of laminin and vitronectin was observed until the third day of healing. The test results may indicate that the apitherapeutic formulation modulates laminin and vitronectin metabolism, most likely influencing keratinocyte migration and proliferation. Moreover, the applied apitherapeutic ointment may also modulate fibroblast activation, thus providing reepithelization and wound closure (Olczyk et al., 2012).

Staniczek et al. validated the regenerative effect of 1% propolis ointment, which accelerated neo-angiogenesis and epithelialization in healing experimentally induced burn wounds. The study demonstrated that an ointment containing propolis at a concentration of 1% decreases healing time, brings an antimicrobial effect, and exhibits a satisfactory impact on the typical means of scar formation (Staniczek et al., 2021). Studies conducted in Poland have compared the therapeutic effectiveness of propolis and silver sulfadiazine (SSD) during experimental burn treatment. Propolis was more effective during burn wounds than SSD, which causes many adverse reactions (among others, argyria, inner organ disorders, erythema multiforme rashes, dermatitis, acute hemolytic anemia, and the exhibition of cytotoxic activity on keratinocytes and fibroblasts). The positive effect of dressings containing propolis on burn regeneration has been confirmed by analyses of blood taken from organisms with thermal injury by electron paramagnetic resonance (Olczyk et al., 2020). Therefore, it should be highlighted that propolis can be significant in treating burns in all healing stages, leading to satisfactory effects among patients suffering from thermal injury. It exhibits healing time–reducing properties. It is worth underlining that propolis exhibits promising photoprotective and regenerative effects, so that it might be used as a potential sunscreen in different formulations. An apitherapeutic agent exerts a beneficial effect on sunburn.

9.8 CONCLUSION

Propolis has been used traditionally for medicinal purposes for ages. It is one of the most prevalent naturally found products, extensively employed for centuries due to its numerous health-promoting and therapeutic properties. Phenolic compounds, polyphenols, flavonoids, phenolic acids, flavanones, flavones, and terpenoids are potential bioactive constituents of propolis and are responsible for possessing significant biological properties. Propolis has been substantially investigated for its numerous health-promoting and therapeutic properties. Moreover, it is further attracting researchers' attention to exploring its biological properties with their modes of action alone or in combination (or encapsulation) with different compounds. The unique properties of propolis have intrigued consumers, drawing their attention toward propolis products, leading to advancements in existing products and the innovation of new products in dietary supplements, therapeutics, and cosmetics. The advancements in propolis formulations and extract preparations suggest that propolis can be a significant, cost-effective, and readily available natural source of effective bioactive components for nutritional supplements, cosmetics, and pharmaceuticals.

REFERENCES

AbdullKadem, M.H.; Al-Mukhtar, E.J.; Madlum, K.N. 2022. Cytotoxicity and immunomodulatory effects of propolis on cancer and normal cells. *Journal of Drug and Alcohol Research*, *11*(8), 1–8. https://doi.org/10.4303/jdar/236191

Abreu, A.M.; Oliveira, D.W.D.; Marinho, S.A.; Lima, N.L.; de Miranda, J.L.; Verli, F.D. 2012. Effect of topical application of different substances on fibroplasia in cutaneous surgical wounds. *International Scholarly Research Notices*. https://doi.org/10.5402/2012/282973

Afata, T.N.; Nemo, R.; Ishete, N.; Tucho, G.T.; Dekebo, A. 2022. Phytochemical investigation, physicochemical characterization, and antimicrobial activities of Ethiopian propolis. *Arabian Journal of Chemistry*, *15*(7), 103931. https://doi.org/10.1016/j.arabjc.2022.103931

Afrouzan, H.; Tahghighi, A.; Zakeri, S.; Es-Haghi, A. 2018. Chemical composition and antimicrobial activities of Iranian propolis. *Iranian Biomedical Journal*, *22*(1), 50. https://doi.org/10.22034/ibj.22.1.50

Akpinar, G.; Canogullari, S.; Baylan, M.; Alasahan, S.; Aygun, A. 2015. The use of propolis extract for the storage of quail eggs. *Journal of Applied Poultry Research*, *24*(4), 427–435. https://doi.org/10.3382/japr/pfv043

Al-Juhaimi, F.Y.; Özcan, M.M.; Mohamed Ahmed, I.A. et al. 2022. Bioactive compounds, antioxidant activity, fatty acid composition, and antimicrobial activity of propolis from different locations in Turkey. *Journal of Apicultural Research*, *61*(2), 246–254. https://doi.org/10.1080/00218839.2021.1898785

Al-Qurashi, A.D.; Awad, M.A. 2018. Postharvest ethanolic extract of propolis treatment affects quality and biochemical changes of 'Hindi-Besennara'mangos during shelf life. *Scientia Horticulturae*, *233*, 520–525. https://doi.org/10.1016/j.scienta.2017.12.030

Alenezi, S.S.; Natto, M.J.; Igoli, J.O. et al. 2020. Novel flavanones with anti-trypanosomal activity isolated from Zambian and Tanzanian propolis samples. *International Journal for Parasitology: Drugs and Drug Resistance*, *14*, 201–207. https://doi.org/10.1016/j.ijpddr.2020.10.011

Ali, A.; Chow, W.L.; Zahid, N.; Ong, M.K. 2014. Efficacy of propolis and cinnamon oil coating in controlling post-harvest anthracnose and quality of chilli (Capsicum annuum L.) during cold storage. *Food and Bioprocess Technology*, *7*(9), 2742–2748. https://doi.org/10.1007/s11947-013-1237-y

Ali, A.; Wei, Y.Z.; Mustafa, M.A. 2015. Exploiting propolis as an antimicrobial edible coating to control postharvest anthracnose of bell pepper. *Packaging Technology and Science*, *28*(2), 173–179. https://doi.org/10.1002/pts.2088

Ali, F.H.; Kassem, G.M.; Atta-Alla, O.A. 2010. Propolis as a natural decontaminant and antioxidant in fresh oriental sausage. *Veterinaria Italiana*, *46*(2), 167–172

Almeida, W.A.d.S.; Sousa, L.R.D.; dos Santos Antunes, A.; et al. 2020. Green propolis: In vitro photoprotective and photostability studies of single and incorporated extracts in a sunscreen formulation. *Revista Brasileira de Farmacognosia*, *30*(3), 436–443. https://doi.org/10.1007/s43450-020-00071-z

American Diabetes Association 2014. Diagnosis and classification of diabetes mellitus. *Diabetes Care*, *37* (Supplement_1): S81-S90. https://doi.org/10.2337/dc10-S062

Anjum, S.I.; Ullah, A.; Khan, K.A.; et al. 2019. Composition and functional properties of propolis (bee glue): A review. *Saudi Journal of Biological Sciences*, *26*(7), 1695–1703. https://doi.org/10.1016/j.sjbs.2018.08.013

Asgharpour, F.; Moghadamnia, A.A.; Zabihi, E. 2019. Iranian propolis efficiently inhibits growth of oral streptococci and cancer cell lines. *BMC Complementary and Alternative Medicine*, *19*(1), 1–8. https://doi.org/10.1186/s12906-019-2677-3

Asma, S.T.; Bobiş, O.; Bonta, V.; et al. 2022. General nutritional profile of bee products and their potential antiviral properties against mammalian viruses. *Nutrients*, *14*(17), 3579. https://doi.org/10.3390/nu14173579

American Diabetes Association. 2014. Diagnosis and classification of diabetes mellitus. *Diabetes Care*, *37*(Supplement_1), S81–S90.

Awad, M.A.; Al-Qurashi, A.D. 2019. Quality and biochemical changes of 'Sukkari'bananas during shelf life as affected by postharvest dipping in ethanolic extract of propolis. *Philippine Agricultural Scientist*, *102*(2), 132–140.

Azemin, A.; Md-Zin, N.; Mohd-Rodi, M.; Kim-Chee, A.; Zakaria, A.; Mohd, K. 2017. Application of metabolite profiling and antioxidant activity in assessing the quality of processed and unprocessed stingless bee's propolis. *Journal of Fundamental and Applied Sciences*, *9*(2S), 637–660.

Bakchiche, B.; Temizer, İ.K.; Güder, A.; et al. 2020. Chemical composition and biological activities of honeybee products from Algeria. *Journal of Applied Biotechnology Reports*, *7*(2), 93–103. https://doi.org/10.30491/jabr.2020.109498

Balica, G.; Vostinaru, O.; Stefanescu, C.; et al. 2021. Potential role of propolis in the prevention and treatment of metabolic diseases. *Plants*, *10*(5), 883. https://doi.org/10.3390/plants10050883

Bankova, V. 2005. Recent trends and important developments in propolis research. *Evidence-Based Complementary and Alternative Medicine*, *2*(1), 29–32. https://doi.org/10.1093/ecam/neh059

Bankova, V.; Popova, M.; Bogdanov, S.; Sabatini, A.G. 2002. Chemical composition of European propolis: Expected and unexpected results. *Zeitschrift für Naturforschung C*, *57*(5–6), 530–533. https://doi.org/10.1515/znc-2002-5-622

Bankova, V.; Popova, M.; Trusheva, B. 2014. Propolis volatile compounds: Chemical diversity and biological activity: A review. *Chemistry Central Journal*, *8*(1), 1–8. https://doi.org/10.1186/1752-153X-8-28

Banzato, T.P.; Gubiani, J.R.; Bernardi, D.I.; et al. 2020. Antiproliferative flavanoid dimers isolated from Brazilian red propolis. *Journal of Natural Products*, *83*(6), 1784–1793. https://doi.org/10.1021/acs.jnatprod.9b01136

Barrera Bello, E.; Gil Loaiza, M.; García Pajón, C.M.; Durango Restrepo, D.L.; Gil González, J.H. 2012. Use of a coating formulated with propolis for postharvest handling of papaya (Carica papaya L. cv. Hawaiian) fruits. *Revista Facultad Nacional de Agronomía, Medellín*, *65*(1), 6497–6506

Batiha, G.E.S.; Hussein, D.E.; Algammal, A.M. 2021. Application of natural antimicrobials in food preservation: Recent views. *Food Control*, *126*, 108066.

Batista, C.; Alves, A.; Queiroz, L. 2018. The photoprotective and anti-inflammatory activity of red propolis extract in rats. *Journal of Photochemistry and Photobiology, B: Biology*, *180*, 198–207. https://doi.org/10.1016/j.jphotobiol.2018.01.028

Belmehdi, O.; Bouyahya, A.; Jekő, J.; et al. 2021. Chemical analysis, antibacterial, and antioxidant activities of flavonoid-rich extracts from four Moroccan propolis. *Journal of Food Processing and Preservation*, *45*(10), e15816. https://doi.org/10.1111/jfpp.15816

Bentham, J.; Di Cesare, M.; BIlano, V.; Boddy, L.M. 2017. Worldwide trends in children's and adolescents' body mass index, underweight and obesity, in comparison with adults, from 1975 to 2016: A pooled analysis of 2,416 population-based measurement studies with 128.9 million participants. *Lancet* *390*(10113), 2627–2642.

Berretta, A.A.; Arruda, C.; Miguel, F.G. 2017. Functional properties of Brazilian propolis: from chemical composition until the market. *Superfood and Functional Food: An Overview of Their Processing and Utilization*, 55–98. https://doi.org/10.5772/65932

Bhargava, P.; Mahanta, D.; Kaul, A.; et al. 2021. Experimental evidence for therapeutic potentials of propolis. *Nutrients*, *13*(8), 2528. https://doi.org/10.3390/nu13082528

Bogdanov, S.; Bankova, V. 2016. The propolis book, chapter 1: Propolis: Origin, production, composition. *Bee Product Science. www. bee-hexagon. net–pristupljeno*, *26*, 2020.

Boke Sarikahya, N.; Varol, E.; Sumer Okkali, G.; Yucel, B.; Margaoan, R.; Nalbantsoy, A. 2022. Comparative study of antiviral, cytotoxic, antioxidant activities, total phenolic profile and chemical content of propolis samples in different colors from Turkiye. *Antioxidants*, *11*(10), 2075. https://doi.org/10.3390/antiox11102075

Bolfa, P.; Vidrighinescu, R.; Petruta, A. 2013. Photoprotective effects of Romanian propolis on skin of mice exposed to UVB irradiation. *Food and Chemical Toxicology*, *62*, 329–342. https://doi.org/10.1016/j.fct.2013.08.078

Boulechfar, S.; Akbulut, Z.; Tepe, H.D. 2022. LC–MS/MS analysis, antioxidant and anticancer effects of phenolic-rich extracts from Algerian propolis: A comparative study. *Journal of Food Measurement and Characterization*, 1–12. https://doi.org/10.1007/s11694-022-01652-3

Boulechfar, S.; Zellagui, A.; Asan-Ozusaglam, M. 2022. Chemical composition, antioxidant, and antimicrobial activities of two essential oils from Algerian propolis. *Zeitschrift für Naturforschung C*, *77*(3–4), 105–112. https://doi.org/10.1515/znc-2021-0028

Braga, V.N.L.; Juanes, C.D.C.; Peres, H.D.S. 2019. Gum arabic and red propolis protecteting colorectal preneoplastic lesions in a rat model of azoxymethane1. *Acta Cirúrgica Brasileira*, *34*(2), e201900207.

Brihoum, H.; Maiza, M.; Sahali, H. 2018. Dual effect of Algerian propolis on lung cancer: Antitumor and chemopreventive effects involving antioxidant activity. *Brazilian Journal of Pharmaceutical Sciences*, *54*(1). https://doi.org/10.1590/s2175-97902018000117396

Busch, V.M.; Pereyra-Gonzalez, A.; Šegatin, N.; Santagapita, P.R.; Ulrih, N.P.; Buera, M.d.P. 2017. Propolis encapsulation by spray drying: Characterization and stability. *LWT*, *75*, 227–235. https://doi.org/10.1016/j.lwt.2016.08.055

Butnariu, M.V.; Giuchici, C.V. 2011. The use of some nanoemulsions based on aqueous propolis and lycopene extract in the skin's protective mechanisms against UVA radiation. *Journal of Nanobiotechnology*, *9*(1), 1–9. https://doi.org/10.1186/1477-3155-9-3

Cai, W.; Xu, J.; Li, G. 2020. Ethanol extract of propolis prevents high-fat diet-induced insulin resistance and obesity in association with modulation of gut microbiota in mice. *Food Research International*, *130*, 108939. https://doi.org/10.1016/j.foodres.2019.108939

Campoccia, D.; Ravaioli, S.; Santi, S. 2021. Exploring the anticancer effects of standardized extracts of poplar-type propolis: In vitro cytotoxicity toward cancer and normal cell lines. *Biomedicine and Pharmacotherapy*, *141*, 111895. https://doi.org/10.1016/j.biopha.2021.111895

Candir, E.E.; Ozdemir, A.; Soylu, E.M.; Sahinler, N.; Gul, A. 2009. Effects of propolis on storage of sweet cherry cultivar Aksehir Napolyon. *Asian Journal of Chemistry*, *21*, 2659–2666

Cardinault, N.; Tourniaire, F.; Astier, J. et al. 2020. Poplar propolis ethanolic extract reduces body weight gain and glucose metabolism disruption in high-fat diet-fed mice. *Molecular Nutrition and Food Research*, *64*(18), 2000275. https://doi.org/10.1002/mnfr.202000275

Cerqueira, P.; Cunha, A.; Almeida-Aguiar, C. 2022. Potential of propolis antifungal activity for clinical applications. *Journal of Applied Microbiology*, *133*(3), 1207–1228. https://doi.org/10.1111/jam.15628

Chen, L.H.; Chien, Y.W.; Chang, M.L. 2018. Taiwanese green propolis ethanol extract delays the progression of type 2 diabetes mellitus in rats treated with streptozotocin/high-fat diet. *Nutrients*, *10*(4), 503. https://doi.org/10.3390/nu10040503

Cheong, A.A.C.K.; Zahid, N. 2014. Composite effect of propolis and gum arabic to control postharvest anthracnose and maintain quality of papaya during storage. *International Journal of Agriculture and Biology*, *16*(6), 1117–1122.

Chien, Y.H.; Yu, Y.H.; Ye, S.R.; Chen, Y.W. 2022. Antibacterial and antioxidant activity of the fruit of Macaranga Tanarius, the plant origin of Taiwanese green propolis. *Antioxidants*, *11*(7), 1242. https://doi.org/10.3390/antiox11071242

Coskun, Z.M.; Ersoz, M.; Gecili, M.; Ozden, A.; Acar, A. 2020. Cytotoxic and apoptotic effects of ethanolic propolis extract on C6 glioma cells. *Environmental Toxicology*, *35*(7), 768–773. https://doi.org/10.1002/tox.22911

Costa Jr A.G.; Yoshida, N.C.; Garcez, W.S.; Perdomo, R.T.; Matos, M.D.F.C.; Garcez, F.R. 2020. Metabolomics approach expands the classification of propolis samples from midwest Brazil. *Journal of Natural Products*, *83*(2), 333–343. https://doi.org/10.1021/acs.jnatprod.9b00783

Cottica, S.M.; Sabik, H.; Bélanger, D.; Giroux, H.J.; Visentainer, J.V.; Britten, M. 2015. Use of propolis extracts as antioxidant in dairy beverages enriched with conjugated linoleic acid. *European Food Research and Technology*, *241*(4), 543–551. https://doi.org/10.1007/s00217-015-2483-1

Cruz, A.I.C.; da Cruz Costa, M.; Mafra, J.F. 2021. A sodium alginate bilayer coating incorporated with green propolis extract as a powerful tool to extend Colossoma macropomum fillet shelf-life. *Food Chemistry*, *355*, 129610. https://doi.org/10.1016/j.foodchem.2021.129610

Cui-ping, Z.; Shuai, H.; Wen-ting, W. 2014. Development of high-performance liquid chromatographic for quality and authenticity control of Chinese propolis. *Journal of Food Science*, *79*(7), C1315–C1322. https://doi.org/10.1111/1750-3841.12510

Davidson, P.M.; Taylor, T.M.; Schmidt, S.E. 2012. Chemical preservatives and natural antimicrobial compounds. In: *Food Microbiology: Fundamentals and Frontiers*, Ed (4): Doyle, M.P.; Buchanan, R.L.; ASM Press Washington, DC, pp. 765–801.

de Oliveira Cardoso, E.; Santiago, K.B.; Conti, B.J. 2022. Brazilian green propolis: A novel tool to improve the cytotoxic and immunomodulatory action of docetaxel on MCF-7 breast cancer cells and on women monocyte. *Phytotherapy Research*, *36*(1), 448–461. https://doi.org/10.1002/ptr.7345

Dezmirean, D.S.; Mărghitaş, L.A.; Chirilă, F. 2017. Influence of geographic origin, plant source and polyphenolic substances on antimicrobial properties of propolis against human and honey bee pathogens. *Journal of Apicultural Research*, *56*(5), 588–597. https://doi.org/10.1080/00218839.2017.1356205

Dezmirean, D.S.; Paşca, C.; Moise, A.R.; Bobiş, O. 2020. Plant sources responsible for the chemical composition and main bioactive properties of poplar-type propolis. *Plants*, *10*(1), 22. https://doi.org/10.3390/plants10010022

El-aziz, E.A.E.D.A.; Elgayar, S.F.; Mady, F.M. et al. 2021. The potential of optimized liposomes in enhancement of cytotoxicity and apoptosis of encapsulated Egyptian propolis on Hep-2 cell line. *Pharmaceutics*, *13*(12), 2184. https://doi.org/10.3390/pharmaceutics13122184

El Adaouia Taleb, R.; Djebli, N.; Chenini, H.; Sahin, H.; Kolayli, S. 2020. In vivo and in vitro anti-diabetic activity of ethanolic propolis extract. *Journal of Food Biochemistry*, *44*(7), e13267. https://doi.org/10.1111/jfbc.13267

El Menyiy, N.; Al-Wali, N.; El Ghouizi, A.; El-Guendouz, S.; Salom, K.; Lyoussi, B. 2019. Potential therapeutic effect of Moroccan propolis in hyperglycemia, dyslipidemia, and hepatorenal dysfunction in diabetic rats. *Iranian Journal of Basic Medical Sciences*, *22*(11), 1331. https://doi.org/10.22038/ijbms.2019.33549.8004

El Sheikha, A.F.; Allam, A.Y.; ElObeid, T. et al. 2022. Impact of a carboxymethyl cellulose coating incorporated with an ethanolic propolis extract on the quality criteria of chicken breast meat. *Antioxidants*, *11*(6), 1191. https://doi.org/10.3390/antiox11061191

Elkhateeb, O.M.; Badawy, M.E.; Noreldin, A.E. et al. 2022. Comparative evaluation of propolis nanostructured lipid carriers and its crude extract for antioxidants, antimicrobial activity, and skin regeneration potential. *BMC Complementary Medicine and Therapies*, *22*(1), 1–13. https://doi.org/10.1186/s12906-022-03737-4

Elmahallawy, E.K.; El Fadaly, H.A.M. et al. 2022. Novel insights on the potential activity of propolis and wheat germ oil against chronic toxoplasmosis in experimentally infected mice. *Biomedicine and Pharmacotherapy*, *156*, 113811. https://doi.org/10.1016/j.biopha.2022.113811

Erler, S.; Moritz, R.F. 2016. Pharmacophagy and pharmacophory: Mechanisms of self-medication and disease prevention in the honeybee colony (Apis mellifera). *Apidologie*, *47*(3), 389–411. https://doi.org/10.1007/s13592-015-0400-z

Feás, X.; Pacheco, L.; Iglesias, A.; Estevinho, L.M. 2014. Use of propolis in the sanitization of lettuce. *International Journal of Molecular Sciences*, *15*(7), 12243–12257. https://doi.org/10.3390/ijms150712243

Fernández-Calderón, M.C.; Navarro-Pérez, M.L.; Blanco-Roca, M.T.; Gómez-Navia, C.; Pérez-Giraldo, C.; Vadillo-Rodríguez, V. 2020. Chemical profile and antibacterial activity of a novel Spanish propolis with new polyphenols also found in olive oil and high amounts of flavonoids. *Molecules*, *25*(15), 3318. https://doi.org/10.3390/molecules25153318

Fokt, H.; Pereira, A.; Ferreira, A.; Cunha, A.; Aguiar, C. 2010. How do bees prevent hive infections? The antimicrobial properties of propolis. *Current Research, Technology and Education Topics in Applied Microbiology and Microbial Biotechnology*, *1*, 481–493

Forma, E.; Bryś, M. 2021. Anticancer activity of propolis and its compounds. *Nutrients*, *13*(8), 2594. https://doi.org/10.3390/nu13082594

Frión-Herrera, Y.; Gabbia, D.; Scaffidi, M. 2020. Cuban brown propolis interferes in the crosstalk between colorectal cancer cells and m2 macrophages. *Nutrients*, *12*(7), 2040. https://doi.org/10.3390/nu12072040

Fu, Y.; Wan, R.; Yang, L. 2022. Propolis inspired sunscreens for efficient UV-protection and skin barrier maintenance. *Nano Research*, 1–10. https://doi.org/10.1007/s12274

Gajger, I.T.; Pavlović, I.; Bojić, M. 2017. Components responsible for antimicrobial activity of propolis from continental and Mediterranean regions in Croatian. *Czech Journal of Food Sciences*, *35*(5), 376. https://doi.org/10.17221/103/2017-CJFS

Galeotti, F.; Maccari, F.; Fachini, A.; Volpi, N. 2018. Chemical composition and antioxidant activity of propolis prepared in different forms and in different solvents useful for finished products. *Foods*, *7*(3), 41. https://doi.org/10.3390/foods7030041

Garcia-Viguera, C.; Ferreres, F.; Tomas-Barberan, F. 1993. Study of Canadian propolis by GC-MS and HPLC. *Zeitschrift für Naturforschung C*, *48*(9–10), 731–735. https://doi.org/10.1515/znc-1993-9-1009

Georgieva, K.; Popova, M.; Dimitrova, L.; et al. 2019. Phytochemical analysis of Vietnamese propolis produced by the stingless bee Lisotrigona cacciae. *PLOS One*, *14*(4), e0216074. https://doi.org/0.1371/journal.pone.0216074

Georgieva, K.; Trusheva, B.; Uzunova, V. 2018. New cycloartane triterpenes from bioactive extract of propolis from Pitcairn Island. *Fitoterapia*, *128*, 233–241. https://doi.org/10.1016/j.fitote.2018.05.024

Gomez-Lopez, V.M.; Pataro, G.; Tiwari, B. 2022. Guidelines on reporting treatment conditions for emerging technologies in food processing. *Critical Reviews in Food Science and Nutrition*, *62*(21), 5925–5949. https://doi.org/10.1080/10408398.2021.1895058

Greenaway, W.; Scaysbrook, T.; Whatley, F. 1988. Composition of propolis in Oxfordshire, UK and its relation to poplar bud exudate. *Zeitschrift für Naturforschung C*, *43*(3–4), 301–304. https://doi.org/10.1515/znc-1988-3-423

Greenaway, W.; Scaysbrook, T.; Whatley, F. 1990. The composition and plant origins of propolis: A report of work at Oxford. *Bee World*, *71*(3), 107–118. https://doi.org/10.1080/0005772X.1990.11099047

Gülbandılar, A. 2022. Antimicrobial activities of propolis samples collected from different provinces of Turkey. *Mas Journal of Applied Sciences*, *7*(2), 433–442. https://doi.org/10.1080/14786410600906277

Gülçin, I.; Bursal, E.; Şehitoğlu, M.H.; Bilsel, M.; Gören, A.C. 2010. Polyphenol contents and antioxidant activity of lyophilized aqueous extract of propolis from Erzurum, Turkey. *Food and Chemical Toxicology*, *48*(8–9), 2227–2238. https://doi.org/10.1016/j.fct.2010.05.053

Hasan, A.E.Z.; Setiyono, A.; Afrian, M. 2022. Hepatoprotective Activity of Propolis Trigona spp., Curcuma Xanthorrhiza, Hibiscus sabdariffa, and myrmeleon sp. in Rats Induced by Paracetamol. *Current Biochemistry*, *9*(1), 38–50. https://doi.org/10.29244/cb.9.1.4

Hossain, R.; Quispe, C.; Khan, R.A. 2022. Propolis: An update on its chemistry and pharmacological applications. *Chinese Medicine*, *17*(1), 1–60. https://doi.org/10.1186/s13020-022-00651-2

Hudz, N.; Korytniuk, O.; Yezerska, O. 2020. Evaluation of the total flavonoid content and antimicrobial activity of the tinctures of propolis of Ukrainian origin. *Acta Poloniae Pharmaceutica – Drug Research*, *77*, 897–907. https://doi.org/10.32383/appdr/131651

Isidorov, V.A.; Szczepaniak, L.; Bakier, S. 2014. Rapid GC/MS determination of botanical precursors of Eurasian propolis. *Food Chemistry*, *142*, 101–106. https://doi.org/10.1016/j.foodchem.2013.07.032

Kaewmuangmoon, J.; Charoonpatrapong, K.; Janebodin, K. 2022. Cytotoxicity of propolis extracts obtained using dichloromethane and hexane solvent on human salivary gland tumor cell line. *Journal of International Society of Preventive and Community Dentistry*, *12*(5), 506. https://doi.org/10.4103/jispcd.JISPCD_303_21

Kontogiannis, T.; Dimitriou, T.G.; Didaras, N.A.; Mossialos, D. 2022. Antiviral activity of bee products. *Current Pharmaceutical Design*. https://doi.org/10.2174/1381612828666220928110103

Kumazawa, S.; Hamasaka, T.; Nakayama, T. 2004. Antioxidant activity of propolis of various geographic origins. *Food Chemistry*, *84*(3), 329–339. https://doi.org/10.1016/S0308-8146(03)00216-4

Kurek-Górecka, A.; Keskin, Ş.; Bobis, O. et al. 2022. Comparison of the antioxidant activity of propolis samples from different geographical regions. *Plants*, *11*(9), 1203. https://doi.org/10.3390/plants11091203

Kurek-Górecka, A.; Olczyk, P. 2022. *Bee Products and Skin Therapy (Bee Products and Their Applications in the Food and Pharmaceutical Industries*. Elsevier, pp. 25–62.

Kurek-Górecka, A.; Rzepecka-Stojko, A.; Górecki, M.; Stojko, J.; Sosada, M.; Świerczek-Zięba, G. 2013. Structure and antioxidant activity of polyphenols derived from propolis. *Molecules*, *19*(1), 78–101. https://doi.org/0.3390/molecules19010078

Kustiawan, P.M.; Yanti, E.N.; Nisa, K.; Zulfa, A.F.; Batistuta, M.A. 2022. Bioactivity of heterotrigona Itama propolis as anti-inflammatory: A review. *Biointerface Research in Applied Chemistry*, *13*(4), 326. https://doi.org/10.33263/BRIAC134.326

Laaroussi, H.; Bakour, M.; Ousaaid, D. et al. 2020. Effect of antioxidant-rich propolis and bee pollen extracts against D-glucose induced type 2 diabetes in rats. *Food Research International*, *138*(B), 109802. https://doi.org/10.1016/j.foodres.2020.109802

Lavinas, F.C.; Macedo, E.H.B.; Sá, G.B. et al. 2019. Brazilian stingless bee propolis and geopropolis: Promising sources of biologically active compounds. *Revista Brasileira de Farmacognosia*, *29*(3), 389–399. https://doi.org/10.1016/j.bjp.2018.11.007

López, A.S.; Subovsky, M.; Maidana, J.; Castillo, A. 2003. Short communication: organoleptic and pshysical characteristics of propolis from Northeastern Argentina. *Spanish Journal of Agricultural Research*, *1*(2), 37–40. https://doi.org/10.5424/sjar/2003012-26

López, B.G.-C.; Schmidt, E.M.; Eberlin, M.N.; Sawaya, A.C. 2014. Phytochemical markers of different types of red propolis. *Food Chemistry*, *146*, 174–180. https://doi.org/10.1016/j.foodchem.2013.09.063

Mahamat, A.A.; Nyemb, J.N.; Gade, I.S. et al. 2020. A New fatty acid and some triterpenoids from propolis of Nkambe (North-West Region, Cameroon) and evaluation of the antiradical scavenging activity of their extracts. *Open Chemistry*, *18*(1), 239–243. https://doi.org/10.1515/chem-2020-0016

Mahdavi-Roshan, M.; Gheibi, S.; Pourfarzad, A. 2022. Effect of propolis extract as a natural preservative on quality and shelf life of marinated chicken breast (chicken Kebab). *LWT*, *155*, 112942. https://doi.org/10.1016/j.lwt.2021.112942

Markham, K.R.; Mitchell, K.A.; Wilkins, A.L.; Daldy, J.A.; Lu, Y. 1996. HPLC and GC-MS identification of the major organic constituents in New Zeland propolis. *Phytochemistry*, *42*(1), 205–211. https://doi.org/10.1016/0031-9422(96)83286-9

Martinello, M.; Mutinelli, F. 2021. Antioxidant activity in bee products: A review. *Antioxidants*, *10*(1), 71. https://doi.org/10.3390/antiox10010071

Martinotti, S.; Ranzato, E. 2015. Propolis: A new frontier for wound healing? *Burns and Trauma*, *3*(1), 1–7. https://doi.org/10.1186/s41038-015-0010-z

Matny, O.N. 2015. Efficacy evaluation of Iraqi propolis against gray mold of stored orange caused by Penicillium digitatum. *Plant Pathology Journal*, *14*(3), 153. https://doi.org/10.3923/ppj.2015.153.157

Mezgabu, E.; Hirpa, E.; Begna, D.; Lama, Y.; Bayan Aa, M.C. 2016. Occurrence and distribution of Varroa mite and antivarroa effect of propolis in Walmara district of Oromia special zone around Finfine, Ethiopia. *Journal of Veterinary Science and Technology*, *7*(5), 10.4172. https://doi.org/10.4172/2157-7579.1000370

Miguel, M.G.; Nunes, S.; Dandlen, S.A.; Cavaco, A.M.; Antunes, M.D. 2010. Phenols and antioxidant activity of hydro-alcoholic extracts of propolis from Algarve, South of Portugal. *Food and Chemical Toxicology*, *48*(12), 3418–3423. https://doi.org/10.1016/j.fct.2010.09.014

Millones-Gómez, P.A.; De la Garza-Ramos, M.A.; Urrutia-Baca, V.H.; Hernandez-Martinez, H.C.; Marín, D.A.H.; Medina, C.A.M. 2022. Cytotoxicity of Peruvian propolis and Psidium guajava on human gingival fibroblasts, PBMCs and HeLa cells. *F1000Research*, *11*(430), 430. https://doi.org/10.12688/f1000research.110352.2

Miyata, R.; Sahlan, M.; Ishikawa, Y.; Hashimoto, H.; Honda, S.; Kumazawa, S. 2019. Propolis components from stingless bees collected on south Sulawesi, Indonesia, and their xanthine oxidase inhibitory activity. *Journal of Natural Products*, *82*(2), 205–210. https://doi.org/10.1021/acs.jnatprod.8b00541

Moise, A.R.; Bobiş, O. 2020. Baccharis dracunculifolia and Dalbergia ecastophyllum, main plant sources for bioactive properties in green and red Brazilian propolis. *Plants*, *9*(11), 1619. https://doi.org/10.3390/plants9111619

Moreno, M.A.; Vallejo, A.M.; Ballester, A.-R. et al. 2020. Antifungal edible coatings containing Argentinian propolis extract and their application in raspberries. *Food Hydrocolloids*, *107*, 105973. https://doi.org/10.1016/j.foodhyd.2020.105973

Nalbantsoy, A.; Sarıkahya, N.B.; Özverel, C.S. et al. 2022. Chemical composition and biological activities of Cypriot propolis. *Journal of Apicultural Research*, *61*(2), 233–245. https://doi.org/10.1080/00218839.2021.1977028

Namae, J. 2022. *The Potential of Propolis as Passion Fruit Juice Preservative*. Makerere University

Nichitoi, M.M.; Josceanu, A.M.; Isopescu, R.D. et al. 2021. Polyphenolics profile effects upon the antioxidant and antimicrobial activity of propolis extracts. *Scientific Reports*, *11*(1), 1–12. https://doi.org/10.1038/s41598-021-97130-9

Nishikawa, S.; Hydo, T.; Aoyama, H.; Miyata, R.; Kumazawa, S.; Tsuda, T. 2020. Artepillin C, a Key component of Brazilian propolis, induces thermogenesis in inguinal white adipose tissue of mice through a creatine-metabolism-related thermogenic pathway. *Journal of Agricultural and Food Chemistry*, *68*(4), 1007–1014. https://doi.org/10.1021/acs.jafc.9b07080

Nna, V.U.; Abu Bakar, A.B.; Ahmad, A.; Eleazu, C.O.; Mohamed, M. 2019. Oxidative stress, NF-κb-mediated inflammation and apoptosis in the testes of streptozotocin–induced diabetic rats: Combined protective effects of Malaysian propolis and metformin. *Antioxidants*, *8*(10), 465. https://doi.org/10.3390/antiox8100465

Nna, V.U.; Bakar, A.B.A.; Lazin, M.R.M.L.M.; Mohamed, M. 2018. Antioxidant, anti-inflammatory and synergistic anti-hyperglycemic effects of Malaysian propolis and metformin in streptozotocin–induced diabetic rats. *Food and Chemical Toxicology*, *120*, 305–320. https://doi.org/10.1016/j.fct.2018.07.028

Olczyk, P.; Komosinska-Vassev, K.; Krzyminiewski, R. et al. 2020. The estimation of blood paramagnetic center changes during burns management with biodegradable propolis-nanofiber dressing. *Oxidative Medicine and Cellular Longevity*, *2020*. https://doi.org/10.1155/2020/3675603

Olczyk, P.; Komosińska-Vassev, K.; Winsz-Szczotka, K. et al. 2012. Propolis modulates vitronectin, laminin, and heparan sulfate/heparin expression during experimental burn healing. *Journal of Zhejiang University. Science. B*, *13*(11), 932–941. https://doi.org/10.1631/jzus.B1100310

Olczyk, P.; Komosinska-Vassev, K.; Winsz-Szczotka, K.; Stojko, J.; Klimek, K.; Kozma, E.M. 2013a. Propolis induces chondroitin/dermatan sulphate and hyaluronic acid accumulation in the skin of burned wound. *Evidence-Based Complementary and Alternative Medicine*, *2013*. https://doi.org/10.1155/2013/290675

Olczyk, P.; Wisowski, G.; Komosinska-Vassev, K. et al. 2013b. Propolis modifies collagen types I and III accumulation in the matrix of burnt tissue. *Evidence-Based Complementary and Alternative Medicine*, *2013*. https://doi.org/10.1155/2013/423809

Olczyk, P.; Komosinska-Vassev, K.; Wisowski, G.; Mencner, L.; Stojko, J.; Kozma, E.M. 2014. Propolis modulates fibronectin expression in the matrix of thermal injury. *BioMed Research International*, *2014*. https://doi.org/10.1155/2014/748101

Oliveira, G.; McManus, C.; Dos Santos, V. 2022. Essential oils and propolis as additives in egg coatings. *World's Poultry Science Journal*, 1–14. https://doi.org/10.1080/00439339.2022.2119914

Oršolić, N.; Jazvinšćak Jembrek, M. 2022. Molecular and cellular mechanisms of propolis and its polyphenolic compounds against cancer. *International Journal of Molecular Sciences*, *23*(18), 10479. https://doi.org/10.3390/ijms231810479

Oršolić, N.; Landeka Jurčević, I.; Đikić, D. et al. 2019. Effect of propolis on diet-induced hyperlipidemia and atherogenic indices in mice. *Antioxidants*, *8*(6), 156. https://doi.org/10.3390/antiox8060156

Oryan, A.; Alemzadeh, E.; Moshiri, A. 2018. Potential role of propolis in wound healing: Biological properties and therapeutic activities. *Biomedicine and Pharmacotherapy*, *98*, 469–483. https://doi.org/10.1016/j.biopha.2017.12.069

Özkök, A.; Keskin, M.; Tanuğur Samancı, A.E.; Yorulmaz Önder, E.; Takma, Ç. 2021. Determination of antioxidant activity and phenolic compounds for basic standardization of Turkish propolis. *Applied Biological Chemistry*, *64*(1), 1–10. https://doi.org/10.1186/s13765-021-00608-3

Payandan, E.; Sayyed-Alangi, S.Z.; Shamloofar, M.; Koohsari, H. 2017. Study of chemical composition and efficacy of different extracts of Iranian propolis on the microbiological and sensory parameters of minced Cyprinus carpio meat at 4 C storage. *Journal of Aquatic Food Product Technology*, *26*(5), 593–603. https://doi.org/10.1080/10498850.2016.1240281

Pereira, F.A.N.; Barboza, J.R.; Vasconcelos, C.C.; Lopes, A.J.O.; Ribeiro, M.N.d.S. 2021. Use of stingless bee propolis and geopropolis against cancer—A literature review of preclinical studies. *Pharmaceuticals*, *14*(11), 1161. https://doi.org/10.3390/ph14111161

Pessolato, A.G.T.; dos Santos Martins, D.; Ambrósio, C.E.; Mançanares, C.A.F.; de Carvalho, A.F. 2011. Propolis and amnion reepithelialise second-degree burns in rats. *Burns*, *37*(7), 1192–1201. https://doi.org/10.1016/j.burns.2011.05.016

Pisoschi, A.M.; Pop, A.; Georgescu, C.; Turcuş, V.; Olah, N.K.; Mathe, E. 2018. An overview of natural antimicrobials role in food. *European Journal of Medicinal Chemistry*, *143*, 922–935. https://doi.org/10.1016/j.ejmech.2017.11.095

Pobiega, K.; Kraśniewska, K.; Gniewosz, M. 2019. Application of propolis in antimicrobial and antioxidative protection of food quality–A review. *Trends in Food Science and Technology*, *83*, 53–62. https://doi.org/10.1016/j.tifs.2018.11.007

Pobiega, K.; Przybył, J.L.; Żubernik, J.; Gniewosz, M. 2020. Prolonging the shelf life of cherry tomatoes by pullulan coating with ethanol extract of propolis during refrigerated storage. *Food and Bioprocess Technology*, *13*(8), 1447–1461. https://doi.org/10.1007/s11947-020-02487-w

Popova, M.; Trusheva, B.; Khismatullin, R. et al. 2013. The triple botanical origin of Russian propolis from the Perm region, its phenolic content and antimicrobial activity. *Natural Product Communications*, *8*(5), 1934578X1300800519. https://doi.org/10.1177/1934578X1300800519

Popova, M.P.; Trusheva, B.S.; Nedialkov, P.T. et al. 2020. New Δ-tocotrienol derivatives from Colombian propolis. *Natural Product Research*, *34*(19), 2779–2786. https://doi.org/10.1080/14786419.2019.1590710

Prata, M.F.; de Carvalho, F.M.; Gonçalves-Júnior, W.D., et al. 2022. Hypolipidemic and anti-obesity effects of hydroalcoholic extract of Brazilian red propolis in a rodent model of dyslipidemia. *European Journal of Lipid Science and Technology*, *124*(9), 2100017. https://doi.org/10.1002/ejlt.202100017

Ripari, N.; Sartori, A.A.; da Silva Honorio, M. et al. 2021. Propolis antiviral and immunomodulatory activity: A review and perspectives for COVID-19 treatment. *Journal of Pharmacy and Pharmacology*, *73*(3), 281–299. https://doi.org/10.1093/jpp/rgaa067

Ristivojević, P.; Dimkić, I.; Guzelmeric, E. et al. 2018. Profiling of Turkish propolis subtypes: Comparative evaluation of their phytochemical compositions, antioxidant and antimicrobial activities. *LWT*, *95*, 367–379. https://doi.org/10.1016/j.lwt.2018.04.063

Rivera-Yañez, N.; Rivera-Yañez, C.R.; Pozo-Molina, G.; Méndez-Catalá, C.F.; Méndez-Cruz, A.R.; Nieto-Yañez, O. 2020. Biomedical properties of propolis on diverse chronic diseases and its potential applications and health benefits. *Nutrients*, *13*(1), 78. https://doi.org/10.3390/nu13010078

Rivera-Yañez, N.; Rodriguez-Canales, M.; Nieto-Yañez, O. et al. 2018. Hypoglycaemic and antioxidant effects of propolis of Chihuahua in a model of experimental diabetes. *Evidence-Based Complementary and Alternative Medicine*, *2018*. https://doi.org/10.1155/2018/4360356

Rivero-Cruz, J.F.; Granados-Pineda, J.; Pedraza-Chaverri, J. et al. 2020. Phytochemical constituents, antioxidant, cytotoxic and antimicrobial activities of the ethanolic extract of Mexican brown propolis. *Antioxidants*. https://doi.org/10.3390/antiox9010070.

Rouibah, H.; Kebsa, W.; Lahouel, M. et al. 2018. Algerian propolis potentiates doxorubicin mediated anticancer effect against human pancreatic PANC-1 cancer cell line through cell cycle arrest, apoptosis induction and P-glycoprotein inhibition. *Anti-Cancer Agents in Medicinal Chemistry (Formerly Current Medicinal Chemistry-Anti-Cancer Agents)*, *18*(3), 375–387. https://doi.org/10.2174/1871520618666180110143239

Roy, S.; Priyadarshi, R.; Rhim, J.-W. 2021. Development of multifunctional pullulan/chitosan-based composite films reinforced with ZnO nanoparticles and propolis for meat packaging applications. *Foods*, *10*(11), 2789. https://doi.org/10.3390/foods10112789

Rufatto, L.C.; Luchtenberg, P.; Garcia, C. et al. 2018. Brazilian red propolis: Chemical composition and antibacterial activity determined using bioguided fractionation. *Microbiological Research*, *214*, 74–82. https://doi.org/10.1016/j.micres.2018.05.003

Russo, A.; Longo, R.; Vanella, A. 2002. Antioxidant activity of propolis: Role of caffeic acid phenethyl ester and galangin. *Fitoterapia*, *73*, S21–S29. https://doi.org/10.1016/S0367-326X(02)00187-9

Saewan, N.; Jimtaisong, A. 2015. Natural products as photoprotection. *Journal of Cosmetic Dermatology*, *14*(1), 47–63. https://doi.org/10.1111/jocd.12123

Salatino, A.; Salatino, M.L.F.; Negri, G. 2021. How diverse is the chemistry and plant origin of Brazilian propolis? *Apidologie*, 1–23. https://doi.org/10.1007/s13592-021-00889-z

Santos, M.F.C.; Oliveira, L.C.; Ribeiro, V.P. et al. 2021. Isolation of diterpenes from Araucaria sp Brazilian brown propolis and development of a validated high-performance liquid chromatography method for its analysis. *Journal of Separation Science*, *44*(16), 3089–3097. https://doi.org/10.1002/jssc.202100374

Sarıkahya, N.B.; Gören, A.C.; Okkalı, G.S. et al. 2021. Chemical composition and biological activities of propolis samples from different geographical regions of Turkey. *Phytochemistry Letters*, *44*, 129–136. https://doi.org/10.1016/j.phytol.2021.06.008

Sato, T.; Mello, D.; Vasconcellos, L.; Valente, A.J.; Borges, A. 2020. Chitosan-based coacervate polymers for propolis encapsulation: Release and cytotoxicity studies. *International Journal of Molecular Sciences*, *21*(12), 4561. https://doi.org/10.3390/ijms21124561

Sehn, E.; Hernandes, L.; Franco, S.; Gonçalves, C.; Baesso, M. 2009. Dynamics of reepithelialisation and penetration rate of a bee propolis formulation during cutaneous wounds healing. *Analytica Chimica Acta*, *635*(1), 115–120. https://doi.org/10.1016/j.aca.2009.01.019

Siheri, W.; Ebiloma, G.U.; Igoli, J.O. et al. 2019. Isolation of a novel flavanonol and an alkylresorcinol with highly potent anti-trypanosomal activity from Libyan propolis. *Molecules*, *24*(6), 1041. https://doi.org/10.3390/molecules24061041

Silva-Beltrán, N.P.; Galvéz-Ruíz, J.C.; Ikner, L.A.; Umsza-Guez, M.A.; de Paula Castro, T.L.; Gerba, C.P. 2022. In vitro antiviral effect of Mexican and Brazilian propolis and phenolic compounds against human coronavirus 229E. *International Journal of Environmental Health Research*, 1–13. https://doi.org/10.1080/09603123.2022.2110576

Simone-Finstrom, M.; Borba, R.S.; Wilson, M.; Spivak, M. 2017. Propolis counteracts some threats to honey bee health. *Insects*, *8*(2), 46. https://doi.org/10.3390/insects8020046

Skowron, K.; Kwiecińska-Piróg, J.; Grudlewska, K. et al. 2019. Antilisterial activity of polypropylene film coated with chitosan with propolis and/or bee pollen in food models. *BioMed Research International*, *2019*. https://doi.org/10.1155/2019/7817063

Socha, R.; Gałkowska, D.; Bugaj, M.; Juszczak, L. 2015. Phenolic composition and antioxidant activity of propolis from various regions of Poland. *Natural Product Research*, *29*(5), 416–422. https://doi.org/10.1080/14786419.2014.949705

Sosnowski, Z.M. 1983. *Method for Extracting Propolis and Water Soluble Dry Propolis Powder*. Google Patents.

Spinelli, S.; Conte, A.; Lecce, L.; Incoronato, A.L.; Del Nobile, M.A. 2015. Microencapsulated propolis to enhance the antioxidant properties of fresh fish burgers. *Journal of Food Process Engineering*, *38*(6), 527–535. https://doi.org/10.1111/jfpe.12183

Spivak, M.; Goblirsch, M.; Simone-Finstrom, M. 2019. Social-medication in bees: The line between individual and social regulation. *Current Opinion in Insect Science*, *33*, 49–55. https://doi.org/10.1016/j.cois.2019.02.009

Staniczek, J.; Jastrzębska-Stojko, Ż.; Stojko, R. 2021. Biological activity of propolis ointment with the addition of 1% nanosilver in the treatment of experimentally-evoked burn wounds. *Polymers*, *13*(14), 2312. https://doi.org/10.3390/polym13142312

Šturm, L.; Črnivec, I.G.O.; Istenič, K. et al. 2019. Encapsulation of non-dewaxed propolis by freeze-drying and spray-drying using gum Arabic, maltodextrin and inulin as coating materials. *Food and Bioproducts Processing*, *116*, 196–211. https://doi.org/10.1016/j.fbp.2019.05.008

Suhag, R.; Kumar, N.; Petkoska, A.T.; Upadhyay, A. 2020. Film formation and deposition methods of edible coating on food products: A review. *Food Research International*, *136*, 109582. https://doi.org/10.1016/j.foodres.2020.109582

Surek, M.; Fachi, M.M.; de Fátima Cobre, A. et al. 2021. Chemical composition, cytotoxicity, and antibacterial activity of propolis from Africanized honeybees and three different Meliponini species. *Journal of Ethnopharmacology*, *269*, 113662. https://doi.org/10.1016/j.jep.2020.113662

Tamfu, A.N.; Sawalda, M.; Fotsing, M.T. et al. 2020. A new isoflavonol and other constituents from Cameroonian propolis and evaluation of their anti-inflammatory, antifungal and antioxidant potential. *Saudi Journal of Biological Sciences*, *27*(6), 1659–1666. https://doi.org/10.1016/j.sjbs.2019.11.035

Tani, H.; Hikami, S.; Takahashi, S. et al. 2019. Isolation, identification, and synthesis of a new prenylated cinnamic acid derivative from Brazilian green propolis and simultaneous quantification of bioactive components by LC-MS/MS. *Journal of Agricultural and Food Chemistry*, *67*(44), 12303–12312. https://doi.org/10.1021/acs.jafc.9b04835

Tavares, L.; Smaoui, S.; Lima, P.S.; de Oliveira, M.M.; Santos, L. 2022. Propolis: Encapsulation and application in the food and pharmaceutical industries. *Trends in Food Science and Technology*. https://doi.org/10.1016/j.tifs.2022.06.003

Touzani, S.; Imtara, H.; Katekhaye, S. et al. 2021. Determination of phenolic compounds in various propolis samples collected from an African and an Asian region and their impact on antioxidant and antibacterial activities. *Molecules*, *26*(15), 4589. https://doi.org/10.3390/molecules26154589

Usman, U.Z.; Bakar, A.B.A.; Mohamed, M. 2018. Propolis improves pregnancy outcomes and placental oxidative stress status in streptozotocin-induced diabetic rats. *BMC Complementary and Alternative Medicine*, *18*(1), 1–6. https://doi.org/10.1186/s12906-018-2391-6

Vargas-Sánchez, R.D.; Torrescano-Urrutia, G.R.; Acedo-Félix, E. et al. 2014. Antioxidant and antimicrobial activity of commercial propolis extract in beef patties. *Journal of Food Science*, *79*(8), C1499–C1504. https://doi.org/10.1111/1750-3841.12533

Viera, V.B.; Piovesan, N.; Moro, K.I.B. et al. 2016. Preparation and microbiological analysis of Tuscan sausage with added propolis extract. *Food Science and Technology*, *36*, 37–41. https://doi.org/10.1590/1678-457X.0045

Wagh, V.D. 2013. Propolis: A wonder bees product and its pharmacological potentials. *Advances in Pharmacological Sciences*, *2013*. https://doi.org/10.1155/2013/308249

Wang, J.; Bhargava, P.; Yu, Y. et al. 2020. Novel caffeic acid phenethyl ester-mortalin antibody nanoparticles offer enhanced selective cytotoxicity to cancer cells. *Cancers*, *12*(9), 2370. https://doi.org/10.3390/cancers12092370

West, C. 2019. Recent trends in chiral supercritical fluid chromatography. *TrAC Trends in Analytical Chemistry*, *120*, 115648. https://doi.org/10.1016/j.trac.2019.115648

Wezgowiec, J.; Wieczynska, A.; Wieckiewicz, W. et al. 2020. Polish propolis—Chemical composition and biological effects in tongue cancer cells and macrophages. *Molecules*, *25*(10), 2426. https://doi.org/10.3390/molecules25102426

Wieczorek, P.P.; Hudz, N.; Yezerska, O. et al. 2022. Chemical variability and pharmacological potential of propolis as a source for the development of new pharmaceutical products. *Molecules*, *27*(5), 1600. https://doi.org/10.3390/molecules27051600

Woźniak, M.; Mrówczyńska, L.; Kwaśniewska-Sip, P.; Waśkiewicz, A.; Nowak, P.; Ratajczak, I. 2020. Effect of the solvent on propolis phenolic profile and its antifungal, antioxidant, and in vitro cytoprotective activity in human erythrocytes under oxidative stress. *Molecules*, *25*(18), 4266. https://doi.org/10.3390/molecules25184266

Xu, X.; Yang, B.; Wang, D.; Zhu, Y.; Miao, X.; Yang, W. 2020. The chemical composition of Brazilian green propolis and its protective effects on mouse aortic endothelial cells against inflammatory injury. *Molecules*, *25*(20), 4612. https://doi.org/10.3390/molecules25204612

Yang, S.; Peng, L.; Cheng, Y.; Chen, F.; Pan, S. 2010. Control of citrus green and blue molds by Chinese propolis. *Food Science and Biotechnology*, *19*(5), 1303–1308. https://doi.org/10.1007/s10068-010-0186-8

Yong, H.; Liu, J. 2021. Active packaging films and edible coatings based on polyphenol-rich propolis extract: A review. *Comprehensive Reviews in Food Science and Food Safety*, *20*(2), 2106–2145. https://doi.org/10.1111/1541-4337.12697

Yousefi, M.; Rahimi-Nasrabadi, M.; Pourmortazavi, S.M. et al. 2019. Supercritical fluid extraction of essential oils. *TrAC Trends in Analytical Chemistry*, *118*, 182–193. https://doi.org/10.1016/j.trac.2019.05.038

Zabaiou, N.; Fouache, A.; Trousson, A. et al. 2017. Biological properties of propolis extracts: Something new from an ancient product. *Chemistry and Physics of Lipids*, *207*(B), 214–222. https://doi.org/10.1016/j.chemphyslip.2017.04.005

Zahid, N.; Ali, A.; Siddiqui, Y.; Maqbool, M. 2013. Efficacy of ethanolic extract of propolis in maintaining postharvest quality of dragon fruit during storage. *Postharvest Biology and Technology*, *79*, 69–72. https://doi.org/10.1016/j.postharvbio.2013.01.003

Zheng, Y.; Wu, Y.; Tao, L. et al. 2020. Chinese Propolis prevents obesity and metabolism syndromes induced by a high-fat diet and accompanied by an altered gut microbiota structure in mice. *Nutrients*, *12*, 959. https://doi.org/10.3390/nu12040959

10 Beeswax

Muhammad Shahbaz, Hammad Naeem, Ali Hamza, Shamas Murtaza, Noor-ul-Huda, Gulzar Ahmad Nayik, Aijaz Ahmad Wachkoo, Rattan Singh and Vikas Nanda

10.1 INTRODUCTION

In 1975, H. Bennett gave a detailed explanation of what wax was, calling it "an unctuous solid with various degree levels of flexibility, smoothness and gloss, which melts swiftly" (Bennett, 1975). Waxes can be characterized in a variety of ways. The fatty molecules known as waxes can be found in plants, insects, animal skins, and minerals. Additionally, they could be referred to as medium- to long-chain-length organic hydrophobic compounds (Post-Beittmiller, 1996). In addition to flavonoids, sterols, triterpenes, alcohols, aldehydes, ketones, and hydrocarbons, wax frequently contains fatty acids with extremely long chains (>C18) (Post-Beittmiller, 1996). Wax is insoluble in most organic solvents and in water. As complex mixtures with a variety of physical properties, waxes are frequently used to describe particular goods based on their sources.

Wax is produced by some insects in nature, but Apoidea, especially bees, produce wax that is widely prized and utilized by mankind (Kaluza et al., 2018). The most common type of wax produced by humans is beeswax, which is created by *Apis cerana* and *Apis mellifera* species. This multifunctional bee product is therefore simple to acquire.

Naturally, beeswax is a product produced by honey bees. When exposed to air, it solidifies into translucent white scales in certain wax-producing glands in the abdomen of young, i.e. 12- to 18-days-old, worker bees (Bogdanov, 2016). These wax scales are used to form wax combs, which are then molded by the mandibles of honey bees. As wax accumulation is not the goal of beekeeping, beekeepers often feed created wax foundation sheets to bee colonies. Recycled substances from older brood combs/honeycombs, and capping waxes, are used to make the raw materials for wax production. Almost all of the wax used for capping is pure wax. Esters of higher fatty acids comprise the majority of the complex mixture that makes up beeswax (Tulloch, 1980). Beeswax is a useful substance to evaluate the history of exposure for lipophilic chemicals inside the hive, since the majority of acaricides are readily fat-soluble as well as non-volatile, because it includes a high concentration of fatty acids (Wallner, 1999). It can survive in the hive for a very long time, and it is recycled by many beekeepers into wax foundations for comb development, which leads to a better concentration of various chemical residual matters used in agriculture as well as beekeeping (Lozano et al., 2019; Ravoet et al., 2015). It also has the lowest replacement rate of all beehive products (Mullin et al., 2010). Beeswax can act as a final sink or pollutant reservoir (Bommuraj et al., 2019). The majority of residual materials remain within wax, and the movement of residues from wax to larvae and beebread is a significant component that may influence the colony's evolution (Morales et al., 2020). Bee longevity, colony performance, development of worker honey bees and queens, and residue accumulation are all possible effects (Haarmann et al., 2002).

10.2 CLASSIFICATION

Waxes are classified into two types: natural and artificial. Synthetic waxes are often made by polymerizing feedstocks including methylene, ethylene, and alpha olefins to create low–molecular weight polymers. There are many different kinds of natural waxes, including mineral waxes generated

 DOI: 10.1201/9781003354116-10

from peat, coal, and oil, as well as animal and plant waxes. Wax sources significantly affect the composition of waxes. The majority of natural wax group esters derived from plants and animals are made up of long-chain fatty acids (LCFA), hydrocarbons, fatty alcohols, and fatty acids, while waxes derived from minerals contain hardly any long-chain hydrocarbons. Many times, waxes are misidentified as gum or fat. Wool wax (or wool grease) made using lanolin is widely used to evaluate its qualities and is thought of as a fat (Tinto et al., 2017).

10.2.1 Animal Wax

Bees (*Apis mellifera*), sheep (*Ovis aries*), snails (*Physeter macrocephalus*), and other animals are regularly used to produce animal wax. Bees (*Apis mellifera*) produce beeswax, a naturally occurring wax, principally for the construction of honeycombs (Tulloch, 1970a).

The construction of the hive uses propolis, sometimes known as "bee glue," a sticky adhesive derived from plants. Propolis can be combined with beeswax; however, beeswax-free propolis is the best. The wax gland naturally secretes beeswax, which is utilized to construct the hive's cell walls (Bogdanov, 2009).

Depending on the age and nutrition of the bees, beeswax can range from yellow to virtually black in color. It also has a mild balsamic taste and an aroma that is somewhat reminiscent of honey. It is composed primarily of free myricyl palmitate (myricin) and cerotic acid, with various high-carbon paraffins, and is soft–brittle, having a specific gravity of approx. 0.95 and melting point slightly higher than 60°C (140°F). It is insoluble in water, but it can be dissolved in chemicals such as warmed ether, carbon tetrachloride, and chloroform.

Non-hydrolyzed beeswax mostly consists of esters, with hydrocarbons making up 15 percent of it, free fatty acids 8 percent, and other components 6 percent (Tuloch, 1970b). When temperatures drop below 62–65°C, beeswax melts and becomes brittle naturally. Beeswax is incredibly flexible and safe. It provides protection from abrasion, corrosion, and evaporation of moisture. Beeswax-based accessible waxes rank among the most practical commercial products (Ogutcu et al., 2015). Commercially, beeswax is utilized for a variety of products, including candles, metal casting, cosmetics, textiles, varnishes, and food.

10.3 HISTORY

Let's first consider the significance of bees to human health and existence in general. Did you realize that over a third of the food we consume is dependent on bees? They pollinate plants that produce foodstuffs and medications. This, and the list goes on, produces great appreciation for our bees. More than simply our food and medications are impacted by bees. Several crucial issues for the wellbeing of people and the environment, such as biodiversity, food security, nutrition, and sustainable agriculture, are related to honey bees using land. There is no way a world without bees could exist, because the health of the honey bee is essential for sustainable agriculture. Beeswax is thought to have been used in the 18th and 19th centuries by ladies who had smallpox scars to smooth out their skin. There is a theory that if someone looked at these marks too closely or for too long, "Mind your own beeswax," or "Stop," could be a woman's command to mind one's own business rather than looking at her beeswax. However, it is said that this tale's genesis is a myth. It appears that the "Little History Lesson" chain email, which went viral in 2000, is where the story first gained popularity. The song "Mind Your Own Beeswax" was initially mentioned in a children's book in 1929. Other early recordings followed in 1934 and 1939, several years after it was claimed that women were putting wax on their faces and making fashionable remarks about it.

However, its use as a dental filling was a novel discovery. Neolithic-era ancient dental artefacts have been found, including a 6500-year-old Slovenian human mandible with a left molar. The beeswax filling may still be seen in the canine crest. Despite the lack of complete information, it is assumed that if the filling was completed at the time of the intervention, it was done to alleviate the victim's ongoing dental sensitivity brought on by either pain or exposed enamel and dentine as a result of nibbling on a chipped tooth, as exciting as it would be to have the oldest direct evidence of a documented medicinal tooth filling.

Beeswax is the beehive's structural support, according to research. Beeswax is used by honey bees to construct their comb, which is then filled with honey and brood to create the hexagonal shape. The bottom of the abdomen of worker bees has eight specialized wax glands that create wax. When released, the wax is in liquid form, but it cools down and solidifies rapidly. The worker bees scrapes the produced wax from her abdomen using her legs. The bee chews wax till it becomes malleable, at which point the bee shapes the wax into specific hexagonal shapes that take the shape of the final honeycomb of the cell walls.

Beeswax was used to bind reeds together to create the musical instrument known as Pan's Pipes, which was devised by the ancient god Pan, the protector of bees. Beeswax sheets were rolled by the famed Greek physician Dioscorides and used to create fake flowers. The skill of creating delicate items from wax and casting them afterward in precious metals was used by ancient jewelers and craftspeople. Ancient wall murals and icons used beeswax to create their colors, which haven't changed in over 2000 years (Birshtein et al., 1976). Beeswax was employed in a soothing ointment by the Greek-Roman physician Galen in the second century AD. In the tenth century AD, the renowned Iranian physician Avicenna suggested using beeswax as medication. Beeswax was acknowledged as a key component of medicine in China's best-known medical text, *The Shen Nong Book of Herbs*, which was written in the first or second century BC, according to Rit and Behrer (1999). It is acknowledged for having a positive impact on the body's general balance, blood and energy systems, and metabolism. Beeswax, according to the author, has qualities that can improve appearance and fight aging. It is used on the skin in combination with other components to cure wounds and as a healthy food for dieting. This trustworthy source of traditional Chinese medicine offers a vast resource of recipes. Ge Hong (284–364, Jin dynasty) and Sun Simiao (58–682) advocated "beeswax therapy," which involves heating skin patches with cloths that have been dipped in molten beeswax. Beeswax therapy was described in detail by Liu Yuxi in 841, more than 1000 years before the Barthe de Sandford wrote his paraffin wax therapy. *The Sages' Prescriptions*, published by the Song Dynasty Imperial Hospital in 992, contains dietary treatment, health improvement, and prescriptions for anti-aging effects using honeycomb, beeswax and honey.

In ancient Egypt, beeswax was recycled to create figurines and religious artefacts. Figures made of stingless beeswax have been produced in Australia and other South American nations that have stingless bee populations. An essential component of early seals was beeswax. Beeswax was initially utilized in its purest form, and resin and color were added only afterward (Crane, 1999).

In the 19th century paraffin wax was introduced; up until then, the manufacturing and reselling of beeswax products, beeswax candles, etc. was a very good business. Currently, beeswax has lost its exclusivity; nonetheless, it stands out as the most expensive wax compared with other waxes.

10.4 HONEY BEE PRODUCTS

Important and practical products produced by honey bees include venom, royal jelly, honey, pollen, bee bread, and beeswax. The nutritional, physiological, and therapeutic benefits of bee products in the ancient Greek, Egyptian, and Chinese cultures have been highlighted in earlier studies (El-Seedi et al., 2020). Additionally, nutritional supplements for improved health can be made from bee products. The biologically active qualities of bee products include antioxidant, anti-inflammatory, anti-cancer, and antimicrobial effects (Nainu et al., 2021; Ranneh et al., 2021). Biologically active components discovered in bee products include peptides, minerals, proteins, flavonoids, fatty acids, terpenes, and phenolic compounds (Thakur and Nanda, 2020; Huang et al., 2014; Carpena et al., 2020).

10.5 BEESWAX

Beeswax is a natural animal wax produced by the honey bee species *Apis mellifera*. Wax is produced by four pairs of glands on the ventral side of the abdomen and is released by worker bees.

Bees manufacture capping and cells out of wax, which they primarily use to store honey pollen and raise their young. Beeswax is used to build the honeycomb's structure. The wax gland secretes beeswax, which is needed to construct the hive's cell walls (Bogdanov, 2016).

Beeswax is a composite material that is secreted as a liquid by wax-producing glands found in the abdomens of young worker bees (belonging to the age group of 12 to 18 days, or during the end period when the bee is acting as a nursing bee). When air contacts the liquid, it transforms it into scale (which is used by the bees to form honeycombs, adding pollen and propolis) (Bogdanov, 2004). Pure beeswax is almost white when released by the bee; not until it comes into contact with honey and pollen does it begin to change color, becoming variously brilliant and yellowish. After about four years, it turns brown, since it contains the cocoon. Boiling alcohol causes it to partially dissolve, whereas carbon disulfide, chloroform, and hot turpentine essence cause it to completely dissolve. Acidic substances and honey bee stomach secretions have no effect on it. Additionally, water and cold alcohol cannot dissolve it (Bogdanov, 2004). Boiling alcohol is used to cure wax. Cerotic acid, either alone or in combination with trace amounts of melissic acid, makes up the component that melts, while melissyl palmitic ether and trace amounts of ether compounds of palmitic and stearic acid make up the component that dissolves. It has a density of 0.960 to 0.970 kg/m^{-3} at 15°C and melts at 63.5 to 64.5°C. Wax oil is created when it undergoes a dry distillation, which results in a buttery mass (Puleo, 1991)

A bee is thought to eat from 6 to 10 lb (or 3 to 4.5 kg) of honey to produce 1 lb of beeswax. The bee secretes wax from wax glands present on the underside of its abdomen in the form of tiny flakes. After the honey is removed, the beeswax is extracted by carefully melting the honeycomb, sifting the beeswax to make it free from impurities, and then pressing the residue to gather the remaining beeswax. After that, the purified wax is put into molds to solidify. Instead of using direct heat to melt the wax, it can be melted in water to preserve color and quality. Wax can also be bleached.

Beeswax is used in the pharmaceutical sector, in addition to being used in candles and cosmetics, since it is thought to be safe for consumption and has no adverse effects on the digestive system of people. Beeswax is used to encapsulate chemicals that when consumed, slowly release their contents into the body. Beeswax is utilized in many different products and procedures, including packaging, processing, and preservation. It serves as an essential component in numerous industrial technology applications.

10.5.1 Crude Beeswax

Wax is extracted from honeycombs after the honey has been extracted. To obtain honey and wax, the honeycomb is subjected to preliminary processing such as melting, molding, and decantation.

10.5.2 Refined Beeswax

Beeswax is purified using purification techniques such as melting in hot water and fine filtration.

10.5.3 Composition of Beeswax

Over 300 distinct components make up the incredibly complex compound known as beeswax. Higher fatty acid esters and alcohols make up the majority of its composition. Beeswax also includes trace amounts of hydrocarbons, acids, and other compounds besides esters. There have also been shown to be about 50 different fragrance constituents.

The various systems of medicine employ the ratio of ester values to acids to describe pure beeswax. Prolonged or severe heating dramatically alters this ratio. The ester to acid ratio is changed beyond the limits defined for pure beeswax at 100°C for 24 hours. Longer heating times or higher temperatures result in increased ester breakdown and loss. The physical characteristics of the wax are also affected by these changes. The structural characteristics of the wax, in addition to the positive effects of many of its minor constituents, not just the volatile and aromatic ones, are altered by excessive heat during rendering or further processing. Aside from the lipophilic compounds that comprise wax, bees also add certain proteins.

The place of production (Europe or Asia), the type of honey bees, and the age of the wax produced all influence the composition of beeswax (Bonvehi and Bermejo, 2012). About 14–15 percent of total unhydrolyzed bee wax is made up of hydrocarbons, 7–8 percent of it is free acid, 69–71 percent of it is esters, and 4–6 percent of it is made up of miscellaneous substances (Tulloch AP, 1970; Bonvehi and Bermejo, 2012). Early research revealed that beeswax contains a complicated mixture of n-alkanes, acids, and alcohols. According to Indian beeswax studies, the wax is mostly composed of hydrocarbons, saturated and unsaturated, and hydroxyl acids (Tinto et al., 2017). The correlation among the characteristics and the chemical structure of Japanese bee wax revealed that higher alcohols and hydroxyl acids are necessary for flexibility. High-temperature gas chromatography was used to evaluate the composition of beeswax, which was subsequently treated with diazomethane and acetylation. There were hydroxyl acids, hydrocarbons (C21–C41), acids, esters, and free alcohols found in the wax. The alkyl esters of beeswax, isolated from unhydrolyzed wax using preparative thin liquid chromatography, were then studied using gas chromatography, revealing that wax contains carbons ranging from C36 to C54 (Tinto et al., 2017). According to research on Polish beeswax, the dark- and light-colored combs have an influence on the hydrocarbons present in beeswax, because the darker comb has a higher percentage of n-alkane, as well as odd-numbered and even-numbered alkanes.

10.6 PROPERTIES OF BEESWAX

The chemical makeup of wax generated by several *Apis mellifera* species and wax from the African *adansonii* species is identical, although a few components are present in varied levels in both. The Asian bee species *Apis cerana*, *Apis dorsata*, and *Apis florae* produce waxes known as ghedda. Asian honey bee wax has fewer components in different quantities and a far simpler makeup. The numerous ghedda waxes resemble one another more than any *mellifera* wax. The use of ghedda waxes in place of *Apis mellifera* wax is therefore not permitted. Despite claims to the contrary, ghedda waxes from Asian species of honey bee are reported to be softer and more flexible. However, there is no significant variation in their melting points.

Aside from beeswax, the following waxes are available. The three that are most frequently used are lanolin, which is obtained from lamb wool, carnauba, which is derived from the leaves of the carnauba plant, and jojoba, which is obtained from the jojoba plant. Beeswax often has a lower melting point than other waxes, ranging from 10 to 20 degrees Celsius. Tulloch explains this discrepancy by pointing to the enormous variety of chemical elements present in beeswax. This characteristic is quite helpful in many crafts and allows the bees to utilize softened materials in the beehive.

Beeswax originally has a white appearance before gradually changing to shades of yellow, deep yellow, and brown in color. In contrast to the brown hue, which is brought on by the pigments in the larvae's feces, the usual yellow color is created by pollen colorants and propolis. Therefore, due to the approximate amounts of various pollen pigments and propolis, the color of wax can differ. Bees, honey, propolis, pollen, and other ingredients all contribute to the distinctive scent of beeswax. Beeswax typically has a nice, neutral flavor; any disagreeable flavor denotes a drop in quality brought on by foreign substances.

Beeswax naturally has a crystalline structure. Conditions for storage have an impact on how beeswax crystallizes. Wax's stiffness and elasticity increase as the crystallization process becomes more rapid after being kept for three to four months. Fresh "scale wax" has mechanical features that are important for its use as a "bee home," since it is stronger, reaches farther, and is less rigid than comb wax. These variations in wax's physical structure are what produce these variations. A crucial aspect of beeswax quality is its hardness; the harder the wax, the higher the quality of the wax.

At a fairly low temperature (about 32°C), beeswax is a solid that is inert and highly flexible. Furthermore, the majority of plant waxes have a crystalline nature and are substantially harder at this temperature. Wax changes physically when it is heated. It becomes plastic at 30 to 35°C, loses its rigid body structure at 46 to 47°C, and starts to melt between 60 and 70°C. Surface foam forms at temperatures between 95 and 105°C, while volatile components start to evaporate above 140°C. Beeswax shrinks by roughly 10 percent after cooling. In order to remove the remaining water, the mixture must be heated for at least 30 minutes at 120°C. This causes the hardness to increase.

Aside from being resistant to various acids, beeswax is also insoluble in water. The majority of organic solvents, including acetone, ether, benzene, xylol, toluene, chloroform, and tetrachloromethane, are soluble in it. All of these solvents, along with ethanol, dissolve wax quickly when heated above its melting point, although none of them do so at ambient temperature.

Monoesters, hydroxylated esters, hydrocarbons, diesters, and free fatty acids (FFAs) are only a few of the different types of compounds that make up pure beeswax. Beeswax is said to include approximately 284 distinct compounds, of which 21 main compounds make up 56 percent of the wax's composition (Tulloch, 1970). The intrinsic structural chain lengths of beeswax determine its physical qualities in terms of its overall structure. Since it is made up of esters with one long chain and one short chain, beeswax has a slightly softer texture than waxes with two long-chain esters, such as plant waxes, which are often stiffer and crystalline. It is widely acknowledged that the melting point of beeswax varies slightly depending on where it was produced. Values in the middle of 62°C and 65°C are frequently stated. The relative density of beeswax at 15°C is reported to be between 0.958 Mg m^{-3} and 0.970 Mg m^{-3}, while its thermal conductivity is stated as 0.25 $Wm^{-1}K^{-1}$ (compared with polyethylene 0.33 $Wm^{-1}K^{-1}$, copper 390 $Wm^{-1}K^{-1}$, etc.) (Koedam et al., 2002).

When it gets cold, natural beeswax becomes brittle. Beeswax is extremely plastic and inert. In addition to being insoluble in water and acid, beeswax is also hydrophobic for most organic solvents. Beeswax has a melting point between 62°C and 65°C. Beeswax's relative density ranges from 0.958 to 0.970 at 15°C. Beeswax has an acid content of 17 to 24 mg KOH/g and a saponification content of 85 to 100 mg KOH/g (Bayer et al., 2011).

10.7 PROCESSING OF BEESWAX

10.7.1 Production

Beeswax is a natural wax produced by *Apis mellifera* and *Apis cerana* honey bees and afterward utilized to make honeycomb. Bees have a life duration of 35 days. Worker bees produce wax during the 10th–16th days of life through four pairs of wax-producing glands. Carbohydrates, including honey sugar, glucose, sucrose, and fructose, are the raw material for wax formation. Bees ingest sugar, which is then transformed into wax by wax-producing glands. Wax is then expelled from tiny pores and found as little flakes on the abdomen of bees. When the translucent flakes are chewed and combined with salivary secretions, they turn white. Then, the bees use this as a building material to create combs (Bogdanov, 2016).

10.7.2 Manufacturing Methods

In the 19th century, the manufacturing of industrial wax began. The German company Mehring began manufacturing comb foundations on an industrial scale in 1857 (Ferber and Nursten, 1977). Beeswax is primarily produced by specialized beeswax producers around the world. Beekeepers supply either shabby wax or used combs.

The production processes have a significant impact on the beeswax's high quality. Melting and chemical extraction are the two processes used to extract wax. The process that is most usually utilized is melting. Wax can be melted using any of the following: steam, electricity, solar power, or boiling water. Chemical extraction with solvents is only practical in a laboratory when a limited amount of wax production is required. Xylene and gasoline are both effective wax solvents. The downside of this approach is that all the impurities of organic wax, pollen, pupae, and propolis, are dissolved. As a result, the quality of the wax may suffer. This process is only practical in laboratories when wax manufacturing is required on a smaller scale. The recovery of wax is determined by the combs and the procedure utilized. Recovery from used combs often ranges between 50 and 60 percent. It may be higher if newer combs and more capping are used. After pure wax has been separated, some wax remains (approximately 30 percent of the comb cake). Solvents can be used to remove the remainder, but the wax won't be of the highest quality. There are two states of beeswax in combs: free and bound. Only free wax will be extracted from combs when heated in sun melters at temperatures below 100°C. The only ways to release the bonded wax are via pressing or solvent extraction. Water emulsions can frequently be constructed while wax formation is being manufactured. There are two sorts of emulsions: the first one contains water-into-wax particles, while the second has wax-into-water particles (Bogdanov, 2016). Emulsifiers are used to create these emulsions. The proteins and dextrin found in pollen, wax fatty acid salts, honey, and other natural items are used as emulsifiers for the first class of emulsions. Wax fatty acid salts comprising calcium, copper, and iron cations cause the second type of emulsion. Cations can be present in hard water or can diffuse out of the containers used to make wax. The use of stainless steel pots and soft water is therefore recommended. If emulsions do form, they can be eliminated by keeping wax in the water bath for a longer period of time while the temperature is between 75 and 80°C. According to general quality standards, wax made by comb capping is of the highest grade. However, this wax still contains the same amount of pesticide residue as regular beeswax.

Beekeepers have an easy and affordable technique to manufacture raw beeswax. The sun's heat rapidly melts the combs that are placed on the sun melter. At least twice a day, the melter should face the sun directly. This technique works well, and the energy is provided "free of charge." As comb storage may be avoided, this approach is preferred for the manufacture of raw wax.

The production method mainly affects the quality of beeswax. The method can be chemical extraction or melting. The following techniques can be used to melt materials: forced immersion in hot water for extraction, boiling water for extraction, pressing of wax, steam extraction, joint steam and press extraction, and centrifugal extraction. Chemical extraction employing solvents is only a practical choice in a laboratory when only a limited amount of wax production is required. Forced immersion is a frequent method for hot water extraction. In this method, the honeycomb is immersed in the water for approximately 4–5 hours and then washed with water to remove the impurities that are water soluble. After repeating this procedure two or three times, the wax is melted by boiling the combs in a container made of aluminum with 20 to 30 liters of water. After boiling, the mixture is filtered and then cooled. When cool, wax rises to the water surface, where it can be collected. Wax rises on water because it is lighter than water. By cooling, wax can be obtained in the form of cakes, blocks, and sheets, which are easy to transfer (Bogdanov, 2016).

10.7.2.1 Solar Melter

Solar energy can be used to melt the wax. For this procedure, we require a rectangular box (slanted), which is covered by any transparent plastic or a piece of glass. At the bottom, a heat-absorbing plate is placed; when the sun's rays pass through the glass, the plate absorbs them. This plate transforms the sun's rays into heat, which increases the temperature in the box. When the wax melts, it drips drop by drop and can be collected in a catch tray. In order to melt the beeswax capping and empty combs, a very suitable instrument is the sun wax burner.

10.7.2.2 Hot Water

Pieces of comb and wax capping can be bundled in a jute bag or cotton for melting wax. For this purpose, a bag is put in a cooking pot containing water and put under a press weight. The wax begins to melt, filters through the surface, and floats on the surface. The molten wax gets hard in water and can be extracted. The water should not be very hot, because the bag should be directly over the fire; we can put the bag on a wooden bar to avoid this situation.

10.7.2.3 Steam Burner

The steam is obtained by another boiling pot. After that, the steam is transferred to a bag attached to the wax melting chamber. After dripping to the bottom, it can be separated. The steam master is very efficient for the extraction of different types of wax and can work on enormous amounts also. It can be a bit difficult to run and set a wax steamer by yourself.

10.7.3 Processing and Marketing

After the procedure, there may be some unwanted substances left in the apparatus or in sheets, and these can be removed. Once the wax is extracted, it is further purified by steam or water at a high temperature. It is then put in small plastic or steel units, but the temperature should not be very high. Then, let it cool for 24 hours. The wax obtained after this whole procedure will have no organic wastes or impurities.

We can use chemicals for bleaching if we want. But, bleaching can also be done naturally, e.g., by keeping in the sunlight, and this is more recommended to beekeepers. The wax pieces can be spread on a mat, or one can also make sheets by using a wet board. For this, we have to put the board in liquid wax, and when the wax sticks to it, we pull it out.

When the procedure is finished, and cleaning actions have also been performed, we can collect the wax in blocks (weighing 20 to 25 kg). The mixing of wax with substitutes like paraffin is not useful (Reybroeck et al., 2014).

10.8 APPLICATIONS

Wax is used in a variety of applications, from cosmetics, candles, and chewing gum to industrial ones such as hot melt adhesives for sealing cartons, plastic resin flow modifiers, and suspension polymerization process additives. Wax is most frequently used in food applications to coat food products in order to preserve quality and lengthen shelf life. It is less commonly employed in chewing gum today. Employing wax crystal networks for oleo gelation, which selected waxes display, is one of the novel applications with great abilities to bind oil. Delivery mechanisms, such as lipid carriers having a nanostructure, are used in other applications.

10.8.1 Use as Edible Coating

Food quality degradation is largely brought on by mass transfer, which includes the transfer of oxygen, flavors, and water between the food and the external environment. To prevent this transfer and preserve the integrity of food products, edible films or coatings are being created. Wax-based compounds in particular are the most effective substances to stop loss of moisture and dehydration due to their strong hydrophobicity (Morillon et al., 2002). Waxes have previously been employed as edible coatings or films for a wide range of culinary products, such as eggs, eggshells, cheese, confectionery, vegetables, and fruits.

Beeswax utilized in edible films is specifically interesting due to the chemical configuration of beeswax, which comprises a high percentage of esters of long-chain fatty alcohols and acids, and a smaller percentage of long-chain hydrocarbons and free fatty acids (FFAs). An alternative substitute is the production of composite films, in which the benefit of each constituent can be utilized to enhance the mechanical, barrier, and physical properties (Pastor et al., 2010).

10.8.1.1 Advantages of Edible Coating

- **Retardation of moisture permeability**
  One of the most noticeable advantages waxes in coatings provide is the prevention of moisture loss and desiccation. Dehydration can cause fresh goods like fruits and vegetables to lose weight and lose their marketability. Water loss can cause unfavorable alterations in processed food products. Given its apolar nature, wax protects food goods from moisture evaporation and keeps their quality high. Numerous citrus products have been effectively protected against moisture evaporation using wax, wax emulsions, and wax mixtures, including, but not limited to, blue-eyed grass (Hagenmaier and Baker, 1994; Kellerman et al., 2018), eggplants, berries, pineapples, and pineapple juice (Li et al., 2018).
- **Regulation of internal atmosphere**
  Waxing also modifies the internal environments of fruits, which affects their metabolism, particularly their respiration, and changes the quality of the inside of the fruit. Fruits that are taken before they are fully ripe may occasionally have their environment gently modified to lengthen their life span (Baldwin et al., 1997). The type of waxes used determines how well the atmosphere may be changed, because different waxes have varying degrees of gas permeability depending on their chemical composition. Reports indicated that as compared with fruits, polyethylene wax- and carnauba-coated grapefruits and oranges had lower carbon dioxide and higher oxygen concentration than shellac-coated equivalents (Hagenmaier and Baker, 1995). The interior browning of pineapple under chilling stress has also been reported to be reduced by wax treatment; however, the causes are unclear (Hu et al., 2012). Waxing also lessens fruit and vegetable deterioration and spoiling. In a hypothetical commercial context, a carnauba wax microemulsion on fresh cucumbers reduced degradation by 15 percent when compared with uncoated cucumbers, yet there was no change in weight loss (Baldwin et al., 1997).
- **Overall appearance enhancement**
  Food goods' overall appearance is enhanced by waxing, which increases their marketability. When applied to fresh fruits, vegetables, and confectionery products, several waxes give them a glossy appearance. Wax coatings are also used to retain the resilience of peel tissue and prevent surface scratch when managing fresh food, which lowers spoilage from infections at the locations of injuries. According to studies, coating avocado with candelilla films reduced the noticeable discoloration produced by browning (Saucedo-Pompa et al., 2009). Another study discovered that after ten days in cold storage, cherry tomatoes wrapped with rice bran–based wax took longer to soften (Zhang et al., 2017).

10.8.1.2 Novel Applications in Oleo Gelation

- **Gelation of oil by wax**
  A large amount of liquid oil (>95%wt) has recently been trapped by waxes acting as effective structural agents to form semisolid, thermo-reversible soft materials known as oleo gels. By creating three-dimensional crystal networks out of higher-melting, solid triacylglycerol (TAGs) (greater than 0.2), oleo gelation can be used to create meals based on lipids with little to no saturated fat and no trans fats instead of traditional hard stock fats. A certain quantity of wax is first combined with a desired liquid oil, typically at a higher temperature, to generate an oleo gel. The wax flakes disintegrate entirely. High–melting point wax gelators generate crystals with distinctive size and shape after cooling. Surface tension and capillary force cause the three-dimensional colloidal network formed by these crystals to join noncovalently and form a trap for the liquid oil (Co and Marangoni, 2012; Singh et al., 2017).
- **Natural wax gelators**
  Natural waxes stand out among the various oleo gelators that have been studied so far because of their exceptional oil-binding abilities, capacity to gel oil at slightly low

concentrations (as low as 0.5%wt), variety of tailored properties, effectiveness and cost-effectiveness (abundance of resources and relatively low cost), natural origin, and accessibility of some waxes having generally regarded as safe (GRAS status). The majority of naturally occurring waxes that may oleo gel currently come from plants, including berry wax, candelilla wax, rice bran wax, sugar cane wax, sunflower wax, etc. (Mattice and Marangoni, 2018). Beeswax is the best-known of the wax-based gelators of animal origin (Ogutcu et al., 2015).

The first study on the rice bran wax–induced gelation of edible oil was published by the Ueno group at Japan's Hiroshima University (Dassanayake et al., 2009). It has a critical gelator concentration (CGC) of 1–2 percent in canola oil and 0.5–1 percent in soybean oil, making it a very efficient gelator (Dassanayake et al., 2009; Hwang et al., 2012). Another excellent gelator is sunflower wax, which has a CGC in canola oil of 1 percent (w/w) and a specified cooling rate in soybean oil of 0.5 percent (w/w) (Hwang et al., 2012). Due to its tendency to form spherical, grain-like crystals with a poor aspect ratio and ineffective ability to entrap oil, carnauba wax is recognized as one of the least effective gelators. In terms of gelation in canola oil, candelilla wax only gelates with a modest 2 percent (w/w) efficiency (Blake and Marangoni, 2015). Nevertheless, the CGC contains 4 percent (w/w) canola oil (Blake and Marangoni, 2015). These natural wax gelators have great promise for large-scale industrial applications, especially when considering the cost-effectiveness and the minimal number of waxes (1–4 percent w/w) required to structure liquid oil as a basis for scalability.

- **Physiochemical properties of wax oleo gels**

  Each wax has a unique set of physicochemical characteristics because of its intricate chemical makeup. Wax-based oleo gels are not all the same. In order to fully comprehend how the following gels operate in certain ways, it is crucial to determine the precise makeup of each wax. Doan et al. have carried out a thorough analysis of the chemical composition of various gels. There is a known relationship between the chemical makeup of natural waxes and their ability to form a gel. The type of wax utilized also has an impact on the microstructure of wax-based oleo gels. It has been shown that thinner, more needle-like fibers or platelets are present in waxes with additional homogeneous compositions, e.g., sunflower wax and rice bran wax (Limpimwong et al., 2017). It is conceivable for many molecules to co-crystallize in a more complex system, such as the components of candelilla wax and carnauba wax, which would impede the formation of an organized crystal structure and would also make tighter molecular packing difficult (Morales-Rueda et al., 2009). Typically, these systems lead to the structural characteristics of a bigger scale. The rheological features of both flocculated suspensions and semi-diluted polymer solutions are particularly fascinating and complex in wax-based oleo gels. Additionally, the rheological profile is impacted by the chemical makeup of distinct wax types. Wax was found to be composed of saturated fatty acids and FAs (fatty acids) with higher melting temperatures. Oleo gels made by acids are viscous and durable (Dassanayake et al., 2012).

10.8.1.3 Health Benefits

- **Treatment of eczema, psoriasis, and diaper dermatitis**

  Beeswax is crucial for treating a variety of skin issues. Treatment for psoriasis, diaper dermatitis, and eczema benefits from a blend of beeswax, olive oil, and honey. An experiment was conducted at Medical Research Labs Islamic Establishment for Education and Dubai Specialized Medical Center to check the effect of this combination on the development of *Staphylococcus aureus* and *Candida albicans* that were isolated from human specimens. The results concluded that this mixture can inhibit the growth of harmful bacteria that cause skin diseases (Al-Waili, 2005).

- **Skin moisturizer**

  Frequently included in skincare products and various cosmetics, beeswax is a superb natural skin moisturizer. It locks all the moisture in the skin that can be used for the protection and repair of dry, rough, and chapped skin. Beeswax is rich in vitamin A and shows emollient properties that can rehydrate and soften the skin. It aids in the development of cellular restoration (Ashburner, 2014).

- **Liver protection**
  The *Korean Journal of Internal Medicine* published research on the antioxidant properties of honeycomb and their potential to protect the liver in 2013. For testing their impact on those with fatty liver disease, researchers combined beeswax and alcohol. The study was completed in 24 weeks. At the end, they concluded that this mixture can help in normalizing liver function (Illnait et al., 2013).
- **Lowers cholesterol levels**
  According to one study, long-chain fatty alcohol present in the waxes can lower the cholesterol level. The nutritional or health beneficial effects due to wax ester and alcohol present in beeswax cause low-density lipoprotein (LDL) (bad cholesterol) to be reduced by 20–28 percent and increase high-density lipoprotein (HDL) (good cholesterol) by 7–14 percent (Hargrove et al., 2004).
- **Reduction of stretch marks**
  One study found that stretch marks are dermally atrophic. The scars have thinned the epidermis due to a reduction in collagen and elastic fiber. According to the study, collagen is a significant extracellular matrix and a crucial component in the healing of wounds. Because beeswax includes vitamin A, which aids in the synthesis of collagen, it can be really beneficial for stretch mark removal. A mixture of grapeseed oil, shea or cocoa butter, royal jelly, and beeswax can be used. Coconut oil offers a natural alternative for preventing and reducing stretch marks and increasing collagen levels at the same time.
- **Treats fungal skin infections and jock itch**
  Although jogger's itch and fungus skin diseases are uncomfortable, beeswax can be used to cure them. Due to its anti-inflammatory properties, it can help lessen the discomfort caused by jock itch and fungus skin infections while also hydrating the skin itchiness. Early studies indicate that using a blend of beeswax and honey and olive oil three times each day for four weeks on the afflicted area can considerably enhance the treatment of jock itch and fungus skin diseases.
- **Encourages relaxation and reduces stress**
  When you think about beeswax, you probably picture candles. This is advantageous since the inhalable soot from paraffin candles can be hazardous to your health and the health of your family. Additionally, it might seriously harm the interior of your computers, electrical devices, and other components. Choose beeswax stress-relieving natural candles as an alternative. Since organic beeswax is free of chemicals, tags should be checked to make sure you get it if you're making your own candles. When you're purchasing ready-made beeswax candles, select 100 percent. The best option is probably beeswax candles with cotton wicks.
- **Pain relief and anti-inflammatory**
  Beeswax has been used for centuries to ease discomfort and inflammation, and it also has anti-inflammation and anti-swelling qualities. Inflammation occurring due to osteoarthritis can be relieved by wax, as reported in the *Korean Journal* (Puente et al., 2014).

10.8.1.4 Food and Non-Food Applications

Beeswax can be used as an edible coating for food. Beeswax is also utilized for the production of waxy foundations. On a commercial scale, beeswax has various applications, e.g., metal castings, modeling, and candle making; it is also utilized in cosmetics, food technology and processing, various industrial technologies, the textile sector, and the production of polishes and varnishes (Bogdanov, 2004).

In the United States, beeswax is recognized as safe for consumption and is frequently used in human food (Bogdanov, 2016). Because it is non-reactive, it does not react with the components of the human digestive system, and thus, it passes through the entire body almost unchanged. Substances that are dissolved or contained in wax, on the other hand, are slowly released. This feature is used in a variety of therapeutic formulations. These characteristics, however, might be troublesome when wax is kept close to dangerous chemicals like insecticides or during internal medical treatment. Whenever this wax is ingested as food, applied to skin as foundation, or given to bees as foundation sheets, any fat-soluble poisons may be absorbed and are then released later on. In the European Union, beeswax is approved as a food preservative under the trade name E901 (EFSA, 2020). Beeswax can be consumed daily up to an amount of 1290 g per person. As a colorant carrier in food technology, it is also used to coat chocolate candies, treat the surface of chocolate, and coat fruits, nuts, and coffee beans. Using beeswax to shield containers from the impact of acidic liquids like fruit juice or honey is a typical, easy, and small-scale application. It is true that honey must be treated to prevent corrosion and iron breakdown in the steel barrels used to store and transport the honey. The remedy may consist of a costly food-grade plastic liner or paint made up of a food-grade plastic film, or a very thin beeswax coating.

Beeswax is the most harmless honey product. Beeswax-related skin allergies have only been reported in one study (Lucente et al., 1996). Using beeswax on the skin increases its smoothness and gives it a smooth, fresh appearance, while also having antimicrobial characteristics. It is used to coat medications and pills, making them easier to swallow but delaying their disintegration. Medication can be combined with beeswax to delay release (Basa et al., 2016). Chewing beeswax can strengthen the gums, boost saliva production, and increase stomach acid (Aslani and Rostami, 2015). When used to treat muscular, nerve, and joint inflammation, warm beeswax offers remarkable warming capabilities. Because of this, the Bulgarian physician Pochinkova recommends using beeswax as the primary bee product for thermotherapy (Bogdanov, 2016). For this purpose, repeated applications are suggested for nerve, tendon, and muscle inflammation due to lumbago, colds, arthrosis, arthritis, and neuritis.

10.8.1.5 Pharmaceutical and Traditional Use of Beeswax

According to the Ebers Papyrus, ancient Egypt employed beeswax as a medicine (1550 BC). Beeswax was a crucial ingredient in various recipes for lotions and ointments used to extract plugs, heal burns and wounds, and reduce joint ache. Hippocrates, known as the "father of medicine," recommended utilizing beeswax to treat purulent tonsillitis. Physicians in ancient Rome used a cream known as a "cold-cream," which contained beeswax, rose water, and olive oil, to cure burns, wounds, cuts, bruises, and fractures. The first cosmetic cream, made of beeswax, olive oil, and an emulsion of water (or rose water) about 150 BC by the great Greek physician Galen, included beeswax as one of its ingredients (Stacey, 2001). In Ayurvedic medicine, the traditional and age-old form of Indian medicine, beeswax is a key component known as Madhuchishtha (Patwardhan, 2014). The "rediscovery" of Ayurvedic medicine in Western nations is a result of the search for natural remedies that can be used instead of or in addition to prescription medications. Madhuchishtha (beeswax) is used topically to cure wounds caused by abrasion or even burns (Gokani, 2014). Heel cracking has been demonstrated to be a condition for which it is especially effective (Sarojini, 2013). According to some accounts, Madhuchishtha can be used in mixes or in addition to other natural ingredients like Guda (jaggery), Madhu (honey), or Taila (oil) (Dnyandeo and Manda, 2014).

10.8.1.6 Antimicrobial Activity of Crude Beeswax

Beeswax has only lately been investigated, in contrast to other bee products, and the antibacterial action of natural materials, particularly products of the hive, has gained prominence in recent years. The yeast *Candida albicans* and various bacterial strains were both resistant to crude beeswax's antibacterial properties (Ghanem, 2011). The sample of beeswax was found to be effective against the Gram-positive bacteria *Staphylococcus epidermidis* ATCC12228, *S. aureus* ATCC25923, *Bacillus subtilis* ATCC27853 and *S. pyogenes* ATCC19615, and also effective against the Gram-negative bacteria *Pseudomonas aeruginosa* ATCC27853, *Escherichia coli* ATCC25922, and a complete inhibitory effect was found against *C. albicans* NCTC2708; no effect was found against *Salmonella typhimurium* ATCC14028 and *Proteus mirabilis* ATCC14153 (Ghanem, 2011). Beeswax methanol and ethanol extracts generated a great deal of curiosity. The effects were quite successful, and they were assessed in millimeters by determining the inhibition zone caused by the extracts' impact on the organisms tested. Beeswax extracts with ethanol and methanol have distinct antibacterial properties, for unknown reasons, although it is hypothesized that these differences result from the different extraction processes, which inhibit or block certain molecules (Kacaniova et al., 2012).

10.8.1.6.1 Antimicrobial Activity Combined with Other Natural Components

There has been additional research and testing on the antibacterial properties of beeswax when combined with other beehive products and natural items. *Staphylococcus aureus* and *Candida albicans* isolated from human patients can develop more gradually when combined with olive oil, honey, and beeswax (1:1:1 v/v) (Al-Waili, 2005). An inhibitory zone of 4 mm and 3.5 mm, respectively, was seen for *S. aureus* and *C. albicans* after the bacteria (for 24 hours) and yeast (for 48 hours) were incubated (Al-Waili, 2005). Following the initial incubation, the two organisms were incubated individually to assess the efficiency of the various components. Patients with pityriasis versicolor, tinea corporis, tinea faciei, and tinea cruris were examined for the in vivo antifungal activity of a honey, beeswax, and olive oil mixture (Al-Waili, 2004). Pityriasis versicolor, a superficial skin condition caused by 4 of the 11 *Malassezia* species, might be linked to pregnancy, conditions like diabetes, chronic malnutrition, or poor personal hygiene. It is characterized by changes in skin pigmentation and uneven, well-defined patches (Nenoff et al., 2014). *Trichophyton rubrum* and *Trichophyton tonsurans* are the two bacteria that cause the exceedingly uncommon infection known as tinea faciei, which affects the face (Nenoff et al., 2014; Fu et al., 2013). For four weeks, the mixture was applied topically every eight hours to 37 patients (14 with *Pityriasis versicolor*, 14 with *Tinea cruris*, 8 with *Tinea corporis*, and 1 with *Tinea faciei*) (Al-Waili, 2004). The average S.D. (standard deviation) for subjects with *Pityriasis versicolor* at the start of therapy was 7.10 to 1.35, but by the end of the treatment (fourth week), it was 1.00 to 1.20, with clinical signs arising in 12 of 14 individuals. According to Al-Waili (2004), the average score for patients with *Tinea corporis* dropped from (8.70 to 2.40) to (1.30 to 1.50) with clinical sign in 6/8 of the patients and from (8.50 to 2.40) to (1.00 to 1.30) with clinical sign in 11/14 of the patients who had *Tinea cruris*. Four infants with diaper dermatitis were used as test subjects for the honey, beeswax, and olive oil mixture's antibacterial activity. A significant erythema with pustules and/or ulceration was present in patients with candidiasis. *C. albicans*–related diaper dermatitis was treated topically for seven days using a honey, beeswax, and olive oil mixture. Following the treatment period, clinical advances were observed. The severity of the lesions had only marginally decreased with the elimination of ulcers and pustules in 50 percent of the infants with *C. albicans* by the time treatment was completed (Al-Waili, 2004). Propolis and beeswax have recently been investigated for their possible antibacterial effects against *S. aureus* ATCC25923, *B. subtilis* ATCC27853, *S. epidermidis* ATCC12228, and *C. albicans* NCTC270 (Castaldo and Capasso, 2002). Bees extract the resinous substance known as propolis from tree buds, and it is then processed with beeswax, pollen, and enzymes. Bees employ it as glue to seal the cell walls because of its insulating qualities, which also increase the antimicrobial effectiveness of the mixture of honey. Its usage in medicine dates back to ancient times due to its antibacterial, antiviral, antifungal, and anti-inflammatory effects (Wulandari et al., 2013; Freires et al., 2016). The organisms most sensitive to the combination inhibition zones were *S. aureus* and *C. albicans* (20 and 22 mm, respectively) (Castaldo and Capasso, 2002). Despite showing significantly higher resistance to the combination (13.5 mm and 10.5 mm, respectively), *S. epidermidis* and *B. subtilis* were still able to demonstrate the bactericidal effects of propolis and wax (Al-Waili, 2003).

10.8.1.7 Uses of Beeswax in OMFS (Oral and Maxillofacial Surgery)

According to research, interesting uses of beeswax have been mentioned in different parts of the world, with its usage starting in 1880. Beeswax can be used in the surgical field, as described here:

- **Augmentation material:**

Beeswax was used for the first time as soft tissue augmentation for enhancing macro facial esthetics, according to documentation. Later, it was discarded due to different undesired reactions (Dastoor et al., 2007).

- **Impression material:**

In 1945, before the innovation and usage of imaging scanners or light body elastomers, beeswax mixed with Vaseline was used for the recording of skull defects and also for reproducing the details of vomerine bone (Saralaya et al., 2021).

- **Hemostatic agent:**

Beeswax is still used most effectively as a local hemostatic in the form of bone wax. Like other cells or tissues, bone in the body also bleeds on cutting; hemostasis is used to maintain the visibility of cutting, clotting, and bleeding places on bone during surgery. The most effective method is to use a tampon consisting of bone wax containing beeswax (88 percent) and iso-propylpalmitate (12 percent) (Saralaya et al., 2021).

- **For treatment of myiasis:**

Myiasis of the oral cavity can be treated by using beeswax for suffocation. In the process, beeswax blocks the orifice, and the aerobic larva is removed with the help of tweezers (Kumar P and V. Singh, 2014).

- **Coating material in sutures:**

Beeswax can be used for coating material like silk that is non-resorbable. The main function of beeswax as a coating is to facilitate the twisted structure during handling; this is mostly useful when the tissue is dragged or knotted during passage from the needle tract and the structure slides (Srinivasulu K and N.D. Kumar, 2014).

- **Non-adhesive dressing:**

Silk is used as a non-adhesive wound dressing. For this, a silk fibroin fabric is glazed with biocompatible beeswax. Wax reduces the surface energy and makes it more hydrophobic (Kanokpanont et al., 2013). Another use of beeswax is its ability to turn necrotic skin color yellow or leathery, having sufficient resistance to infection to help the surgeon excise the wound at different stages where excision is not possible without using beeswax (Kajjari et al., 2014).

- **In cosmetic products:**

Since ancient times, the standard formula for creams and ointments has been a mixture of beeswax and oil in varied quantities depending on the desired consistency. When used properly in cosmetic formulas, beeswax won't clog pores or result in any other problems, because it has a comedogenicity rating of 0 to 2 and a zero-irritation potential. However, it has a number of very useful characteristics, such as widespread healing and softening, antibacterial activity, and moisturizing capabilities, which make it a superb addition to cosmetic products. Beeswax has special qualities that make it a superior cosmetic ingredient because it helps to make stable emulsions, improves the way ointments and creams bind to water, gives the skin a non-occlusive protective action, increases the effectiveness of sunblock as a preventative measure, and increases the elasticity and plasticity of the product to enable thinner films and greater permanence on the skin and lips. Beeswax enhances the function of soap, leaves a protective layer on the skin, and increases its elasticity (Vera et al., 2013). It enhances the protective properties of sunscreens. It has thermo-storing and antibacterial characteristics. It doesn't cause an allergic reaction (Basa et al., 2016). Beeswax concentrations of 1 to 3 percent are frequently sufficient to produce the desired results. It also enhances the look and uniformity of cosmetic creams and various lotions due to its contribution to sheen, consistency, and color stabilization, but this wax is also a desired ingredient for lipsticks. It strengthens sunscreen's defensive properties (Kurek-Gorecka et al., 2020).

Additional cosmetic uses include deodorants (up to 34 percent by weight), depilatories (hair removers, up to 45 percent), hair creams (7–11 percent), hair conditioners (4 percent), mascara (7–10 percent), rouge (8–16 percent), eye shadows (9–15 percent), and others. Vegetable oils were once the preferred choice, but they quickly go rancid, reducing the amount of time that such creams can be used. Nowadays, inorganic oils like liquid paraffin or additional preservatives have largely supplanted plant oils in many products. It is still common practice to utilize vegetable oils selectively from sources including olives, corn, peanuts, jojoba, cacao, palm, coconut, and others, because many of their beneficial properties cannot be duplicated by synthetic mineral oils. Any cream or lotion must have beeswax, oils, and water, all of which are incompatible with one another unless an emulsifier is introduced. The traditional emulsifier, borax, is accessible at most pharmacies. Many different synthetic emulsifiers are used in today's high-chemistry percent cosmetics. Emulsification is a chemical process that is based on saponification, which turns the acids in beeswax into soap. Making use of the associated cleansing activity are so-called cleansing creams, which imitate conventional skin creams in many ways. A unique procedure was created to eliminate the free acids from beeswax, allowing it to be combined with pigments and mineral materials without the requirement for an emulsifier (Burlando and Cornara, 2013)

10.9 CYTOTOXIC EFFECTS OF BEESWAX

Stingless bee products including propolis, bee pollen, and honey have anti-cancer properties as a natural resource. According to the research, bee products from seven types of bees were used. These bee products were tested for cytotoxicity on HeLa, Caco-2, and MCF-7 cancer cells. The data showed that *Homotrigona fimbriata* bee products had higher cytotoxicity than other bee products. Mangiferonic acid is an active compound obtained from ethanol extraction of propolis from *H. fimbriata*, fractionated and separated by biological activity-guided fractionation and chromatography. Its presence in propolis emphasizes its potential use in drugs for cancer treatment. (Arung et al., 2021).

10.10 CONCLUSION

The cosmetics, pharmaceutical, and food sectors all utilize beeswax. It is abundantly used to create a variety of emollients, gels, balms, and moisturizers. In the food business, it is used to seal food containers and as a cheese coating. As acaricide contamination has spread across commercial beeswax, there is a growing need for organic beeswax that is residue free, which is a characteristic that is provided by beeswax manufacturers. A suitable choice for the coming years is acaricide-free African beeswax. (African bees can tolerate varroa; hence, no acaricides are required.)

The majority of beeswax is utilized in beekeeping to create the foundations for combs, making it impossible to collect accurate statistics on beeswax output. Beeswax comes in second place to honey in terms of economic significance among all bee products. According to reports, 1.5–2.5 percent as much beeswax is produced as honey. Based on an estimate from Food and Agriculture Organization (FAO) figures, between 17,850 and 29,750 tons of wax were produced over the same period as 1.19 million tons of honey. China is the main producer in the world, producing 6,000 tons annually.

Beeswax's rising popularity in the medical industry is a result of its natural antibacterial characteristics, inertness, and excellent flexibility at a temperature of 32°C, which is relatively low. The melting point of beeswax is 64°C, and its specific gravity is 0.95. A variety of infectious agents, including *Aspergillus niger*, *Candida* species, *Salmonella enterica*, *Staphylococcus aureus*, *Pseudomonas aeruginosa*, and *Escherichia coli*, have been demonstrated to be sensitive to beeswax when it is used in its raw form. It should be mentioned that because beta carotene and other physiologically active compounds are present, it has the ability to stimulate the regeneration of tissues, leading to rapid healing of the injured tissue. The use of beeswax in assisted tissue regeneration has not been investigated, despite its usage as a material for regeneration of tissue. Beeswax has been well-documented in the literature for its anti-inflammatory, antibacterial, and anti-oxidative properties.

In the coming years, it is projected that beeswax marketing will continue to increase significantly. Honey bees produce beeswax, which is said to be a natural wax. Worker bees in the hive generate honey, which is used to make beeswax. It has a significant number of fatty acids and long-chain alcohols. It serves as a cleansing, seasoning, and lubricating agent. The second factor that supports the growth of the beeswax industry is the cheap production cost.

REFERENCES

Al-Waili, N. S. 2003. Topical application of natural honey, beeswax and olive oil mixture for topic dermatitis or psoriasis: Partially controlled, single-blinded study. *Complement. Ther. Med.* 11(4):226–234.

Al-Waili, N. S. 2004. An alternative treatment for pityriasis versicolor, tinea cruris, tinea corporis and tinea faciei with topical application of honey, olive oil and beeswax mixture: An open pilot study. *Complement. Ther. Med.* 12(1):45–47.

Al-Waili, N. S. 2005. Clinical and mycological benefits of topical application of honey, olive oil and beeswax in diaper dermatitis. *Clin. Microbiol. Infect.* 11(2):160–163.

Al-Waili, N. S. 2005. Mixture of honey, beeswax and olive oil inhibits growth of Staphylococcus aureus and Candida albicans. *Arch. Med. Res.* 36(1):10–13.

Arung, E. T., R. Ramadhan, B. Khairunnisa, Y. Amen, M. Matsumoto, M. Nagata and K. Shimizu. 2021. Cytotoxicity effect of honey, bee pollen, and propolis from seven stingless bees in some cancer cell lines. *Saudi J. Biol. Sci.* 28(12):7182–7189.

Ashburner, G. 2014. *How to Use Beeswax and Honey to Cure Skin Problems Paperback*. Createspace Independent Publishing Platform. South Carolina, United States.

Aslani, A. and F. Rostami. 2015. Medicated chewing gum, a novel drug delivery system. *J. Res. Med. Sci.* 20(4):403–408.

Baldwin, E. A., M. O. Nisperos, R. D. Hagenmaier and R. A. Baker. 1997. Use of lipids in edible coatings for food products. *Food Technol.* 51:57–63.

Basa, B., W. Belay, A. Tilahun and A. Teshale. 2016. Review on medicinal value of honeybee products: Apitherapy. *Advan. Biol. Res.* 10:236–247.

Bayer, I. S., D. Fragouli, P. J. Martorana, L. Martiradonna, R. Cingolani and A. Athanassiou. 2011. Solvent resistant superhydrophobic films from selfemulsifying carnauba wax-alcohol emulsions. *Soft Matter* 7(18):7939–7934.

Bennett, H. 1975. *Industrial Waxes*, second ed. Chemical Publishing Co., Inc., New York.

Birshtein, V. Y., V. M. Tulchinskii and A. V. Troitskii. 1976. A study of organic components in ancient Central Asian and Crimean wall paintings. *Vestn. Moskosvkogo Univ.* 31:33–38.

Blake, A. I. and A. G. Marangoni. 2015. The effect of shear on the microstructure and oil binding capacity of wax crystal networks. *Food Biophys.* 10(4):403–415.

Bogdanov, S. 2004. Beeswax: Quality issues today. *Bee World*. 85(3):46–50.

Bogdanov, S. 2009. Beeswax: Uses and trade. In *The Beeswax Book, Bee Product Science*. 1–16.

Bogdanov, S. 2016. Beeswax: History, uses and trade. *Bee Product Science*, 1–18.

Bogdanov, S. 2016. Beeswax: Production, properties composition and control bee product science. *Bee Product Science*. 1–17.

Bommuraj, V., Y. Chen, H. Klein, R. Sperling, S. Barel and J. A. Shimshoni. 2019. Pesticide and trace element residues in honey and beeswax combs from Israel in association with human risk assessment and honey adulteration. *Food Chemistry*: 123–132.

Burlando, B. and L. Cornara. 2013. Honey in dermatology and skin care: A review. *J. Cosmet. Dermatol.* 12(4):306–313.

Carpena, M., B. Nunez-Estevez, A. Soria-Lopez and J. Simal-Gandara. 2020. Bee venom: An updating review of its bioactive molecules and its health applications. *Nutrients* 12(11):3360–3386.

Castaldo, S. and F. Capasso. 2002. Propolis, an old remedy used in modern medicine. *Fitoterapia* 73:1–6.

Co, E. D. and A. G. Marangoni. 2012. Organogels: An alternative edible oil-structuring method. *J. Am. Oil Chem. Soc.* 89:749–780.

Crane, E. 1999. *The World History of Beekeeping and Honey Hunting.* Routledge, London.

Dassanayake, L. S. K., D. R. Kodali, S. Ueno and K. Sato. 2009. Physical properties of rice bran wax in bulk and organogels. *J. Am. Oil Chem. Soc.* 86(12):1163–1168.

Dassanayake, L. S. K., D. R. Kodali, S. Ueno and K. Sato. 2012. Crystallization kinetics of organogels prepared by rice bran wax and vegetable oils. *J. Oleo Sci.* 61(1):1–9.

Dastoor, S. F., C. E. Misch and H. L. Wang. 2007. Dermal fillers for facial soft tissue augmentation. *J. Oral Implantol.* 33(4):191–204.

Dnyandeo, N. Y. and S. Manda. 2014. Treatment of burn wound by Ghruta (Ghee) and Madhu (Honey) – A case report. *IAMJ.* 4:43–46.

El-Seedi, H. R., S. A. M. Khalifa, A. A. Abd El-Wahed, R. Gao, Z. Guo and H. E. Tahir. 2020. Honeybee products: An updated review of neurological actions. *Trends Food Sci. Technol.* 101:17–27.

European Food Safety Authority (EFSA). 2020. Risk assessment of beeswax adulterated with paraffin and/or stearin/stearic acid when used in apiculture and as food (honeycomb) 17(5):1859–1863.

Ferber, C. E. and H. E. Nursten. 1977. The aroma of beeswax. *Journal of the Science of Food and Agriculture* 28(6):511–518.

Freires, I. A., V. C. Queiroz, V. F. Furletti, M. Ikegaki, S. M. de Alencar and M. C. Duarte. 2016. Chemical composition and antifungal potential of Brazilian propolis against Candida spp. *J. Mycol. Med.* 26(2):122–132.

Fu, M., Y. Ge, W. Chen, S. Feng, X. She and X. Li. 2013. Tinea faciei in a newborn due to Trichophyton tonsurans. *J. Biomed. Res.* 27(1):71–74.

Ghanem, N. 2011. Study on the antimicrobial activity of honey products and some Saudi Folkloric substances. *Res. J. Biotech.* 6:38–43.

Gokani, T. 2014. Ayurveda-the science of healing. *Headache* 54(6):1103–1106.

Haarmann, T., M. Spivak, D. Weaver, B. Weaver and T. Glenn. 2002. Effects of fluvalinate and coumaphos on queen honey bees (Hymenoptera: Apidae) in two commercial queen rearing operations. *Journal of Economic Entomology* 95(1):28–35.

Hagenmaier, R. D. and R. A. Baker. 1994. Wax microemulsions and emulsions as citrus coatings. *J. Agric. Food Chem.* 42(4):899–902.

Hagenmaier, R. D. and R. A. Baker. 1995. Internal gases, ethanol content and gloss of citrus fruit coated with polyethylene wax, carnauba wax, shellac or resin at different application levels. *Proceedings of the Florida State Horticultural Society 107*: 261–265.

Hargrove, J. L., P. Greenspan and D. K. Hartle. 2004. Nutritional significance and metabolism of very long chain fatty alcohols and acids from dietary waxes. *Exp. Biol. Med.* 229(3):215–226.

Hu, H., X. Li, C. Dong and W. Chen. 2012. Effects of wax treatment on the physiology and cellular structure of harvested pineapple during cold storage. *J. Agric. Food Chem.* 60(26):6613–6619.

Huang, S., C. P. Zhang, K. Wang, FL Hu and GQ Li. 2014. Recent advances in the chemical composition of propolis. *Molecules* 19(12):19610–19632.

Hwang, H. S., S. Kim, M. Singh, J. K. Winkler-Moser and S. X. Liu. 2012. Organogel formation of soybean oil with waxes. *J. Am. Oil Chem. Soc.* 89(4):639–647.

Illnait, J., I. Rodriguez, S. Mendoza, Y. Fernandez, R. Mas, M. Miranda. 2013. Effects of D-002, a mixture of high molecular weight beeswax alcohols, on patients with non-alcoholic fatty liver disease. *Korean J. Intern. Med.* 28(4):439–448.

Kacaniova, M., N. Vukovic, R. Chlebo and P. Hascik, K. Rovna and J. Cubon. 2012. The antimicrobial activity of honey, bee pollen loads and beeswax from Slovakia. *Arch. Biol. Sci. Bel.* 64:927–934.

Kajjari, P. B., L. S. Manjeshwar and T. M. Aminabhavi. 2014. Novel blend microspheres of cellulose triacetate and bee wax for the controlled release of nateglinide. *J. Ind. Eng. Chem.* 20(2):397–404.

Kaluza, B. F., H. M. Wallace, T. A. Heard, V. Minden, A. Klein and S. D. Leonhardt. 2018. Social bees are fitter in more biodiverse environments. *Scientific Reports* 8(1):23–53.

Kanokpanont, S., S. Damrongsakkul, J. Ratanavaraporn and P. Aramwit. 2013. Physico-chemical properties and efficacy of silk fibroin fabric coated with different waxes as wound dressing. *Int. J. Biol. Macromol.* 55:88–97.

Kellerman, M., E. Liebenberg, N. Njombolwana, A. Erasmus and P. H. Fourie. 2018. Postharvest dip, drench and wax coating application of pyrimethanil on citrus fruit: Residue loading and green mould control. *Crop Prot.* 103:115–129.

Koedam, D., H. Jungnickel, J. Tentschert, G. R. Jones and E. D. Morgan. 2002. Production of wax by virgin queens of the stingless bee Melipona bicolor (Apidae, Meliponinae). *Insectes sociaux* 49:229–233.

Kumar, P. and V. Singh. 2014. Oral myiasis: Case report and review of literature. *Oral Maxillofac. Surg.* 18(1):25–29.

Kurek-Górecka, A., M. Górecki, A. Rzepecka-Stojko, R. Balwierz and J. Stojko. 2020. Bee products in dermatology and skin care. *Molecules* 25(3):556–560.

Li, X. P., X. Y. Zhu, H. L. Wang, X. F. Lin, H. W. Lin and W. X. Chen. 2018. Postharvest application of wax controls pineapple fruit ripening and improves fruit quality. *Postharvest Biol. Technol.* 136:99–110.

Limpimwong, W., T. Kumrungsee, N. Kato, N. Yanaka and M. Thongngam. 2017. Rice bran wax oleogel: A potential margarine replacement and its digestibility effect in rats fed a high-fat diet. *J. Funct. Foods* 39:250–256.

Lozano, A., M. D. Hernando, S. Uclés, E. Hakme and A. R. Fernández-Alba. 2019. Identification and measurement of veterinary drug residues in beehive products. *Food Chem.* 274:61–70.

Lucente, P., M. Cavalli, C. Vezzani, C. Orlandi and C. Vincenzi. 1996. Contact cheilitis due to beeswax. *Contact Derm.* 35:1–5.

Mattice, K. D. and A. D. Marangoni. 2018. New insights into wax crystal networks in oleogels. In Patel, A. R. (Ed.), *Edible Oil Structuring: Concepts, Methods and Applications.* CPI Group, Ltd, Croydon. 71–78

Morales, M. M., M. J. G. Ramos, P. P. Vázquez, F. J. D. Galiano, M. G. Valverde, V. G. Lopez and A. R. Fernández-Alba. 2020. Distribution of chemical residues in the beehive compartments and their transfer to the honeybee brood. *Science of the Total Environment* 710:136–148.

Morales-Rueda, J. A., E. Dibildox-Alvarado, M. A. Charó-Alonso, R. G. Weiss and J. F. Toro-Vazquez. 2009. Thermo-mechanical properties of Candelilla wax and dotriacontane organogels in safflower oil. *Eur. J. Lipid Sci. Technol.* 111(2):207–215

Morillon, V., F. Debeaufort, G. Blond, M. Capelle and A. Voilley. 2002. Factors affecting the moisture permeability of lipid-based edible films: A review. *Crit. Rev. Food Sci. Nutr.* 42(1):67–89.

Mullin, C. A., M. Frazier, J. L. Frazier, S. Ashcraft, R. Simonds, D. VanEngelsdorp and J. S. Pettis. 2010. High levels of miticides and agrochemicals in North American apiaries: implications for honey bee health. *PloS one* 5(3):54–67.

Nainu, F., A. Masyita, M. A. Bahar, M. Raihan, S. R. Prova and S. Mitra. 2021. Pharmaceutical prospects of bee products: Special focus on anticancer, antibacterial, antiviral, and antiparasitic properties. *Antibiotics (Basel)* 10(7):822–858.

Nenoff, P., C. Kruger, J. Schaller, G. Ginter-Hanselmayer, R. Schulte-Beerbuhl and H. J. Tietz. 2014. Mycology an update. Part 2: dermatomycoses: Clinical picture and diagnostics. *JDDG* 12(9):749–777.

Ogutcu, M., N. Arifoglu and E. Yilmaz. 2015. Preparation and characterization of virgin olive oil-beeswax oleogel emulsion products. *J. Am. Oil Chem. Soc.* 92(4):459–471.

Pastor, C., L. Sánchez-González, M. Cháfer, A. Chiralt and C. González-Martínez. 2010. Physical and antifungal properties of hydroxypropylmethylcellulose based films containing propolis as affected by moisture content. *Carbohydrate Polymers* 82(4):1174–1183.

Patwardhan, B. 2014. Bridging Ayurveda with evidence-based scientific approaches in medicine. *EPMA J.* 5(1):5–19.

Post-Beittmiller, D. 1996. Biochemistry and molecular biology of wax production in plants. *Annu. Rev. Plant Physiol. Plant Mol. Biol.* 47:405–430.

Puente, R., J. Illnait, R. Mas, D. Carbajal, S. Mendoza, J. C. Fernandez. 2014. Evaluation of the effect of D-002, a mixture of beeswax alcohols, on osteoarthritis symptoms. *Korean J. Intern. Med.* 29(2):191–202.

Puleo, S. L. 1991. Beeswax minor components: a new approach. *Cosmetics and Toiletries* 106(2):83–89.

Ranneh, Y., A. M. Akim, H. A. Hamid, H. Khazaai, A. Fadel and Z. A. Zakaria. 2021. Honey and its nutritional and antiinflammatory value. *BMC Complement Ther.* 21:30–46.

Ravoet, J., W. Reybroeck and D. C. de Graaf. 2015. Pesticides for apicultural and/or agricultural application found in Belgian honey bee wax combs. *Bull. Environ. Contam. Toxicol.* 94(5):543–548.

Reybroeck, W., J. W. Van Veen and A. Gupta. 2014. Beekeeping for poverty alleviation and livelihood security. Springer, New York.

Rit, T. and R. Behrer. 1999. *Beeswax Through the Ages.* Koster Keunen Holland BV Publication, AG Bladel.

Saralaya, S., N. S. Thomas and S. SM. 2021. Bee wax and honey-a primer for OMFS. *Oral Maxillofac. Surg.* 25(1):1–6.

Sarojini, B. 2013. A comparative clinical study on Shala niryasa and madhuchishtha in the management of padadari. *IAMJ* 1:1–7.

Saucedo-Pompa, S., R. Rojas-Molina, A. F. Aguilera-Carbó, A. Saenz-Galindo, H. de La Garza, D. Jasso-Cantú and C. N. Aguilar. 2009. Edible film based on candelilla wax to improve the shelf life and quality of avocado. *Food Res. Int.* 42(4):511–515.

Bonvehi, S.J. and F. O. Bermejo. 2012. Detection of adulterated commercial Spanish beeswax. *Food Chem.* 132(1):642–648.

Singh, A., F. I. Auzanneau and M. A. Rogers. 2017. Advances in edible oleogel technologies–A decade in review. *Food Res. Int.* 97:307–317.

Srinivasulu, K. and N. D. Kumar. 2014. A review on properties of surgical sutures and applications in medical field. *Int. J. Res. Eng. Technol.* 2:85–96.

Stacey, R. J. 2001. Composition of some roman medicines: Evidence for Pliny's Punic wax? *Anal. Bioanal. Chem.* 401(6):1749–1759.

Thakur, M. and V. Nanda. 2020. Composition and functionality of bee pollen: A review. *Trends Food Sci. Technol.* 98:82–106.

Tinto, W. F., T. O. Elufioye and J. Roach. 2017. Waxes. In S. Badal and R. Delgoda (Eds), *Pharmacognosy.* Academic Press, Massachusetts. 443–455.

Tulloch, A. P. 1970. The composition of beeswax and other waxes secreted by insects. *Lipids* 5(2):247–258.

Tulloch, A. P. 1980. Beeswax—Composition and analysis. *Bee World* 61(2):47–62.

Vera, A., H. Weatherburn and K. Malaker. 2013. A case of post varicella recurrent presternal butterfly keloid. *Int. J. Med. Phys. Clin. Eng. Radiat. Oncol.* 2:88.

Wallner, K. 1999. Varroacides and their residues in bee products. *Apidologie* 30(2–3):235–248.

Wulandari, R., S. Amin and A. Silviana. 2013. Tinea faciei caused by Trichophyton rubrum melanoid type and Pityriasis versicolor. *Indian J. Dermatol. Venereol. Leprol.* 2:59–65.

Zhang, L. F., F. S. Chen, P. L. Zhang, S. J. Lai and H. S. Yang. 2017. Influence of rice bran wax coating on the physicochemical properties and Pectin nanostructure of cherry tomatoes. *Food Bioprocess Technol.* 10(2):349–357.

11 Bee Venom
Composition and Therapeutic Potential

Umer Bin Farook, Shahnawaz Ahmad Dar, Uzma Arifie, Ruqqaya Javid, Rattan Singh, Bilal Ahmad Pandit and Nadeya Khaliq

11.1 INTRODUCTION

Apis spp. (honey bees) are the most economically important arthropods that provide key ecological services to mankind. They provide pollen, royal jelly, honey, beeswax, and propolis (resin) (Mizrahi and Lensky, 2013). They also help pollinate numerous agriculturally important crops (Hoover and Oinge, 2018). Bees are docile and rarely threaten humans unless provoked, inflicting painful and toxic stings (Chen et al., 2016). Female workers of the colony secrete the venom from a well-organised venomous gland networked to the venom reservoir sac in the abdominal chamber. The sting is lethal for the female workers, as their sting possesses barbs, such that on secreting venom, it does not come quickly from the skin and causes fatality to the bees. However, the queen can sting multiple times, since her stinger has no barb. Bee venom contains mainly water (80%), and the rest consists of bioactive compounds such as enzymes (hyaluronidase and phospholipase A2), peptides (apamin, melittin, mast cell degranulating [MCD] peptide, and adolapin), and amino acids, besides some other volatile compounds (Szókán et al., 1994). Apitoxin is an evolved defence for defending a colony (Baracchi et al., 2011). Globally, various studies on bee venom have shown its potential anti-inflammatory (Sobral et al., 2016), antibacterial (Han et al., 2016), anti-cancerous (Chu et al., 2007), and radioprotective properties (Gajskii and Garaj-Vrhovac, 2009) and antiviral potential (Hood et al., 2013). In some instances, bee venom causes severe allergic reactions (skin, cardiovascular, gastric, and respiratory systems) and sometimes severe hypotensive shock that could lead to ischaemia of the brain or heart (Bilo and Bonifazi, 2009; Golden, 2007). These allergic reactions are activated due to the occurrence of several protein antigens within the venom possessing high enzymatic activity (Moreno and Giralt, 2015). Bee venom contains PLA2, melittin, and hyaluronidase, which are immunoglobulin E (IgE) inducers. The presence of melittin in bee venom induces a non-immune response (Mingomataj and Bakiri, 2012).

Studies indicate that in addition to IgE-mediated pathways, allergens may also cause IgE antibodies to react independently, causing complications like bradykinin (BK) mediator, which can result in a variety of anaphylactic symptoms (Kinins & Kaplan 2010). Melittin, a PLA2 activator that resembles the effects of BK on the air constriction of the tracheal tube, can stimulate the synthesis of this non-immune mediator (Ziai et al., 1990).

Considering the benefits and marvels of bee venom, it is relevant to compile the existing knowledge regarding the uses of bee venom and its bioactive components, which may be used to create therapeutic plans.

11.2 HISTORY OF BEE VENOM USES

Apitoxin as a therapeutic is not a new concept; it was used to treat rheumatoid arthritis and relieve joint discomfort during the time of Hippocrates (Kim, 2013). The most comprehensive use of bee

DOI: 10.1201/9781003354116-11

venom is deployment with acupuncture to combat inflammatory disorders in Asia (Zurier et al., 1973). The bee venom is directly injected or applied with the help of acupuncture needles. The utilisation of apitherapy is gaining importance in Europe and the United States, as it has shown significant results in treating pathogenic disorders based on various studies on animals both in vitro and in vivo. The contemporary trend of using apitoxin is in wound healing (Amin and Abdel-Raheem, 2014) and curing rheumatism, dermatitis, and backache (Robert, 1988). The current medical system uses apitoxin for treating multiple sclerosis (Hauser et al., 2001), arthritis, and Parkinsonism (Alvarez-Fischer et al., 2013).

11.3 BEE VENOM VARIABILITY AMONG DIFFERENT BEE SPECIES (*APIS* SPP.)

Bees belong to the order Hymenoptera and possess a stinger as a defence system. The bee venom of this order contains dopamine, adrenaline, noradrenaline, serotonin, histamine, phospholipases A and B, and hyaluronidase (Moreno and Giralt, 2015). In contrast, the bee venom of honey bees contains a peptide that releases histamines, melittin, and apamin (Pucca et al., 2019). The variability among apitoxin compositions in different bee species is attributed primarily to species, age, geographical conditions (flora and fauna), hive conditions, and colony coherence (El-Wahed et al., 2019). In the colony, only females and queens can sting. The drones are male and lack a stinger. The freshly emerged bees do not sting, as venom synthesis begins after 2–3 days. The female bees at age 15–21 days have maximum bee venom concentration. An average sting produces approximately 100 μg of dry bee venom, and the LD_{50} dose ranges between 2.8 and 3.5 mg of venom per kg of normal human body weight (Schumacher et al., 1994; Ali, 2012). An allergic person stung by fewer than 500 bees may die, whereas non-allergic persons can survive 1500 bee stings (França et al., 1994). Overall, the resistance to bee venom is characterised by the victim's age, immune status, number of stings, and body weight.

11.4 COMPOSITION OF BEE VENOM

Honey bee venom is a yellowish to colourless opalescent liquid. It tastes like honey, is sweet, acidic, and hot, and dissolves in water but not in alcohol. The pH varies from 4.50 to 5.50 with a specific gravity of 1.130. Apitoxin/bee venom is a mixture of bioactive compounds encompassing peptides, enzymes, and other macromolecules (Dotimas et al., 1987).

11.4.1 Enzymes

Many enzymes in apitoxin make it a potent candidate in different therapeutic models. It includes phospholipase A2 (PLA2), phospholipase B (PLB), hyaluronidase, acid phosphatase (Api m 3), and dipeptidyl peptidase IV (Api m 5).

11.4.1.1 Api m 1 or Phospholipase-A2 (PLA2)

PLA2 or Api m 1 is the main component of bee venom, possessing immunogenic and allergenic potential. It is essential in many cellular processes and possesses antibacterial properties (Dennis et al., 2011). Api m 1 has been categorised into three classes: cytosolic (c-PLA2), secreted (s-PLA2), and calcium (Ca^{2+})-independent PLA2 (iPLA2). Api m 1 can break down phospholipids and produce free fatty acids and lysolipids. It also modulates the T cells (regulatory) to bring into play an anti-inflammatory effect, besides reducing the progression of brain diseases and easing neuropathic pain (Kim et al., 2015; Li et al.,2016; Ye et al., 2016).

11.4.1.2 Phospholipase B

Phospholipase B is a thermostable compound found in organisms ranging from bacteria to mammals (Doery and Pearson, 1964). The remodelling of fatty acids by the removal of both sn-1 and sn-2 ester bonds of diacyl-phospholipids is catalysed by phospholipase B. Phospholipase B is thought to have more abundant and broader roles than PLA2 in vivo (Saito, 2014).

11.4.1.3 Hyaluronidase or Api m 2

Api m 2, an enzyme and potent allergen in apitoxin (1–3%), shares 3% of its peptide sequence with a human hyaluronidase enzyme (Fox, 2013; Horta et al., 2014). It is widely known as a "spreading factor" because of its potency to disperse toxins quickly. The hyaluronidase enzyme releases hyaluronan from the intercellular matrix (Elieh Ali Komi et al., 2018; Kreil, 1994), resulting in a more rapid and pervasive local irritation by obstructing tissues (e.g., in the synovial bursae of people with atrophic arthritis) and is hence employed as an anti-inflammatory (Markovi'c-Housley et al., 2000).

11.4.1.4 Phosphatase

The presence of the enzyme phosphatase in apitoxin prevents the progression and spread of sarcoma by preventing phosphorylation of a few critical factors, such as vascular endothelial growth factor (Huh et al., 2010), thus signifying its role in many therapeutic problems. It was characterised over two decades ago (Barboni et al., 1987). The apitoxin acid phosphatase is believed to be responsible for a robust discharge of histamine from stimulated human white blood cells (Grunwald et al., 2006). The bee venom of honey bees contains acid phosphatase Api m 3, considered to be one of the allergens. Using recombinant technology, more secure and effective vaccines/therapies against bee venom allergies can be developed (Georgieva et al., 2009).

11.4.2 Peptides

11.4.2.1 Melittin

Melittin, which contains 26 amino acid residues, is the primary lethal constituent of apitoxin; the first 20 are predominantly hydrophobic, whereas the C-terminal end of the molecule is mainly positively charged (Bazzo et al., 1988). Melittin constitutes 50–60% of the venom (Raghuraman and Chattopadhyay, 2007). The discomfort brought on by a bee sting is attributed to melittin and is incited *via* actions on principal nociceptor cells, both direct and indirect (Chen et al., 2016). The main nociceptors become more sensitive as a result of melittin's activation of the heat nociceptors on ion channels of transient receptor potential (TR.) vanilloid 1 (TRPV1) through the PLA2 signalling pathway (Ding et al., 2011; Hao et al., 2008). Melittin's pore-forming properties allow the release from mast cells of chemicals that produce pain, including H^+, ATP, and 5-hydroxytryptamine (5-HT). Additionally, melittin damages tissue, which activates pain receptors. Melittin can also cause pores to develop and release substances, including bradykinin, histamine, and ATP, that trigger G-protein-coupled receptors (GPCRs) and phosphorylate phospholipase C (PLC).

11.4.2.2 Apamin

Apamin is another vital constituent of apitoxin and constitutes 1–3% of crude venom (Moreno and Giralt, 2015). The apamin in bee venom consists of only 18 amino acids, making it the simplest neurotoxic peptide. It has four cysteine residues with two disulphide bonds. Apamin acts centrally and crosses the blood–brain barrier to cause hyperkinetic motor excitation (Strong, 1990). A few apamin-based drug delivery systems have been created because of its potent penetration and precise distribution throughout the central nervous system (CNS) (Lamy et al., 2010).

11.4.2.3 Mast Cell Degranulating (MCD) Peptides

MCD peptide, a 22-amino-acid positively charged peptide residue with two disulphide bridges, is also a crucial ingredient in apitoxin, analogous to the secondary structure of apamin (Hider and Ragnarsson,1981). It is accountable for the dispersion of histamine in the human body on exposure to bee venom. Mast cells' secretion of histamine is crucial in allergies and inflammatory processes (Habermann, 1972). The researchers showed that at concentrations lower than 0.1 mg/mL, MCD causes histamine release. Still, it prevents inflammation at higher concentrations, thus acting as an anti-inflammatory (Kondo et al., 1990).

11.4.2.4 Tertiapin

Tertiapin, like melittin or apamin, is not much studied or widely described. The compound is prone to oxidation and comprises nearly 21 amino acids. It comprises 0.1% of total apitoxin. This apitoxin peptide blocks potassium channels, mainly inward-rectifier, in humans (Cornara et al., 2017). Various studies on tertiapin have confirmed its usefulness for treating ailments via atrioventricular transmission (Drici et al., 2000). Currently, it is mainly used and recognised as a modifier of K^+ channels.

11.4.2.5 Adolapin

Adolapin consists of 103 amino acid residues, comprising 1% of the total dry weight of apitoxin. It was named and isolated by Shkenderov and Koburova in 1982 (Shkenderov and Koburova, 1982). Most studies on adolapin suggest its role as anti-inflammatory and analgesic, besides inhibiting PLA2 (7 nM/mL inhibits ca. 80% of 2.5 nM/mL PLA2) and lipoxygenase from human platelets (Koburova et al., 1985). Adolapin, like non-steroid analgesics, displays antipyretic activity by suppressing the rise in average temperature.

11.5 COLLECTION OF APITOXIN/BEE VENOM

Bee venom is collected from the worker caste only using different methods and other apparatus. In one method, the bees are killed, and their stinging apparatus is extracted and dissected. Another way is by using a bee venom extractor that uses electricity, producing 33 V for 5 minutes; this yields 1 g of bee venom from 20 hives (Urtubey, 2005). The bee venom extractor method is commercially used in America and European countries. The process of extracting bee venom is done three to four times per season, but research showed decreased brood production and a reduction of up to 15% in honey yield (Krivtsov and Lebedev, 1995). The bee venom harvested from different *Apis* spp. varies only slightly and shows overall similar activity. Bee venom from warm zones is more potent than that of temperate areas. It was discovered that *Apis mellifera* and *A. dorsata* bee venom showed comparable toxicity, but *A. dorsata* venom includes significantly more alarm pheromones.

11.6 THERAPEUTIC POTENTIAL AND APPLICATION OF APITOXIN

11.6.1 Anti-Inflammatory Potential of Apitoxin

The globally best-documented benefit of bee venom is its powerful anti-inflammatory effect, primarily attributed to the presence of melittin. Although melittin in high concentration can cause pain, itching, and inflammation, its administration in low doses can significantly reduce inflammation, as when it is used in treating rheumatoid arthritis (RA) (Lee et al., 2016). Additionally, melittin decreases signal transduction pathways that stimulate inflammatory cytokines, such as nuclear factor-kappa B (NF-B), protein kinase (Akt), and exogenous signal-regulated kinases (ERK1/2), in human keratinocytes treated with *Porphyromonas gingivalis* lipopolysaccharide (PgLPS). The results suggest that apitoxin containing melittin reduces inflammatory cytokines by blocking their primary signalling pathways, resulting in a decrease in swelling of dermal, hepatic, joint, and neural tissue (Kim et al., 2018).

Bee-venom acupuncture, which involved 5–15 bee stings every other day, was found to have symptom-relieving effects that were comparable to those of conventional rheumatoid arthritis drugs like Methotrexate and Celecoxib in an 8-week study of 120 RA patients (Lee and Bae, 2016) Several skincare companies have begun to incorporate bee venom into products such as serums and moisturisers. This ingredient may benefit skin health in various ways, including inflammation reduction, antibacterial effects, and wrinkle reduction. Kim and co-workers discovered that apitoxin prevents atopic dermatitis, the most prevalent long-lasting allergic inflammatory skin disease (Ong and Leung, 2016).

11.6.2 Antibacterial and Antiviral Potential of Bee Venom

Apitoxin contains antibacterial properties due to the presence of PLA2 and melittin. The venom from *Apis mellifera* has been documented to have antimicrobial activity against bacteria (Gram-negative and Gram-positive), including coagulase-negative bacteria, *Salmonella* spp., *Enterobacter cloacae*, *Citrobacter freundii*, *Staphylococcus aureus*, and *Escherichia coli* (Zolfagharian et al., 2015). These compounds work against microbes by abrasion of their membrane, causing them to cleave and then lyse (Leandro et al., 2015). The antibacterial and antiparasitic nature of apitoxin is attributed to PLA2. It possesses strong antiprotozoal action against *Trypanosoma brucei*, which cause African human trypanosomiasis. An in vitro study reported that even at a low concentration (1 mg/mL), it disrupts the calcium-mediated signal transduction pathway in parasitic cells 30 min after administration, causing unevenness and misregulation of this pathway (Boutrin et al., 2008). Additionally, they verified that bee venom PLA2 possesses bactericidal doses of 0.01–0.001 g/mL against *C. freundii*, *E. cloacae*, and *E. coli*.

Luciano and co-workers investigated the antimalarial properties of PLA2 in mice and concluded that it disrupts the interaction between *Plasmodium* (a parasite that causes malaria) and the intestinal cells of mosquito mid-gut (acts as a vector) (Moreira et al., 2002).

A recent investigation explored the capability of antiviral effects of apitoxin and discovered some interesting results both in vitro and in vivo. Bee venom and melittin were found to have promising antiviral efficacy against a variety of capsulated viruses (herpes simplex virus, influenza-A virus, and vesicular stomatitis virus) as well as unencapsulated viruses (coxsackie virus and enterovirus-71) (Uddin et al., 2016). The same study showed that melittin shielded test organisms (mice) from fatal doses of the H1N1 virus. Despite the actual mode of action of apitoxin and melittin as antiviral factors being unknown, it has now been demonstrated that apitoxin interacts precisely with the viral coat. Furthermore, apitoxin and its components can trigger type I interferon (IFN), which inhibits the multiplication of virus cells in the host organism (Bachis et al., 2010).

The protective viral envelopes of human immunodeficiency virus and other viruses can be breached by the potent toxin melittin contained in bee venom. Nanoparticles containing bee venom destroy the human immunodeficiency virus without destroying the surrounding cells. Free melittin can severely harm the body. The researchers demonstrated that melittin-loaded nanoparticles do not affect healthy, normal cells. When normal cells, which are typically much more numerous, come into contact with nanoparticles that have protective bumps on their surface, the nanoparticles prefer to bounce off rather than attach. Compared with nanoparticles, HIV is significantly smaller and can fit between the bumps. Melittin fuses with the viral envelope on the nanoparticles, breaks apart the envelope, and creates tiny pore-like assault complexes, tearing it off the virus. While most anti-HIV drugs aim to stop the virus from multiplying, this one targets a crucial component of its structure and kills it (Hood et al., 2013).

A vaginal gel to stop the progression of AIDS infection and AIDS medication, especially drug-resistant models, are two potential uses for melittin-loaded nanoparticles, according to researchers. The administration of melittin-loaded nanoparticles could be used to target and eliminate several viruses, including hepatitis B and C viruses. The gel may also target sperm, according to the researchers, making it a viable contraceptive drug.

11.6.3 Anticoagulant Potential of Bee Venom

Enzymes, proteins, peptides, and a range of smaller compounds make up the venoms of most stinging insects, including honey bees. Venoms from different stinging insects exhibit striking biochemical and pharmacological convergence. Most venoms can kill red blood cells, cause severe pain immediately, and contain other enzymes like hyaluronidase and phospholipases (Schmidt et al., 1986).

An in vitro study conducted on the anticoagulant properties of apitoxin showed anticoagulation factors, such as melittin and phospholipase-A2 (PLA2), which cause blood clotting time to increase. Toxins from many species of bees from the Apidae family, particularly honey bees, contain abundant novel chemicals with uses in biochemistry and medicine. In contrast, toxin/venom from Iranian bees is beneficial in slowing blood coagulation. Prothrombin and other coagulation tests have demonstrated that bee venom generally possesses anticoagulant characteristics. In one study, plasma samples from four groups of six test animals each were used to examine the antithrombotic effects of bee venom using prothrombin tests. The four groups each received venom (apitoxin) infusions at doses of 1, 2, 3, and 4 mg/mL, respectively. The Iranian crude bee venom's ability to prevent coagulation is highlighted by the mean length of the prothrombin test in the said instance (P.T. = more than 300 seconds).

11.6.4 Therapeutic Potential Against Neurodegenerative Diseases

The central and peripheral nervous systems are affected differently by bee venom, which is used to treat a variety of neurological disorders, mainly multiple sclerosis (MS), amyotrophic lateral sclerosis (ALS), Alzheimer's disease (AD), and Parkinson's disease (Hwang et al., 2015). Glutamate, the primary excitatory neurotransmitter in the CNS, changes the activity of glutamate transporters in many neurodegenerative diseases, including Parkinson's disease, AD, and ALS (Hauser et al., 2001). However, bee venom is beneficial for lowering glutamatergic cell toxicity in neurodegenerative disorders because it protects against cell death and significantly inhibits cellular glutamate toxicity. Therapy with bee venom altered MAP kinase activation after glutamate exposure (Lee et al., 2011).

A study conducted on mice showed that bee venom acupuncture in the spinal cord of symptomatic ALS mice at ST36 inhibits neuroinflammation by significantly lowering the levels of inflammatory proteins such as TLR4, CD14, and tumour necrosis factor (TNF) (Cai et al., 2015); on the other hand, the presence of PLA2 in bee venom has shown promising results in preventing the progression of AD in transgenic mice (Ye et al., 2016). According to the research, bee venom PLA2 can increase glucose metabolism in the brain and prevent neuroinflammatory signs, thus restricting AD pathogenesis.

11.6.5 Anticancer Potential of Bee Venom

Current cancer treatments have increased survival rates, but the overall quality of life has been reduced due to the negative side effects of chemotherapy. Several potent molecules are being researched as potential cancer treatments, and apitoxin is one of them (Jung et al., 2018). The study showed melittin and Api m as possible candidates for cancer treatment. Their mechanisms of anticancer effects included cell lysis, programmed cell death, cell targeting, and regulation of gene expression.

Melittin is thought to cause cell death non-specifically. It can disrupt the entry to the lipid bilayer and is toxic when delivered intravenously (Hong et al., 2019). Api m can inhibit tumour growth in multiple cancer cells (lung, breast, liver, and prostate). In vivo and in vitro research concludes that apitoxin can inhibit cancer cells via NF-kB inhibition signalling and triggering the caspase pathways (3 and 9). Suppression of Rac1-dependent pathways inhibited hepatocellular tumour cell death in vivo and in vitro (Sabaratnam et al., 2013).

Furthermore, breast cancer, which accounts for nearly 30% of all newly diagnosed malignancies in women, is one of the most prevalent diseases (Berek, 2019; Ahmad, 2019). The two types of breast cancer that are most prevalent are triple-negative and HER2-positive. Patients with early

HER2-positive cancer have a higher survival percentage using anti-HER2 therapy. However, it develops drug resistance and is challenging to treat once it reaches the final stage. The effectiveness of bee venom in preventing the spread of breast tumour cells and limiting cell viability in response to dosage and timing has recently been examined and proven (Yeo et al., 2003). Other ways of inhibiting breast cancer cells discovered by researchers are cytostatic cell death, expression of gene modulation, and cell breakdown (Kamran et al., 2020; Jung et al., 2018). Studies have shown that bee venom dramatically reduces the growth of pancreatic carcinoma cells by arresting the cell cycle in S-phase.

Additionally, the effects of apitoxin therapy on cell apoptosis and cell migration have been explored. The results showed that apitoxin treatment significantly increased cell death and prevented the spread of colon cancer cells. All the mentioned facts point to the anti-tumoural properties of apitoxin as a panacea for pancreatic cancer (Liu et al., 2002; Zheng et al., 2015; Kim et al., 2020). The most dangerous side effect of bee venom, the haemolytic impact (Duffy et al., 2020), can be reduced by improving selectivity and reducing the dose to the right level. Melittin and bee venom are beneficial against human malignant hepatocellular carcinoma, prostate cancer, and ovarian cancer (Cho et al., 2010). However, integrative research must be done, because cell lines, vehicles, and outcomes vary.

11.6.6 Radioresistance Potential of Bee Venom

Radioprotectors, often known as preventative anti-radiation techniques, are used to lessen the harmful effects of radiation on people (Orsoloc et al., 2007; Son et al., 2007). However, the efficacy/tolerability standards are not usually met by current radioprotectors. In this context, research is ongoing, domestically and abroad, regarding new radioprotectors derived from numerous kinds of synthetic compounds, that is, chemicals with natural sources. Numerous research results have shown that venom injection is associated with longer radioresistance, which lessens the impact of ionising therapy on the lifespan of experimental mice after single gamma-irradiation. The survival time of irradiation-exposed experimental animals can be extended by administering bee venom concurrently with radiation exposure and in the immediate post-irradiation phase. Expanding the concept of vaguely defined irradiation resistance and suggesting the development of new products based on substances with biological activity from animal sources that increase the radioresistance of living organisms is made possible by research into apitoxin doses via mono fractionated gamma-irradiation.

11.7 COVID-19 AND BEE VENOM

The family of viruses known as coronaviruses can affect a diverse range of organisms and produce minor to severe respiratory illnesses in humans. The evolving coronaviruses have become a new concern for public health in the 21st century as a result of the emergence of two contagious viral zoonotic coronaviruses, Middle East respiratory syndrome coronavirus (MERS-CoV) and severe acute respiratory syndrome coronavirus (SARS-CoV), both of which caused chronic lung human illness in 2002 and 2019, respectively (Cui et al., 2019: Wu et al. 2020).

The severe acute respiratory syndrome-Covid-2 belongs to the β-coronavirus family. The overall structural integrity of the virus is maintained by protein molecules, which form an insulating sheath around its nucleic acid (RNA) (Mousavizadeh and Ghasemi,2020). The coronavirus envelope consists of three or four viral proteins. Melittin's presence in bee venom can penetrate the defensive barrier sheath surrounding viruses, including the human immune deficiency virus (Dawood, 2020). Several viral pathogens, such as severe acute respiratory virus, rely on their barrier protection coating and may be susceptible to bee venom therapy with melittin as a guide. Besides melittin, the PLA2 element of apitoxin possesses antiretroviral properties (Fenard et al., 1999). Melittin-loaded nanoparticles supplied a considerable amount of melittin internally to K14-HPV16 mice with epidermal dysplasia and malignancy harbouring human papillomavirus (HPV) recombinant factors, focusing on and eliminating precancerous lesions (E6 and E7 oncogenes) (Soman et al., 2009).

In a survey conducted by the regional beekeepers' union from February 23 to March 8, 2019, 5,115 apiarists were sampled in Hubei Province, the hotspot of China's SARS epidemic, and it was discovered that none of them had experienced any of the symptoms seen in Covid-19 victims. (Yang et al., 2020). The researchers interviewed 5 victims in Wuhan and 121 of their patients who had undergone apitherapy between October and December 2019. Two radiotherapists had been in direct contact with suspected or confirmed Covid-19 victims without protective gear (Hui et al. 2020). Despite being exposed to Covid-infected cases, none of the beekeepers displayed any signs of complications associated with Covid-19.

11.8 CONCLUSION AND FUTURE ASPECTS

Bee venom has been used for treating multiple diseases as sketched in Table 11.1 since time immemorial, dating back to Hippocrates (the father of medicine), who used honeybee stings for healing rheumatoid arthritis (RA). Physicians are currently using bee venom worldwide, predominantly in Asian countries, to treat patients with inflammatory diseases by directly injecting

TABLE 11.1
Bee Venom Components and Their Biological Attributes

Substance group	Component	% of dry weight	Therauptic potential
Proteins (Enzymes)	Phospholipase A2	10–12	• Antiprotozoal, antimicrobial, neuronal protection, anticancer properties
	Phospholipase B	1	
	Hyaluronidase enzyme	1–2	
	Phosphatase enzyme	1	
	α-Glucosidase	0.6	
Peptides	Melittin	40–50	• Anti-inflammatory, antimicrobial, antiviral, fungicidal, and anticancer properties
	Apamin	2–3	
	MCD peptide	2–3	• Exerts neurotoxic effects such as severe local pain, hypersensitivity, seizures, jerks, spasms. Potential treatment for neurological disabilities such as learning deficit disorder, Parkinson's disease by activating inhibitory muscarinic receptors of motor nerve terminals
	Secapine	0.5–2	
	Pamine	1–3	
	Minimine	2	
	Adolapine	0.5–1	
	Procamine A, B	1–2	
	Protease inhibitor	0.1–0.8	• Can ease and manage lumbar disc disease, osteoarthritis of the knee, rheumatoid arthritis, adhesive capsulitis, and lateral epicondylitis
	Tertiapine, cardiopep, melittin F	1–2	
Volatiles (pheromones)	Complex ethers	4–8	• They alert bees to danger, causing the stinging response. Among sting pheromones, the volatile component is the main element that induces aggression
Sugars	Glucose, fructose	2–4	–
Minerals	P, Ca, Mg	3–4	–
Biogenic amines	Histamine	0.5–2	• Mediators of inflammatory and allergic reactions. can trigger allergic reactions, which can occasionally result in brain or cardiac ischaemia. Can induce discomfort by altering neurons or by releasing chemicals that produce pain
	Dopamine	0.2–1	
	Noradrenaline	0.1–0.5	

(Al-Samie and Ali, 2015; Banks and Shipolini, 1986; Beck, 1935; Urtubey, 2005)

venom via the acupuncture method. The biological activities of numerous apitoxin constituents have been studied in vitro and in vivo, providing insights into their potential to treat autoimmune disorders, highly contagious and microbial infections, neurological diseases, metabolic disorders, and malignancy. The optimised strategies mainly concentrate on the future use of nanostructures in the delivery of apitoxin to avoid its generalised cytotoxic effect. The fact that apitoxin shows a considerable harmful impact on various enveloped viruses, such as complex AIDS viruses, and just a few unencapsulated viruses makes the antiviral activity of apitoxin also promising. Last but not least, even though there is more work to be done before bee venom therapy can be employed in healthcare situations, experts are encouraged that their ongoing research will hopefully make bee venom and its components good contenders for inclusion in a range of therapies in the foreseeable future. Many new compounds and possible therapeutic modalities have been discovered thanks to tremendous advancements in different domains of multiomics and chromatographic techniques; many more are expected to be found. To fully leverage the promise that bee venom offers, additional in-depth investigations are required to investigate particular structure–function correlations and biological activities. However, there have only been a few pertinent clinical investigations on the uses of honey bee products, particularly apitoxin, in managing inflammatory and autoimmune conditions. If novel products are to be employed to produce the best scientific and economic results in drug discovery development, severe allergic responses and unselective toxicity still present problems. Therefore, it's critical to perform additional clinical research to examine the advantages of bee venom utilisation in developing new medicines.

REFERENCES

Abd El-Wahed, A.A., Khalifa, S.A.M., Sheikh, B.Y., Farag, M.A., Saeed, A., Larik, F.A. (2019). Bee venom composition: From chemistry to biological activity. *Stud. Nat. Prod. Chem.* 60:459–484. doi: 10.1016/B978-0-444-64181-6.00013-9.

Ahmad, A. (2019). *Breast Cancer Metastasis and Drug Resistance*, 2nd ed. New York: Springer Nature.

Ali, M.A.A.S.M. (2012). Studies on bee venom and its medical uses. *Int. J. Adv. Res Technol.* 1:69–83.

Ali, M.A.A.S.M. (2015). Studies on bee venom and its medical uses. *Int.J.Adv.Res.Techn.* 1:1–11.

Alvarez-Fischer, D., Noelker, C., Vulinovi´c, F., Grünewald, A., Chevarin, C., Klein, C. (2013). Bee venom and its component apamin as neuroprotective agents in a Parkinson disease mouse model. *PLOS One.* 84:61–70. doi: 10.1371/journal.pone.0061700.

Amin, M.A., Abdel-Raheem, I.T. (2014). Accelerated wound healing and anti-inflammatory effects of physically cross linked polyvinyl alcohol–chitosan hydrogel containing honey bee venom in diabetic rats. *Arch. Pharm. Res.* 37(8):1016–1031. doi: 10.1007/s12272-013-0308-y.

Bachis, A., Cruz, M.I., Mocchetti, I. (2010). M-tropic H.I.V. envelope protein gp120 exhibits a different neuropathological profile than T-tropic gp120 in rat striatum. *Eur. J. Neurosci.* 32:234–254.

Banks, B.E.C., Shipolini, R.A. (1986). Chemistry and pharmacology of honeybee venom. In Piek, T. (ed.) *Venoms of the Hymenoptera.* London: Academic Press; pp. 330–416.

Baracchi, D., Francese, S., Turillazzi, S. (2011). Beyond the antipredatory defence: Honey bee venom function as a component of social immunity. *Toxicon* 58(6–7):550–557.

Barboni, E., Kemeny, D.M., Campos, S., Vernon, C.A. (1987). The purification of acid phosphatase from honey bee venom (*Apis mellifica*). *Toxicon* 25(10):1097–1103, ISSN 0041-0101. doi: 10.1016/0041-0101(87)90266-2.

Bazzo, R., Tappin, M.J., Pastore, A., Harvey, T.S., Carver, J.A., Campbell, I.D. (1988). The structure of melittin a ^{1}H-NMR study in methanol. *Eur. J. Biochem.* 173(1):139–146.

Beck, B.F. (1935). *Bee Venom Therapy.* New York and London: D. Appleton-Century Company.

Berek, J.S. (2019). *Berek and Novak's Gynecology*, 16th ed. Philadelphia, PA: Lippincott Williams and Wilkins; p. 2796.

Bilò, M.B., Bonifazi, F. (2009). The natural history and epidemiology of insect venom allergy: Clinical implications. *Clin. Exp. Allergy* 39(10):1467–1476.

Boutrin, M.-C., Foster, H., Pentreath, V. (2008). The effects of bee (*Apis mellifera*) venom phospholipase A2 on *trypanosoma brucei brucei* and *enterobacteria. Exp. Parasitol.* 119(2):246–251.

Cai, M.D., Choi, S.M., Yang, E.J. (2015). The effects of bee venom acupuncture on the central nervous system and muscle in an animal hSOD1G93A mutant. *Toxins* 7(3):846–858.

Chen, J., Guan, S.M., Sun, W., Fu, H. (2016). Melittin, the major pain producing substance of bee venom. *Neurosci. Bull.* 32(3):265–272. doi: 10.1007/s12264-016-0024-y.

Cho, H.J., Jeong, Y.J., Park, K.K., Park, Y.Y., Chung, I.K., Lee, K.G., Yeo, J.H., Han, S.M., Bae, Y.S., Chang, Y.C. (2010). Bee venom suppresses PMA-mediated MMP-9 gene activation via J.N.K./p38 and NF-kappaB-dependent mechanisms. *J. Ethnopharmacol.* 127(3):662–668.

Chu, S.-T., Cheng, H.-H., Huang, C.-J., Chang, H.-C., Chi, C.-C., Su, H.-H ., Hsu, S.-S., Wang, J.-L., Chen, I.-S., Liu, S.-I., Lu, Y.-C., Huang, J.-K., Ho, C.-M., Jan, C.-R. (2007). Phospholipase A2-independent Ca2+ entry and subsequent apoptosis induced by melittin in human MG63 osteosarcoma cells. *Life Sci.* 80(4):364–369.

Cornara, L., Biagi, M., Xiao, J., Burlando, B. (2017). Therapeutic properties of bioactive compounds from different honeybee products. *Front. Pharmacol.* 8:412.

Cui, J., Li, F., Shi, Z.L. (2019). Origin and evolution of pathogenic coronaviruses. *Nat. Rev. Microbiol.* 17(3):181–192.

Dawood, A.A. (2020). Mutated COVID-19 may foretell a great risk for mankind in the future. *New Microbes New Infect.* 35:100673. doi: 10.1016/j.nmni.2020.100673.

Dennis, E.A., Cao, J., Hsu, Y.H., Magrioti, V., Kokotos, G. (2011). Phospholipase A2 enzymes: Physicalstructure, biological function, disease implication, chemical inhibition, and therapeutic intervention. *Chem. Rev.* 111(10):6130–6185.

Ding, J., Xiao, Y., Lu, D., Du, Y.-R., Cui, X.-Y., Chen, J. (2011). Effects of SKF-96365, a TRPC inhibitor, on melittin-induced inward current and intracellular Ca2+ rise in primary sensory cells. *Neurosci. Bull.* 27(3):135–142.doi: 10.1007/s12264-011-1018-4.

Doery, H.M., Pearson, J.E. (1964). Phospholipase B in snake venoms and bee venom. *Biochem. J.* 92(3):599–602.

Dotimas, E.M., Hamid, K.R., Hider, R.C., Ragnarsson, U. (1987). Isolation and structure analysis of bee venom mast cell degranulating peptide. *Biochimica et Biophysica Acta (B.B.A.) Protein Struct. Mol. Enzymol.* 911(3) :285–293, ISSN 0167-4838. doi: 10.1016/0167-4838(87)90069-0.

Drici, M.D., Diochot, S., Terrenoire, C., Romey, G., Lazdunski, M. (2000). The bee venom peptide tertiapin underlines the role of I(KACh) in acetylcholine-induced atrioventricular blocks. *Br. J. Pharmacol.* 131(3):569–577.

Duffy, C., Sorolla, A., Wang, E., Golden, E., Woodward, E., Davern, K., Ho, D., Johnstone, E., Pfleger, K., Redfern, A. (2020). Honeybee venom and melittin suppress growth factor receptor activation in HER2-enriched and triple-negative breast cancer. *N.P.J. Precis. Oncol* 4:24.

Elieh Ali Komi, D., Shafaghat, F., Zwiener, R.D. (2018). Immunology of bee venom. *Clin. Rev. Allergy Immunol.* 54(3):386–396. doi: 10.1007/s12016-017-8597-4.

Fenard, D., Lambeau, G., Valentin, E., Lefebvre, J.-C., Lazdunski, M., Doglio, A. (1999). Secreted phospholipases A2, a new class of H.I.V. inhibitors that block virus entry into host cells. *J. Clin. Invest.* 104(5):611–618. doi: 10.1172/ JCI6915.

Fox, J.W. (2013). A brief review of the scientific history of several lesser-known snake venom proteins: l-amino acid oxidases, hyaluronidases and phosphodiesterases. *Toxicon* 62:75–82. doi: 10.1016/j.toxicon.2012.09.009.

França, F.O., Benvenuti, L.A., Fan, H.W., Dos Santos, D.R., Hain, S.H., Picchi- Martins, F.R. (1994). Severe and fatal mass attacks by "killer" bees (Africanized honey bees–Apis mellifera scutellata) in Brazil: Clinicopathological studies with measurement of serum venom concentrations. *Q. J. Med.* 87(5):269–282.

Gajski, G., Garaj-Vrhovac, V. (2009). Radioprotective effects of honeybee venom (*Apis mellifera*) against 915-MHz microwave radiation–induced DNA damage in Wistar rat lymphocytes: *In vitro study. Int. J. Toxicol.* 28(2):88–98.

Georgieva, D., Greunke, K., Genov, N., Betzel, C. (2009). 3-D Model of the bee venom acid phosphatase: Insights into allergenicity. *Biochem. Biophys. Res. Commun.* 378(4):711–715, ISSN 0006-291X. doi: 10.1016/j.bbrc.2008.11.101.

Golden, D.B.K. (2007). Insect sting anaphylaxis. *Immunol. Allergy Clin. North Am.* 27(2):261.

Grunwald, T., Bockisch, B., Spillner, E., Ring, J., Bredehorst, R., Ollert, M.W. (2006). Molecular cloning and expression in insect cells of honeybee venom allergen acid phosphatase (Api m 3). *J. Allergy Clin. Immunol.* 117(4):848–854.

Habermann, E. (1972). Bee and wasp venoms. *Science* 177(4046):314–322. doi: 10.1126/science.177.4046.314.

Han, S.M., Kim, J.M., Hong, I.P., Woo, S.O., Kim, S.G., Jang, H.R., Pak, S.C. (2016). Antibacterial activity and antibiotic-enhancing effects of honeybee venom against methicillin-resistant Staphylococcus aureus. *Molecules* 21(1):79.

Hao, J., Liu, M.-G., Yu, Y.-Q., Cao, F.-L., Li, Z., Lu, Z.-M (2008). Roles of peripheral mitogen-activated protein kinases in melittin-induced nociception and hyperalgesia. *Neuroscience* 152(4):1067–1075. doi: 10.1016/j.neuroscience.2007.12.038.

Hauser, R.A., Daguio, M., Wester, D., Hauser, M., Kirchman, A., Skinkis, C. (2001). Bee-venom therapy for treating multiple sclerosis: A clinical trial. *Altern. Complement. Ther.* 7(1):37–45. doi: 10.1089/107628001300000714.

Hider, R.C., Ragnarsson, U. (1981). A comparative structural study of apamin and related bee venom peptides. *Biochimica et Biophysica Acta (B.B.A.) - Protein Struct.* 667(1):197–208, ISSN 0005-2795. doi: 10.1016/0005-2795(81)90080-5.

Hong, J., Lu, X., Deng, Z., Xiao, S., Yuan, B., Yang, K. (2019). How melittin inserts into cell membrane: Conformational changes, inter-peptide cooperation, and disturbance on the membrane. *Molecules* 24(9):1775. doi: 10.3390/molecules24091775.

Hood, J. L., Jallouk, A. P., Campbell, N., Ratner, L., Wickline, S.A. (2013). Cytolytic nanoparticles attenuate HIV-1 infectivity. *Antivir. Ther.* 18(1):95–103. doi: 10.3851/IMP2346. Epub 2012 Sep 6. PMID: 22954649.

Hoover, S.E., Ovinge, L.P. (2018). Pollen collection, honey production, and pollination services: Managing honey bees in an agricultural setting. *J. Econ. Entomol.* 111(4):1509–1516.

Horta, C.C.R., Magalhães, B.de F., Oliveira-Mendes, B. B. R., do Carmo, A. O., Duarte, C. G., Felicori, L. F. (2014). Molecular, immunological, and biological characterisation of Tityus serrulatus venom hyaluronidase: New insights into its role in envenomation. *PLOS Negl. Trop. Dis.* 8:2693. doi: 10.1371/journal.pntd.0002693.

Huh, J.E., Baek, Y.H., Lee, M.H., Choi, D.Y., Park, D.S., Lee, J.D. (2010). Bee venom inhibits tumor angiogenesis and metastasis by inhibiting tyrosine phosphorylation of VEGFR-2 in LLC-tumor-bearing mice. *Cancer Lett.* 292(1):98–110.

Hui, D.S., I. Azhar, E. I., Madani, T.A., Ntoumi, F., Kock, R., Dar, O., Ippolito, G., Mchugh, T.D., Memish, Z.A., Drosten, C., Zumla, A., Petersen, E. (2020). The continuing 2019-nCoV epidemic threat of novel coronaviruses to global health - The latest 2019 novel coronavirus outbreak in Wuhan, China. *Int. J. Infect. Dis.* 91:264–266. doi: 10.1016/j.ijid.2020.01.009.

Hwang, D.S., Kim, S.K., Bae, H. (2015). Therapeutic effects of bee venom on immunological and neurological diseases. *Toxins* 7(7):2413–2421.

Jung, G.B., Huh, J.-E., Lee, H.-J., Kim, D., Lee, G.-J., Park, H.-K (2018). Anticancer effect of bee venom on human MDA-MB-231 breast cancer cells using Raman spectroscopy. *Biomed. Opt. Express* 9(11):5703. doi: 10.1364/BOE.9.005703.

Jung, G.B., Huh, J.E., Lee, H.J., Kim, D., Lee, G.J., Park, H.K., Lee, J.D. (2018). Anticancer effect of bee venom on human MDA-MB-231 breast cancer cells using Raman spectroscopy. *Biomed. Opt. Express* 9(11):5703–5718.

Kamran, M.R., Zargan, J., Keshavarzalikhani, H., Hajinoormohamadi, A. (2020). The comparative cytotoxic effects of Apis mellifera crude venom on MCF-7 breast cancer cell line in 2D and 3D cell cultures. *Int. J. Pept. Res. Ther.* 26(4):1819–1828.

Kinins and Kaplan, A.P. (2010). Airway obstruction, and anaphylaxis. *Chem. Immunol. Allergy* 95:67–84.

Kim, C.M.H. (2013). Apitherapy – Bee Venom Therapy. In: Grassberger, M., Sherman, R., Gileva, O., Kim, C., Mumcuoglu, K. (eds) In: *Biotherapy - History, Principles and Practice.* Heidelberg: Springer; pp.– 77–112.

Kim, D. H., Lee, H. W., Park, H. W. (2020). Bee venom inhibits the proliferation and migration of cervical-cancer cells in an HPV E6/E7-dependent manner. *B.M.B. Rep.* 53(8):419–424.

Kim, H., Lee, H., Lee, G., Jang, H., Kim, S.S., Yoon, H., Kang, G.H., Hwang, D.S., Kim, S.K., Chung, H.S. (2015). Phospholipase A2 inhibits cisplatin-induced acute kidney injury by modulating regulatory T cells by the CD206 mannose receptor. *Kidney Int.* 88(3):550–559.

Kim, W.H., An, H.J., Kim, J.Y., Gwon, M.G., Gu, H., Jeon, M., Kim, M.K., Han, S.M., Park, K.K. (2018). Antiinflammatory effect of melittin on Porphyromonas gingivalis LPS-stimulated human keratinocytes. *Molecules* 23(2):332.

Koburova, K.L., Michailova, S.G., Shkenderov, S.V. (1985). Further investigation on the anti-inflammatory properties of adolapin--Bee venom polypeptide. *Acta Physiol. Pharmacol. Bulg.* 11(2):50–55.

Kondo, T., Ikenaka, K., Kato, H., Ito, K.-I., Aimoto, S., Hojo, H., Mikoshiba, K. (1990). Long-term enhancement of synaptic transmission by synthetic mast cell degranulating peptide and its localisation of binding sites in hippocampus. *Neurosci. Res.* 8(3):147–157, ISSN 0168-0102. doi: 10.1016/0168-0102(90)90017-9.

Kreil, G. (1994). Hyaluronidases–a group of neglected enzymes. *Protein Sci.* 4(9):1666–1669. doi: 10.1002/pro.5560040902.

Krivtsov, N., Lebedev, V. (1995). *The Bee Products* (In Russian). Niwa Niwa: Editing House.

Lamy, C., Goodchild, S.J., Weatherall, K.L., Jane, D.E., Liégeois, J.-F., Seutin, V (2010). Allosteric block of KCa2 channels by apamin. *J. Biol. Chem.* 285(35):27067–27077. doi: 10.1074/jbc.M110.110072.

Leandro, L.F., Mendes, C.A., Casemiro, L.A., Vinholis, A.H., Cunha, W.R., De Almeida, R., Martins, C.H. (2015). Antimicrobial activity of apitoxin, melittin and phospholipase A2 of honey bee (*Apis mellifera*) venom against oral pathogens. *An. Acad. Bras. Cienc.* 87(1):147–155.

Lee, G., and Bae, H. (2016). Anti-inflammatory applications of melittin, a major component of bee venom: Detailed mechanism of action and adverse effects. *Molecules* 21(5):616. doi: 10.3390/molecules21050616. PMID: 27187328. PMCID: PMC6273919.

Lee, S.M., Yang, E.J., Choi, S.M., Kim, S.H., Baek, M.G., Jiang, J.H. (2011). Effects of bee venom on glutamate-induced toxicity in neuronal and glial cells. *Evid. Based Complement. Alternat. Med.*:368196–368205.

Li, D., Kim, W., Shin, D., Jung, Y., Bae, H., Kim, S.K. (2016). Preventive effects of bee venom derived phospholipase A2 on oxaliplatin-induced neuropathic pain in mice. *Toxins* 8(1):27.

Liu, X., Chen, D., Xie, L. (2002). Effect of honey bee venom on proliferation of K1735M2 mouse melanoma cells in-vitro and growth of murine B16 melanomas in-vivo. *J. Pharm. Pharmacol.* 54(8):1083–1089.

Markovi'c-Housley, Z., Miglierini, G., Soldatova, L., Rizkallah, P.J., Müller, U., Schirmer, T. (2000). Crystal structure of hyaluronidase, a major allergen of bee venom. *Structure* 8(10):1025–1035. doi: 10.1016/S0969-2126(00)00511-6.

Mingomataj, E.Ç., Bakiri, A.H. (2012). Episodic hemorrhage during honeybee venom anaphylaxis: Potential mechanisms. *J. Investig. Allergol. Clin. Immunol.* 22(4):237–244.

Mizrahi, A., Lensky, Y. (2013). *Bee Products: Properties, Applications, and Apitherapy.* Tel Aviv: Springer Science and Business Media.

Moreira, L.A., Ito, J., Ghosh, A., Devenport, M., Zieler, H., Abraham, E.G., Crisanti, A., Nolan, T., Catteruccia, F., Jacobs-Lorena, M. (2002). Bee venom phospholipase inhibits malaria parasite development in transgenic mosquitoes. *J. Biol. Chem.* 277(43):40839–40843.

Moreno, M., Giralt, E. (2015). Three valuable peptides from bee and wasp venoms for therapeutic and biotechnological use: Melittin, apamin and mastoparan. *Toxins* 7(4):26–50. doi: 10.3390/toxins7041126.

Moreno, M., Giralt, E. (2015). Three valuable peptides from bee and wasp venoms for therapeutic and biotechnological use: Melittin, apamin and mastoparan. *Toxins* 7(4):1126–1150. doi: 10.3390/toxins7041126.

Mousavizadeh, L., Ghasemi, S. (2020). Genotype and phenotype of COVID- 19: Their roles in pathogenesis. *J. Microbiol. Immunol. Infect.* 5:202. doi: 10.1016/j.jmii.2020.03.022.

Ong, P.Y., Leung, D.Y.M. (2016). Bacterial and viral infections in atopic dermatitis: A comprehensive review. *Clin. Rev. Allergy Immunol.* 51(3):329–337.

Orsolic, N., Benkovic, V., Horvat-Knezevic, A. (2007). Assessment by survival analysis of the radioprotective properties of propolis and its polyphenolic compounds. *Biol. Pharm. Bull.* 30(5):946–451.

Pucca, M.B., Cerni, F.A., Oliveira, I.S., Jenkins, T.P., Argemí, L., Sørensen, C.V., Ahmadi, S., Barbosa, J.E., Laustsen, A.H. (2019). Bee updated: Current knowledge on bee venom and bee envenoming therapy. *Front. Immunol.* 10:2090. doi: 10.3389/fimmu.2019.02090.

Raghuraman, H., Chattopadhyay, A. (2007). Melittin: A membrane-active peptide with diverse functions. *Biosci. Rep.* 27(4–5):189–223. doi: 10.1007/s10540-006-9030-z.

Sabaratnam, V., Gurunathan, S., Raman, J., Abd Malek, S.N., John, P. (2013). Green synthesis of silver nanoparticles using Ganoderma neo-japonicum Imazeki: A potential cytotoxic agent against breast cancer cells. *Int. J. Nanomedicine* 8:4399. doi: 10.2147/IJN.S51881.

Saito, K. (2014). Reminiscence of phospholipase B in *Penicillium notatum. Proc. Jpn Acad. B* 90(9):333–346.

Schmidt, J. O., Blum, M. S., Overal, W. L. (1986). Comparative enzymology of venoms from stinging hymenoptera. *Toxicon* 24(9):907–921.

Schumacher, M.J., Tveten, M.S., Egen, N.B. (1994). Rate and quantity of delivery of venom from honeybee stings. *J. Allergy Clin. Immunol.* 93(5):831–835. doi: 10.1016/0091-6749(94)90373-5.

Shkenderov, S., Koburova, K. (1982). Adolapin - A newly isolated analgetic and anti-inflammatory polypeptide from bee venom. *Toxicon* 20(1):317–321, ISSN 0041-0101. doi: 10.1016/0041-0101(82)90234-3.

Sobral, F., Sampaio, A., Falcão, S., Maria João, R.P., Queiroz, M. J. R., Calhelha, R. C., Vilas-Boas, M., & Ferreira, I. C. (2016). Ferreira Chemical characterisation, antioxidant, anti-inflammatory and cytotoxic properties of bee venom collected in Northeast Portugal. *Food Chem. Toxicol.* 94:172–177.

Soman, N.R., Baldwin, S.L., Hu, G., Marsh, J.N., Lanza, G.M., Heuser, J.E., Arbeit, J.M., Wickline, S.A., Schlesinger, P.H. (2009). Molecularly targeted nanocarriers deliver the cytolytic peptide melittin specifically to tumor cells in mice, reducing tumor growth. *Journal of Clinical Investigation* 119(9):2830–2842. doi: 10.1172/JCI38842. Epub 2009 Aug 10. PMID: 19726870; PMCID: PMC2735896.

Son, D.J., Lee, J.W., Lee, Y.H., Song, H.S., Lee, C.K., Hong, J.T. (2007). Therapeutic application of anti-arthritis, pain-releasing and anticancer effects of bee venom and its constituent compounds. *Pharmacol. Ther.* 2(2):246–270.

Strong, P.N. (1990). Potassium channel toxins. *Pharmacol. Ther.* 46(1):137–162.

Szókán, G.Y., Horvath, J., Almas, M., Saftics, G.Y., Palocz, A. (1994). Liquid chromatographic analysis and separation of polypeptide components from honey bee venoms. *J. Liq. Chromatogr.* 17(16):3333–3349.

Uddin, M.B., Lee, B.H., Nikapitiya, C., Kim, J.H., Kim, T.H., Lee, H.C., Kim, C.G., Lee, J.S., Kim, C.J. (2016). Inhibitory effects of bee venom and its components against viruses in vitro and in vivo. *J. Microbiol.* 54(12):853–866.

Urtubey, N. (2005). *Apitoxin: from Bee Venom to Apitoxin for Medical Use.* Argentina: Termas de Rio Grande Santiago del Estero.

Wu, J.T., Leung, K., Leung, G.M. (2020). Nowcasting and forecasting the potential domestic and international spread of the 2019-nCoV outbreak originating in Wuhan, China: A modelling study. *Lancet* 395(10225):689–697.

Yang, W., Hu, F., Xu, X. (2020). Bee venom and SARS-CoV- 2. *Toxicon* 181:69–70. doi: 10.1016/j.toxicon.2020. 04.105.

Ye, M., Chung, H.S., Lee, C., Yoon, M.S., Yu, A.R., Kim, J.S., Hwang, D.S., Shim, I., Bae, H. (2016). Neuroprotective effects of bee venom phospholipase A2 in the 3xTg A.D. mouse model of Alzheimer's disease. *J. Neuroinflammation* 13:10.

Yeo, S.W., Seo, J.C., Choi, Y.H., Jang, K.J. (2003). Induction of the growth inhibition and apoptosis by beevenom in human breast carcinoma MCF-7 Cells. *J. Korean Acupunct. Mox. Med. Sci.* 20:45–62.

Zheng, J., Lee, H.L., Ham, Y.W. (2015). Anticancer effect of bee venom on colon cancer cell growth by activation of death receptors and inhibition of nuclear factor kappa B. *Oncotarget* 6(42):44437–44451.

Ziai, M.R., Russek, S., Wang, H.C., Beer, B., Blume, A.J. (1990). Mast cell degranulating peptide: A multifunctional neurotoxin. *J. Pharm. Pharmacol.* 42(7):457–461.

Zolfagharian, H., Mohajeri, M., Babaie, M. (2015). Honey bee venom (*Apis mellifera*) contains anticoagulation factors and increases the blood-clotting time. *J. Pharmacopuncture* 18(4):007–011. doi: 10.3831/KPI.2015.18.031.

Zolfagharian, H., Mohajeri, M., Babaie, M. (2016). Bee venom (*Apis mellifera*) an effective potential alternative to gentamicin for specific bacteria strains: Bee venom an effective potential for bacteria. *J. Pharmacopuncture* 19(3):225–230. doi: 10.3831/KPI.2016.19.023. PMID: 27695631. PMCID: PMC5043086.

Zurier, R.B., Mitnick, H., Bloomgarden, D., Weissmann, G. (1973). Effect of bee venom on experimental arthritis. *Ann. Rheum. Dis.* 32(5):466–470.

Index

C

D

E

F

G

H

Q

R

S

T

V

W

Made in the USA
Las Vegas, NV
19 September 2026

286c5769-de3a-445a-9f89-68be1cd2a419R01